GANGA : A WATER MARVEL

GANGA : A WATER MARVEL

A.C. Shukla and Vandana, A.
Biopollution Study Centre and Environmental Studies Unit
Christ Church College, Kanpur
Kanpur University, Kanpur, India

ASHISH PUBLISHING HOUSE
8/81, Punjabi Bagh
New Delhi-110 026

Published by:
S.B. Nangia
for **Ashish Publishing House**
8/81, Punjabi Bagh, New Delhi-110 026
Tel.: 500581, 5410924

Showroom: 5, Ansari Road, Delhi 110 002
Tel.: 3274050, 3285807

ISBN 81-7024-681-4

1995

Typeset:
RADIANT PRINTERS
New Delhi-110008

Printed at:
EFFICIENT OFFESET PRINTERS
Delhi-110035

Preface

Ganga has served as an elixir of life—the water from times immemorial to millions of people of Indo-Gangetic Plains. The river is known for its pristine purity of water and has been linked with religious sentiment and fervour. Ganga signifies the tradition, culture and civilization of India. The river is looked upon with religious sentiment and is rightly revered as 'Mother Ganga'. Indo-Gangetic Plain civilization is the symbol of India. The river flows through a string of cities. Each marvels a uniqueness, serves as basic building block and seat of Indian polity. Ganga has been a mute spectator of Indian composite culture but its message to people remained perceptible. The river shines with a lustre of its own and finds expression in social, political and economic canvas of the nation. Socio-economic and political linkages interalia attributes of river as a biological marvel are of greater relevance today than ever before.

The growing influx of population has exerted tremendous pressure and stress on the river. Urbanisation and industrialisation has bred enormous pollution into the riverine system. The alarming deterioration in water quality has grown beyond comprehension. Ganga eco-system faces severe imbalances, and pollutant spills into the river exceed its self-purification capacity. Potability of water is in jeopardy and public health hazards looming large. Life of fish, cattle and human beings forming intermeshing fabric of Ganga eco-system are threatened. Recognizing the magnitude of the problem aspects and prospects of this natural resource need greater attention. The importance of water quality, is a cardinal element of river management for optimum exploitation to benefit dwellers along

Indo-Gangetic Plains. How best Ganga Basin could be harnessed and tailored conducive to maintain and even improve its quality is a million dollars question. Cleaning Ganga of its pollutants and ressurecting its sanctity has in store prosperity and development of Indian people. Ganga is not merely a source of water for drinking, domestic, agricultural and industrial purposes but also linked with futuristic production of proteins, vitamins, fertilizers, antibiotics and fish. Broad spectrum of its biological gamut marvels genetic stock of organisms required in fundamental and applied research par excellence. The significance of the river in transport is another dimension which needs further development in consonance with changing times.

The accompanying text provides a conceptual synthesis of physico-chemical and biological attributes of Ganga-ecosystem fabric, implicit use and re-use of this natural resource, causes and control of spread of diseases, monitoring, abatement of pollution and maintenance of river in healthy state. The data and assumptions presented are either average of results or explicit observations on maximum polluted belt in the heartland of Indo-Gangetic Plain. Ensuing treatise is not only aimed to cater to the needs of the environmentalists or researchers alone but also to arouse curiosity and interest of avid readers to know Ganga more closely. The book provides a conceptual synthesis of social, economic and political attributes of Ganga interwoven with environmental tenets. In here science and technology appear harmoniously wedded with social science establishing the interconnection between natural sciences, institutional structures, and ideology, giving the book a truly inter-disciplinary approach. Dr. Vandana A. collaborates efficiency with Dr. A.C. Shukla and the authors combine the scientific and technological knowhow of one with the socio-economic and political thinking of the other expressed in the form of this unique confluence.

How best this riverine system could be exploited to benefit dwellers of the Indo-Gangetic Plains and can Ganga be cleaned is the strength of the book. GANGA: A WATER MARVEL should find a place in institutional libraries and shelves of scholars, administrators, politicians, bureaucrats and common man alike.

Contents

Chapter 1

Ganga Basin

The river Ganga, a life line for millions of people is aptly called the "River of India". Pandit Jawaharlal Nehru has rightly remarked "The story of the Ganga from source to the sea, from old times to new is the story of India's civilization and culture." It has basins spread over eight States: Himachal Pradesh, Punjab, Haryana, Uttar Pradesh, Rajasthan, Madhya Pradesh, Bihar, West Bengal and Union Territory of Delhi. Due to indiscriminate over use and abuse on account of biological and chemical pollution and heavy silt load the river Ganga has lost its pristine purity. A river is a living dynamic and a continuing water resource system rather than a physical entity. The dimensions of pollutional problems of the river are unduly large and complex. Maintenance and upgradation of Ganga is linked with religious sentiment and applied significance for excellence.

The confluence of the two rivers Bhagirathi originating at Gaumukh in Gangotri glacier, 3,129 m above sea level and Alaknanda originating in Sapta Tal Glacier transcends as a perennial river "Ganga". Thereafter gorging a distance of about 220 km in the Himalayas it enters the plains at Hardwar. Finally, it flows into the Bay of Bengal after traversing over a distance of 2290 km in the plains of Uttar Pradesh, Bihar and West Bengal. Ganga flows through wide geographical diversity comprising hilly terrain in North, low lying agricultural alluvial plains in the middle and deltaic region in the East. Snowfed

Yamuna, Kali, Karnali, Ram Ganga, Gandak and Kosi arising in the Himalayas and Chambal, Betwa, Tons, Ken and Sone arising in the Vindhyan highlands of the Central India comprise the tributaries of the river Ganga before it merges into the Bay of Bengal.

The Ganga has been revered since the Vedic times in the Hindu scriptures and regarded as religious, social, cultural and linguistic wealth. Obviously, the Ganga Basin edifice is of great importance as a prelude to the in-depth treatment to the vital life-line of the alluvial agricultural plain of North India.

For the sake of convenience the topic has been divided into various zones through which the river flows as described below:

ZONE ABOVE UPPER GANGA PLAIN

The zone above the Upper Ganga Plain covers the catchment areas of Bhagirathi, Bhillanganga and Yamuna rivers. While the northern side is bound by a high ridge separating the State of Uttar Pradesh from Himachal Pradesh and China, the southern boundary is formed by a ridge separating the Bhagirathi catchment beginning from Devprayag in the East and extending to Mussoorie and Kasli in the West. This region, thus, incorporates within its area the districts of Uttarkashi, Chamoli, Dehradun and Tehri Garhwal. The geological formations of the area belong to Carboniferous, Permian, Archean, Crustaceous and Tertiary ages alongwith varied rock types. The entire area is hilly with steep to precipitous slopes. Apart from the numerous rivers and rivulets that drain into the region there are two main systems of drainage (1) The Ganga (2) The Yamuna. The region falls in the general category of temperate climate with severe winters. The higher reaches of the region are snow bound throughout the year. The entire area is very rich in forests and four different soils *viz.* redloam, brown forest, podzols and meadow. The region is endowed with natural wealth of forest resources, scenic beauty, agricultural, horticultural and livestock resource.

CULTURAL BACKDROP

The population in the Himalayan region is concentrated mainly in the middle zone along lower parts of the Bhagirathi, Alaknanda, Yamuna, Ramganga and Kosi valleys. The forest area has a lesser density of population than in the area of forest clearings. Historically, the region demonstrates a relatively low population growth due to recurring natural calamities like floods, landslides, avalanches, earthquakes alongwith migration of youth to other areas for employment. The river finds four types of sites. Ridges, river terraces, confluences and gate towns occupying the region. In the zone above Upper Ganga plain the river flows through the urban towns and urban agglomerations of Chamoli, Joshimath, Rudraprayag, Karnaprayag, Devprayag, Uttarkashi, Tehri, Narendra Nagar, Pauri, Srinagar, Rishikesh, Hardwar have developed as small towns. A major factor in the urbanisation of the region has been its religious sentiment. The calm, serene and peaceful atmosphere in the Himalayas provides the natural environment for communion with God. Hindu philosophy believes that meditation in sublime atmosphere can lead us to God. Nowhere is this serenity and sublimity more explicit than in the exquisite natural beauty of the Himalayas where devotees perform penance, prayer and meditation. Religious cities have grown as saints and sages installed holy shrines. Devprayag the confluence of Alaknanda and Bhagirathi where the river acquires the name Ganga is highly congested. Joshimath is an important military centre. Between Rishikesh and Hardwar there is a chain of urban settlements. Development has also occurred due to the establishment of the IDPL antibiotic factory at Virbhadra, Bharat Heavy Electricals at Ranipur and a Military Centre at Raiwala.

Forests occupy almost 40 per cent of the area with the largest proportion existing in the districts of Uttarkashi, Tehri and Dehradun. Only a little more than one third of the area is agriculture-based. Cultivation of any kind ceases at 6000 ft. Above it dense subtropical and temperate forests occupy the mountain slopes. Among the major crop combinations millet, wheat and rice occupy the largest area. The cropping and distributional pattern reflects variability in the availability of water, soil, climate and altitude, apart from differences in method of cultivation. Terracing and crop rotation are their main agronomic practices.

Horticulture also occupies a prominent place in the region. The altitude, climate and soil are favourable for cultivation of apples, pears, citrus, walnuts, cherries and apricots. Fruit cultivation is very popular in some areas of the region.

The industrial base of the region comprises of the cheap and abundant manpower and hydroelectric power of the region. The river basins of Alaknanda and Bhagirathi have tremendous power potential. Exploratory mining work has been taking place but there is a conspicuous absence of coal and fossil fuels in the region. There exist small scale and cottage industries like tea processing, wool weaving and rope, leather, carpentry, paper and pulp manufacturing industries. The IDPL at Virbhadra and BHEL at Ranipur are the only major plants of the region.

The region is a great tourist and pilgrim attraction with prominent deities installed in temples in natural scenic beauty with a variety of flora and fauna. Tourists spots like valley of flowers, and holy shrines of Badrinath attracts lakhs of people to the Himalayas. Tourism is facilitated in these spots through the aegis of administration. Accelerated development of transportation and communication in this region is resorted. Another reason contributing to development of transport has also been the strategic importance of the northern frontiers more so, after the Chinese invasion of 1962. Considerable activity has taken place to improve accessibility to the region due to horticultural and commercial importance. Rishikesh marks beginning of the mountain roads to Badrinath and Gangotri, at times blocked by natural hazards of land slides yet accessible otherwise.

Despite exquisitely beautiful and endowed with vast forest resources, live stocks, agricultural products yet the region faces the peril of urbanisation leading to depletion of forest resources. Consumption for natural hearth is part of their process of survival. Moreover, the industrial development led to consequent clearing of forests adds to the problem. It was dawning of this indiscriminate deforestation that led to a powerful movement by the local population know as "Chipko". Deforestation increases the flood severity of the region in the regions of the upper and middle Ganga plains causing immense losses to human lives, property and agriculture.

UPPER GANGA PLAIN

The upper Ganga plain occupies almost two-thirds of the area of Uttar Pradesh and lies between the Yamuna in the West and 100m contour in the East. The international boundary of Nepal limits the region in the East while Yamuna demarcates its border with Bundelkhand in the South. The North is marked by a 300 m contour separating it from Garhkum Himalaya West of river Sarda. The region comprises of great alluvial deposits and is watered by rivers Jamuna, Ganga and their tributaries. Physico, socio and economic features in the region are different from the bordering northern and southern areas but such variations are imperceptible in its counterparts on the East and West.

The region includes divisions of Agra, Rohilkhand, Lucknow, Meerut excluding Dehradun district, Allahabad and Faizabad excluding a few Tehsils of these divisions and also southern part of Haldwani Tehsil of Nainital and Kumaun Division. Politically the region has enjoyed stability except for occasional disturbances and transitions due to change in regimes.

Physical Characteristics

Geologically it forms part of alluvial asymmetrical Indo-Ganga trough with a dividing line formed by subterranean Delhi-Ridge, a protrusion of the Peninsular Block. The river debri of the peninsular along with Upper and Post-tertiary Himalayan debris constitutes by and large the alluvial deposits mainly composed of unconsolidated beds of clay, sand, gravel and their mixture. The alluvial filling decreased gradually southward, thinning out to merge with the irregular edge of the peninsular block. 'Khadar' 'Bhangar' the lithological constituents are variants of alluvium depending upon location and subsequent processes involved in formation.

The region is drained by the two important tributaries of the Ganga, the Ghaghra and the Gomati joining master stream in the Middle Ganga Plain. All streams have a North-West, South-East direction according to the lie of the land. The Ganga and its major tributaries Yamuna, Ram Ganga and Ghaghra are the only Himalayan rivers carrying sufficient water through out the year though varying due to seasonal fluctuations. Chambal is the only important tributary that emerges from the South running parallel with the Yamuna for miles together before merging into it. The Ghaghra and the Gomati join most of the Awadh and the Sarayupar plain while the Ramganga drains through the Rohilkhand region. The Ganga-Yamuna Doab also has a few streams draining the area before joining the rivers.

The region is a subhumid regime, acquiring the characteristics of both the dry Punjab Plain and humid Middle Ganga Plain and a greater incidence of winter rain. January is the coldest part of the year. The winter rainfall occurs between December to February ranging 1" to 4" with the same depressions that cause rainfall to the Himalayan region. But even this rainfall is beneficial to rabi crops. May is the hottest month in the region with heatwaves when temperatures shoots up exceptionally high varying from 102°F to 108°F. Winds blow from the West with violent dust storms in the months of May and June. The rainy months account for over 90 per cent of the total annual rainfall. The onset of the monsoon begins around the middle of June providing rainfall till September to the magnitude varying from 40" in the East to 20" in the West. When cyclonic storms from the Bay of Bengal move North-Westwards across North India, the area gets very heavy rains at places exceeding 10". Winter conditions start settling in November with fall in temperature alongwith dry and chilly westerlies.

There occurs 'micro regional' differences of slope in some areas of this vast alluvial expanse but they are associated with water table variations. The region consists of by and large homogeneous soils varying in texture at places due to conditions of drainage, chemical and mechanical constituents and climate. Variations in properties lead to formation of 'Usar', 'Bhur', 'Khadar' and 'Bhangar'. 'Usar' is found in the tract of the Ganga-Ghaghra Doab from Sultanpur, Pratapgarh through Raebarelli, Lucknow, Hardoi and Shahjahanpur. 'Bhur' is the sandy river deposit localised in Ramganga tract and in the narrow belt along the Ganga. The Ganga 'Khadar' soil is sandy to silty loamy texture with imperfect drainage and immature profile. The 'Bhangar' is more mature as compared to the Ganga and loamy to sandy-loam in texture. Slightly different from these is the clay soil found in the Terai zone tracts and in depressions of the land. The region is potentially rich in ground water resources and the total physical resource base of the area is confined to its rich alluvial soil and abundance of water resources due to its perennial rivers. Both serve as the basic input for development of agriculture.

Thick forestation has been part of the original Upper Ganga Plain with the famous, 'Bansudha' forests, 'Namisaranya and Khandaw' referred to in old Hindu literature. The decline of natural forests in the past has been due to the increase of human population and gradual clearing of natural vegetation for agriculture and subsistence. The existing forests in the plains are uneconomical and today only a few patches of woodland remain. Forests of the region is constituted of mainly tropical, moist, deciduous; tropical wet and subtropical dry forests. The first group is of 'Sal' as the available species. The second group consists of 'Semal' while the third has 'Shisham', 'Babul' and 'Semal' as the main species scattered in places but seldom occupying large tracts of land.

CULTURAL BACKDROP

The Upper Ganga Plain is one of the most densely populated parts of India. Man has been a basic constituent of the ecosystem here. In earlier decades agricultural suitability of the region has been in the 'Bhangar' tracts where population growth rate and density is higher. The 'Khadar' and 'Bhur' tracts are agriculturally less suitable and the population has been less. However, in the later years urbanization has led to tremendous increase in population. About seventy five per cent of its people are engaged in agricultural activity. But increased industrialization and development in the recent decades has reflected a shift from the rural to the urban base. The urban population is very unevenly distributed. The extent of urbanisation is increasing from East to West. In general there is heaviest concentation of urbanization in the Doab and adjoining regions of Kanpur and Lucknow. The region contains mostly Class-I cities of Uttar Pradesh. The two distinct urban belts of the region are (i) In the Doab. There are two conspicuous agglomerations (a) Aligarh-Hathras-Mathura-Agra-Firozabad and (b) Ghaziabad-Hapur-Meerut-Muzaffarnagar-Saharanpur (ii) Through Rohilkhand from

Bijnor to Fatehpur comprising (a) Amroha-Moradabad (b) Rampur (c) Pilibhit-Bareilly-Badaun, and (d) Etah.

Agriculture forms the main source of employment and revenue but the region has a very low percentage of land under forests. Irrigation has contributed largely to the agricultural prosperity of the region. The Upper Ganga Plain is one of the highly irrigated agricultural regions. The Upper and Middle Doab constitutes greater proportion of irrigated areas. The Tarai districts experiencing higher rainfall and lower temperature have low irrigation needs. Surface wells, wells, tubewells and canals are the major source of irrigation throughout the region unlike the Middle Ganga Plain where tanks, jhils, tals and other sources are more important. The Upper Ganga Plain has an extensive canal network. The upper Ganga Canal (1853) is built on the right bank of Ganga at Hardwar where a dam is built across the river. It irrigates around 900 km covering over approximately two and a half million acres mainly in Saharanpur, Muzaffarnagar, Meerut and Bulandshahr districts to Mainpuri alongwith Deoband and Anupshahr branches. Due to Ganga valley tubewell scheme a 36 m fall in the bedslope of the canal from Hardwar to Aligarh has been broken into 13 vertical falls to generate electricity for tubewells. They provide water to Bhur tracts of Moradabad, Bijnor and areas of western Rohilkhand not covered under canal irrigation. The electricity generated by the falls provides power for irrigational, domestic and industrial purposes in the area.

The lower Ganga Canal (1872) is diverted to the right bank of Ganga at Naraura (Bulandshahr) due to construction of a dam across the river. The main branches cover 1,000 km while distributaries cover over 5,000 km covering over 1.5 million acres of Central and Lower Doab in Aligarh, Etah, Etawah, Mainpuri, Farrukhabad, Kanpur, Fatehpur and upto Allahabad districts. Other canals of the region are Sarda Canal, Ramganga Multipurpose Project and Agra Canal.

Cropping is diversified in the region than the Middle Ganga Plain. Food crops cover over 85 per cent of the total cropped area. Cereals cover 3/5th of the cropped land, pulses 1/5th, while about 8 per cent is covered by sugarcane. Oil seeds is another cash crop apart from sugarcane. Wheat is the premier crop of the region covering over 1/5th of the total cropped land mainly in the concentrated districts of Upper Doab and contiguous parts of the Western Rohilkhand Plain. Moving towards West, rice replaces wheat as the first crop in U.P. claiming over 1/6th of the total cropped land. It is an important crop in humid sub-Himalayan Tarai in Saharanpur, Bahraich, and Pilibhit merging into Eastern Awadh and over Doab belt. The Central Doab, South-West Rohilkhand and Central Awadh devotes sizeable portion of the crop-land to rice. Pulses and oilseeds are grown throughout the region. Several Kharif crops (urad, moong, arhar and moth) and rabi (gram, peas and masur) crops, are also grown. Millets and maize claim 16.8 per cent of the total cropped land, with millets grown mainly in the West and South-West region and maize in the belt of the Tarai region. Gram and peas cover 1/6th of the total cropped land in the region. Arhar is grown significantly in Agra and Kanpur districts, while Masur finds prominence in the Tarai districts from Bareilly to Gonda in Shahjahanpur and Bahraich. Groundnut is the most important oil seed and grown in the North-West—South-West belt from Moradabad district to Lucknow in the Upper Central part of the region. It is most suitable for the 'Bhur' tracts of Moradabad, Budaun, sandy soils of Sitapur and Hardoi as it is best grown on light sandy soils. Allahabad is important for linseed. Mustard is also grown in two clustered regions, (a) Tarai districts of Kheri, Bahraich, Gonda, extending Eastward to Gorakhpur and Kanpur (b) Mathura to Kanpur in the South and beyond.

Jute, sugarcane and cotton are the important commercial crops of the region. About 46 per cent of sugarcane is grown in U.P. and Upper Ganga Plain covering 7.5 per cent of the cropland. Sugarcane clusters in (i) Upper Doab and Western Rohilkhand (ii) Central Tarai and neighboring districts. June is a new entrant confined to the region in Kheri, Bahraich and further Eastern U.P. Cotton is rather decreasing as a crop and found in the West Plain between Muzaffarnagar and Mathura.

The Ganga system covering a length of approximately 8,047 km harbours the richest fresh water fish fauna of India ranging from mahseers and the torrential fishes of the hills, hilsa and wide range of fishes of great commercial value.

INDUSTRIAL ECONOMY

The Upper Ganga Plain is the most industralised part of the region. The upper Ganga Plain provides the bulk of sugarcane products. In U.P. largest concentration of sugar factories is located making it largest producer of sugar in the country. The profitability of these units is really high yet they economize through development of produce by industries. The molasses are used in distilleries attached to sugar mills. The largest distillery is, however, outside the sugar belt at Lucknow. Units of wine at Rampur and soft drinks at Kanpur are located.

Kanpur and Agra are world famous leather goods centres with leather industries employing about four fifths of the total workers of the State engaged in this work. Kanpur has majority of the tanneries employing around 90 per cent of the States tannery workers. Agra, Unnao, Farrukhabad and Fatehpur are other tannery centres. Meerut is also an important centre of leather goods.

Textile industries of cotton, wool, jute, silk and others comprise largest single industrial group in the region. They provide largest employment and highest industrial output. Kanpur has nearly two dozen cotton factories employing 2/3rd of the State's cotton workers. The initial cotton mills were established during the British era between 1869 and 1882. It is the leading centre of cotton trade, manufacturing varieties of cloth. Agra, Saharanpur, Modinagar, Hathras, Allahabad also have important units of textile in the region. Silk units are found mainly in Varanasi and Dehradun. Kanpur is the largest knitting centre and has two units producing wearing apparrel.

Food processing industries are also found in the region, mainly related with rice, dal and edible oils; bakeries, confectionery, dairy, hydrogenated oils and fruit processing. They are located in large and small centres mostly in Doab region. Outside this region Lucknow, Bareilly, Shahjahanpur, Rampur are notable for industries. Rice mills and dal mills are also located in this region in small towns instead of East U. P. Plains. Kanpur leads in edible oil industry other centres being Agra, Aligarh and Hathras. Bakery and sugar confectionery units are centred mainly at Kanpur, Aligarh and Saharanpur. Kanpur enjoys the status of having factories of all kinds and paper and pulp industry units of U.P. are also located in this region. Saharanpur and Lucknow paper mills employ around 60 per cent of the total workers of the State in this group. Meerut, Kanpur, Allahabad, Saharanpur prepare paper board and strawboard. Plywood unit of Sitapur is the largest, others are in Bijnor and Saharanpur. Saw milling units are concentrated in Lucknow district and others in Clutterbuckganj (Bareilly), Saharanpur and Haldwani. Governmental units also exist at Bareilly and Allahabad.

Region lacks in mineral wealth so no heavy metallurgical or mineral based industry exists. Basic ferrous metal industries form the largest group in Kanpur as the chief centre engaging nearly half of the State's workers in this group.

Heavy chemical factories in the region also exist in Kanpur, topping the list. Fertilizer units have also been installed at Kanpur, Moradabad, Allahabad and at other small centres like Fatehpur. Nearly half of the States pharmaceutical, chemical units are concentrated in Lucknow and Kanpur. Meerut, Agra and Kanpur manufacture rubber goods.

Agricultural implements are manufactured in Kanpur, Lucknow, Rampur, Bareilly and Ghaziabad while Agra and Kanpur manufacture wood working machinery, machine tools, other tools and textile machinery. Kanpur has about half of U.P.'s bicycle units while Meerut and Ghaziabad are the biggest holders of bicycle units. Kanpur, Bulandshahr, Ghaziabad also make electrical goods and accessories. Insulated wires and cables are made at factories in Agra, Ghaziabad, Mathura, Aligarh and Kanpur. Kanpur also has an aeronautical engineering factory.

Cottage and small scale industries occupy an important place in the region. Handicrafts and handloom weaving is the most important cottage industry doing excellent jobs of chikans, brocades, embroideries, etc. Etawah and Sandila are famous for silk and art-silk weaving. Wool weaving is important in districts of Meerut, Bijnor and Muzaffarnagar. Sports industry exists in Meerut and Agra. Brass and Copperware industry is traditionally centred at Moradabad, Farrukhabad, Hathras, Shamli and Aligarh. Iron and steel goods and utensils are made in Meerut, Agra, Fatehpur, Hathras, Etah, Moradabad, Hardoi, Lucknow, Ghaziabad and Bareilly. Firozabad is an important glass product centre. Ceramics and pottery of Khurja and clay modelling industries of Allahabad and Lucknow are famous.

The region is seared by Northern and North-Eastern railway systems. The district has all types of traffic. Rail traffic flow is of inter State character. Roads cover around 750 km of surface roads per 10,000 km^2 of area. However, there is heavy traffic density on trunk road and other roads running parallel to railways. The Upper Doab region has highest density of both passenger and freight traffic. The cities of Kanpur, Lucknow, Allahabad and Agra are on the air map of the country.

PROBLEMS

The Upper Ganga plain is relatively more developed but is victim to its own environmental problems leading to public health hazards and imbalances in the ecosystem of the region. Soil erosion and ravine formations has been taking place due to overgrazing and deforestation. Vigorous activities and unnatural exploitation has led to accentuating water logging, and inundation due to floods. The increase in human population has led to increased exploitation of forests. Desilting of land, and recurrence of floods in the downstream region are rampant. Large scale destruction of human life, cattle, livestock, property and crops occurs every year due to floods.

Industrial pollution of the river has also increased in the region due to increased urbanization and quick industrial development. Kanpur, Lucknow, Allahabad are the main areas where the problem is very acute. Kanpur is the largest city of U.P. and the most important industrial town in the region. It has a vast industrial and commercial base with many textile and woollen mills, tanneries, thermal power stations, small metallurgical and chemical factories. The main sources of pollution are industrial liquid-waste and effluents, domestic effluents and urban liquid-waste carried through sewage and drains. Surface waste out of cultivated land contains chemical fertilizers, pesticides, manures and insecticides drained into water makes it unsafe for drinking and bathing. Surface runoff from urban and industrial solid wastes also contributes to the inflow of pollutants. The course of the river has been steadily moving towards Unnao polluting the narrow stream of water heavily at the bank of the river on the city side. Festival bathing and consignment of half burnt bodies adds to pollution. Allahabad another urban agglomerate though comparatively less industralized in the region also suffers the brunt of pollution. Pollution load in the Ganga is added as a result of release of effluents from Iffco Complex. These fertilizer wastes further aggravate the problem leading to heavy fish mortality, loss of plankton population and other marine life.

The scale of pollution also depends on the velocity of the flow of water and degree of dilution. For this it is imperative to maintain a minimum discharge in the river, especially, at points of location of urban wastes, and industrial units. This has been disregarded while diverting water for irrigation. Since 43 per cent of the total irrigated land in the country is located in the Ganga Basin the entire dry weather flow is diverted to the Upper Ganga Canal at Hardwar and regenerated flow between Hardwar and Aligarh is again diverted to lower Ganga Canal near Aligarh. Little dry weather flow in Ganga at Kanpur and Kannauj pours very heavy pollutant load into the river. Since Ganga receives more than 60 per cent of its water from Yamuna, Ghaghra, Kosi and Gandak joining the river around Allahabad the Hardwar-Allahabad stretch is vulnerable and problem of pollution is at its peak.

MIDDLE GANGA PLAIN

The Middle Ganga Plain covers the area of Eastern Uttar Pradesh and Bihar Plains in their entirety. This region is on either side of the Ganga and the Saryu (Ghaghra) within the Himalayan and the peninsular ramparts on the North and South respectively. The Indo-Nepal International Boundary cutting across Bhabar and Tarai zone down the Siwaliks forms the northern limit. In the South, 150 m contour marks the boundary and continues down the zone of Vindhyan-Rohtasgarh Chotanagpur plateau excluding most of Mirzapur district, Chakia tehsil of Varanasi district, Kaimur-Rohtasgarh plateau and moving further East, the southern strips of Gaya Monghyr and Bhagalpur districts and sub-divisions of Santhal Parganas except Godda. In the West the eastern boundary of Upper Ganga Plain marks the Western limit of this region while Bihar-Bengal State boundary (except Kishanganj sub-division in N.E. of Purnea) form the area in the East.

The Middle Ganga Plain includes the administrative divisions of Gorakhpur and Varanasi districts, tehsils of Balrampur and Utraula (Gonda), Faizabad, Tanda and Akbarpur (Faizabad), Phulpur, Handia, Karchhana and Meja (Allahabad) in Eastern Uttar Pradesh and divisions of Tirhut, Bhagalpur sub-division, Sultanpur and Kadipur (Sultanpur), Patti (Pratapgarh) and Patna. The maximum length from East to West is about 600 km while its width from North to South approximates 300 km.

The alluvial filling varies widely in the region. The depth increases upto 8-10,000 M as Himalayas are approached while it thins down reaching the peninsular block. These sediments have less 'Kankar' formation due to riverain character of the plain with mainly 'Khandar' lands. The region is mostly homogeneous in nature. Drainage is mostly dendritic with rivers meeting at acute angles. Ganga receives parallel streams in the region. The Ghaghra, Kosi, Gandak, Son, form the regional drainage sub-systems.

The region is transitional in character between relatively drier upper and prehumid Lower Ganga Plains. But summer winds blow from the West all over the region. Winter cyclones persist at times with cold waves. Dust storms are rainier than in the Upper Plain and the monsoon is in full swing here with a great effect in the overall human occupance pattern and economic development. The onset of South-West monsoon begins around 7th of June retreating by the end of September with October being a sticky and sultry month. Most of the region has alluvial soil cover and soil is rich in minerals and organic plant food.

Natural vegetation in the region has been affected due to increase human population and centuries of plough and pastoral culture. The region is a natural habitat of a dense forest cover of Sal and other species like Shisham, Jamun, Mahua, Ber etc. The remnants of Sal forests are found in northern Gorakhpur, Saharsa and Purnea districts. Naturally, growing trees like peepal, banyan, tamarind, neem, palmyra, date-palm, mahua, etc. are found in the 'diaras', 'Tarai' or other low lying areas. The grasses like Bher, Kans, Jhau, moonj, etc., also grow in this area.

Orchards of fruits like mangoes, jackfruit, lemon, plums, mahua, are also planted. Shisham trees are found around and within settlements. The Champaran forests in Bihar are the Siwalik Sal, Gangetic moist deciduous riverain forest and moist Gangetic high alluvial Sal. In the vegetation cover North of Ghaghra and the Ganga, and to their South groves, orchards, marchy vegetation are found. This holds true in North Bihar Plains.

In most of the districts like Ballia, Ghazipur, Jaunpur, Azamgarh, Muzaffarpur, Ganga-Ghaghra Doab and Saharsa in North Bihar, plain forest cover is particularly non-existent, while it is negligible in Pratapgarh, Faizabad, Deoria, Sultanpupr, Basti, Patna. Good forest cover can be seen in pockets of Gorakhpur, Gonda and Bachraich districts and, in districts South of the Ganga, forests occur mostly outside the plains. The Middle Ganga Plain has porus alluvial soil which stores adequate ground water but is needed to be tapped scientifically. However, the main resource base is from the rich fertile soils and surface and ground water resources of the vast plains. A variety of food and industrial crops grow because of adequate annual rainfall and in rain deficient months surface and ground water store is generally supplemented for water needs. Droughts and floods occur alternatively. The region has lots of tubewells due to availability of upper surface water. Due to inadequate forest cover fuelwood resources and timber are in short supply. The natural pockets existing are destroyed by the people for fuelwood and timber supply for their domestic and agricultural needs. Pottery and tiles are made by clay and sand found in this region while 'Kankar' is utilized for making lime and roads.

CULTURAL BACKGROUND

On an area of about 154 thousand km^2 the Middle Ganga Plain occupies a prominent place on the population map of India due to very high density in the area. The high density of population reflects the isotropic nature of the region, its fertility and habitable conditions, well organised drainage system, tapability of potable ground water resource through wells and borings and a high percentage of arable land. There is a general decrease in density towards the West corresponding with decreasing rainfall.

The livelihood in the middle Ganga Plains is provided on agricultural subsistence and engages approximately 80 per cent of the population spread over 45 per cent cultivators and 35 per cent agricultural labourers. The remaining constitutes workers in local industry and other services. There has been of late a tendency of a decline in agrarian employment and a vast exodus to urban areas in search of livelihood. During seventies the migration was to the tune of 2.57 per cent and the trend continues almost to the same extent or even more. The ancient cities found in the region are Ayodhya, Vindhyachal, Chunar, Gadhipur (Ghazipur), Sultanpur in eastern U.P., Pataliputra, Champa, Gaya in Bihar. Varanasi is the oldest city of the country. Due to the sanctity and sacredness of the river and fertile plain the valley has served as a highway for commerce, defence and politics. The Ganga alongwith its tributaries Ghaghra, Rapti, Gomati, Gandak and Son has acquired diverse roles and led to establishment of urban settlements. History has been testimony to the growth of towns as trade centres, and centre of economic activity. During the modern period too, new towns have multiplied to supplement economic factors. Irrigation has an important role in the agricultural operations of the region due to ill distribution of rainfall and other monsoon vagaries. Irrigation is needed to supplement for natural rains of crops like sugarcane or zaid crops. 'Kharif' and 'Rabi' crops are irrigated by tubewells driven by power. The irrigation provided through electrically driven tubewells are considered main agricultural stay, together with irrigation from tanks, jheels and impoundment of rainwater in low lying areas. The ground water knowhow is still scanty and tapping of ground water through use of pumps provides a promise in the region and supplements drinking water to the rural natives. Impounding of rainwater by laying embankments in small rivulets assures irrigation to the area in times of scarcity of irrigation water. The underdeveloped agricultural economy reflects a peculiar cropping pattern. 90 per cent of cultivated area is under food cropping while cash crops like sugarcane occupies 4.8 and oil seeds 1.4 per cent. Introduction of high yielding varieties of seeds and usage of modern technique of production alongwith facilities of irrigation, marketing, transport and storage exists. Crops of rice, wheat and maize, earlier traditional subsistence crops have become 'Cash crops' yielding money and competed with sugarcane, jute, etc.

Industrially, the Middle Ganga Plain can be marked by a few prized industries in East U.P. Plain of chemicals, diesel locomotive and fertilizers. Dalmianagar consists of cement, chemicals, sugar and vegetable oil, complexes and a few petrochemicals fertilizers and Bata Shoe Factory in South Bihar and Barauni. Textiles and other agrobased industries of rice, dal, oil, fruit, fish canning etc. constitute around 50 per cent of the total employment. Sugar factories are concentrated in the region and a large part of the cane production yields 'Gur' and 'Khandsari' particularly in U.P. Varanasi is famous for 'Banarasi' Silk, while Bhagalpur is known for 'Tassar' Silk. Woollen Carpet centres are located in Mirzapur, Varanasi and Bhadohi. Handloom centres are located in Tanda, Gaya, Buxar, Madhubani, Bihar Sharif, Mubarakpur, etc. Cities like Varanasi, Gorakhpur, Patna, Gaya, Muzaffarpur, Mirzapur enjoy industrial status.

The transportation system of the region consists of a network of roads of diverse nature. A complete system of railways is operated mostly by the Northern, Eastern and North-Eastern Railways and couple of cities are on the air map of India. Rivers are also navigated by boats and ferries. Being a densely populated region communications occupies an important place in the region. Movement of enormous quantities of food, raw materials for industries as well as heavy passenger traffic is a regular feature of this belt. Intra regional transport necessitated a great spurt for bridge and road construction in this region. Navigation has decreased due to advent of railways and roads. Water transport is just limited to ferry steamer service, ferry boats and country boat services. Air flights are available in Varanasi and Patna with connections to Kathmandu (Nepal).

The physico-geographical and industrial setting of the region sets forth many problems for the people of the Middle Ganga Plain. Problems arise mainly out of man made or man induced activities and also due to natural causes.

Natural calamities arise in this region due to floods and soil erosion. With no forests, grasslands or pastures, no grazing lands and absence of methodical soil management, the region is a victim to recurrent floods leading to damage in area, population, crops, housing and cattle heads etc.

Problems are also added due to population exodus leading to enormous problems of shortage of food, malnutrition, disease, health, education and employment. Exploitation of land and definite trend towards urbanization has led to industrialization of the region. The increasing industrialization leads to pollution mainly confined to urban agglomerations and Class-I cities but its impact is spread to a larger area. Ganga is the main stream of the region and all industries as well as domestic bodies discharge effluents and untreated waste affecting the quality of the rivers. The Hindu belief of 'Moksha' by dumping dead bodies and unburnt bodies in the river pose a great threat to the quality and purity of the river. The burnt wood and its resultant ash also add to pollution in the river. The wash out of fertilizer and chemical industries exerts a heavy strain on the river. Religious festival and bathing of pilgrims during this time retards water quality. Cattle bathing also causes considerable damage to flora and fauna of the river. Fishing in northern India has also been greatly affected after the construction of Farakka Barrage. Hilsa has become a threatened species at present. Other aquatic lives such as dolphin, crocodile, etc., also are gradually becoming extinct.

LOWER GANGA PLAIN

The Lower Ganga Plain includes the Kishanganj tehsil of Purnea district (Bihar), West Bengal State excluding mountainous parts of Darjeeling district and Purulia district, and most of Bangladesh as well. Its topography can be defined to contain old mud, new mud and marshes. Two thirds of this plain lying South of Rajmahal-Garo alignment constitutes probably the real deltaic region and the largest delta in the world. The region as a whole embraces an area of about 64,468 Km with 21°25′ - 26°50′ N and 86°30′ - 89°58′ E.

PHYSICO GEOGRAPHICAL BACKDROP

The region is a product of fluvial action which emerged due to the tectonic activities associated with the (Tertiary) phase of earth's organic epoch, resulting in the formation of the Trans-Eurasion lofty mountain chains. Himalayas created a trench like trough onto its South (the Sindhu-Ganga trough) while upwarp its counterpart—Arakan Yoma-Andaman-Sumatra are aided the formation of the Bay of Bengal due to subsidence.

The Midnapur coast reveals the effects of the changing sea level and the North-Eastern winds sweep the coast. The altitudinal position of the region is low to the extent that even a 7 m high tide is sufficient to submerge the region upto Calcutta. The recent withdrawal of the sea and occasional march of the tides in this region is testified by the saline water expanses in the vicinity of Calcutta. The general slope of the plain is to the South-East. The Delta proper has a gradient of less than 2 cm per km while the areas in its immediate vicinity show gradients upto 40 cm except for a narrow strip along the Northern border exhibiting gradients upto 100 cm. The following physiographic subdivisions can be found:

(i) The Northern Plain comprising two district land forms
  (a) Duars
  (b) The Barind Tract
(ii) The Delta proper comprising
  (a) Land of dead decaying rivers (Moribund delta) in the North-Murshidabad and Nadia
  (b) The Active delta of the Sunderbans
  (c) Mature delta parts of Birbhum, Burdwan, Midnapur and entire districts of Hooghly and Howrah
(iii) The Western margin of the delta called the 'Rarh' plain where the lateritic alluvial landscape alongwith the coastal scenery at Digha beach has developed.

The region is drained by the tributaries of the Ganga and the Brahmaputra alongwith some insignificant systems (Kosi, the Subarnarekha, etc.) discharging into Bay of Bengal and draining the South-West part of the region. The Mayurakshi, Damodar, Dwarakeshwar, Kasai and Subarnarekha are streams draining the Rarh Plain. The Bhagirathi-Hooghly system, taking off at an obtuse angle, occupies a weak zone resulting from the change in the basin axis. The shifting courses also have their impacts on the human occupance of the region. There is a constant threat to Calcutta port and the city as the siltation of the Hooghly increases. Ancient towns of Sonargaon, Satgaon and Tamralipti famous as centres of trade and commerce have lost their significance.

The climate is hot, humid and monsoonal. The climatic characteristics are determined largely by the alignment of the Himalayas in the North and the Meghalaya Plateau in the North-East alongwith proximity to the Bay of Bengal in the South. The average humidity in the region is generally high over fifty per cent and rainfall is fairly widespread with uneven seasonal and spatial distribution characteristic of monsoonal conditions. The northern and southern parts due to proximity of Himalayas and Bay of Bengal experience more annual rains than the central part. Soils in Bengal have been grouped into (i) Laterites (ii) Red Earths (iii) Tarai soils (iv) Coastal soils (v) Alluvial soils.

The Chotanagpur highlands are covered with lateritic soils of acidic character, deficient in organic matter and low water holding capacity. Sal forests thrive on this kind of soil. Red soil or lateritic alluvium are found in eastern margins of the 'Rarh' Plain and the Barind tract of Malda and West Dinajpur area. These shallow, coarse textured soils are acidic in character and are poor in organic matter and plant nutrients especially nitrogen. Lime concretions have also been observed. Deforestation has brought this land under cultivation accelerating the process of soil erosion. Darjeeling Himalayas are responsible for typical Tarai soil in Jalpaiguri and Siliguri Tehsils with deficiency of plant food and organic matter and acidity. Alluvial soils are very rich agriculturally. The Ganga alluvium found in North Bengal Plain and whole of remaining West Bengal Delta is rich in plant nutrients, organic matter and shows alkaline reaction in some areas and in uplands somewhat neutral. The narrow alluvial strip in parts of districts of Murshidabad, Bankura, Burdwan, Hoogly and Midnapore are neutral and relatively poor in plant nutrients, organic matter and acidic in reaction. The coastal soils are result of rivers and tides and are found in the district of the 24 Parganas and Midnapur. These are saline and alkaline and contain deposits rich in calcium, magnesium and partially decomposed organic matter.

Ground water resources in the region varies seasonally and spatially. During summer, water table goes as low as 30 m, while in rainy season water table rises up to the surface in most areas.

Large scale mineral based industries have mushroomed in the region around Asansol, Durgapur complex, industries of iron, copper lead zinc etc. Raniganj coal fields of national importance are located here and the region is considered to be the birth place of coal industry in India.

Around 9,00,000 hectares of land constitutes natural vegetation in the region. The mangrove and tidal forests in Sundarbans and humid tropical forests in extreme North of the region are preserves of natural vegetation. The western fringe is covered with tropical deciduous forests while tropical evergreen forests are found is Duar/Tarai region. The natural vegetation in the region comprises of gurjan, in tropical moist deciduous forests comprised of Sal and tropical moist forests includes casuarina and mangrove in this region.

The physical resource base is endowed with vast stretch of flat lands of fertile soil becoming a potent resource of the region. The agricultural landscape is dominated by rice-jute culture. Fishes are the common diet of the people. The perennial rivers apart from supplementing the diet of the people through fish catches are also potential sources of irrigation. Water borne connections are facilitated by movements of tides since ancient times. Bay of Bengal provides connection with outside world. Forest based industries for indigenous woodworks and paper mills is provided by 'Sundari trees' of Sundarbans. Wild life is preserved alongwith scenic natural landscape and basic industry minerals have helped stimulate the development of large scale industries.

CULTURAL BACKDROP

The cultural backdrop of this region depicts a very complex picture of population, economy and transport systems. The region is rather densely populated among all the regions of the country but the distributional unevenness is well-marked at the sub-regional level. The main population thrust is concentrated in the Delta region covering only two-fifths of the area. While initially human occupance in the region started with rural agglomerates based on agriculture, the region has had a significant increase in urban population. The two major areas of urban concentration are the industrial and commercial Hooghly side and industrial aim mining Asansol-Durgapur complex. This alone accommodates about four-fifths of the region's urban population. The region has a very low proportion of workers due to lack of technical and skilled labour and high population density. But in spite of urbanization trend, agriculture still contributes substantially to employment of the total working force including cultivators and agricultural labourers alongwith household industrial workers and workers in other services. However, recent trends show a decline in agricultural worker population and an increase in other working classes showing a tendency to become more industrialized and urbanized.

Trade, commerce and industry have contributed to the growth and development of urban centres in this region. Many towns like Arambagh (Hooghly), Murshidabad, Barinpur (24-parganas) and Ondal (Burdwan) have gained importance due to development of trade, commerce and transportation. The economy of the Lower Ganga Valley region has been adversely affected due to decline of trade through Bhagirathi because of the partition of Bengal. The region has around ten urban agglomerations of Class-I cities. Largest area of influence and positive growth are main characteristics of the towns and cities situtated in the Southern region. In Calcutta this influence is overtapped in towns of North Bengal Plain which is very weak due to an underdeveloped ecomomic base and lesser development of resources. Cities of Durgapur, Kharagpur, Kalyani in the Rarh Plain and Delta region reveal a transmissional stage of economic development. Shantipur, Krishnanagar and Nabadip are notable centres on trade and communication routes. Urban centres have grown and emerged in most cases as a consequence of successive accrual of non-agricultural activities around pre-existing agrarian village in an unplanned manner.

Agriculture comprises more than fifty per cent of the total working population as around seventy per cent of the total area of the region is under cultivation. North Bengal Plain is primarily agricultural. Double cropping has been common practice since the 1950's. 'Aus' land is utilised for the purpose for growing vegetables and rabi crops, a practice popular in the delta districts of Hooghly, Nadia and Murshidabad. Irrigation facilities in the region are inadequate in spite of average rainfall being fairly high. Irrigation is necessary for agricultural prosperity and adverse water balance in the Lower Ganga Plain region. The completion of Damodar Valley Corporation and Mayurakshi Projects provide large scale irrigation. Canal irrigation by government is popular in Birbhum, Bankura, Hooghly and Murshidabad. Pond irrigation is more popular in Murshidabad, Midnapur, Howrah followed by Bankura and Birbhum. Well irrigation is not very popular except in Nadia and Malda. The most important crop in the region is rice covering three-fourth of the net sown area and around 85 per cent of the total cropped land. 'Aman' variety of rice dominates throughout the region while 'Aus' is significant in Malda, Nadia, Bankura and Murshidabad. In the marshy tracts of Howrah, Hooghly and 24-Parganas 'Boro' is cultivated on hardly 1 per cent of cropped area. Jute is the cashcrop with 4.3 per cent of cropped area and stands second in importance. Apart from this, pulses and Til in western margin and vegetables in sub-urban zones are other important crops. Groundnut is also grown in 'Danga' land used as a pasture. The cropping pattern in this plain is stable due to level land and rotational pattern. Two types of rotations are followed (i) 'Aman' - fallow - Aman system in lowlying fields and (ii) 'Aus'-Jute-rabi pulses or vegetables system on the upland fields. While there is little scope in 'Aman' fields for crop rotation, several rotational practices can be developed over 'Aus' fields where maize, potato, jute, vegetables and rabi pulses are grown and other irrigation facilities available. The landscape of this region has been rated as fair and medium on the basis of yields of various crops. Fisheries in the region can be classified into culture fisheries in tanks and ponds, estuarine fisheries, riverine fisheries and marine fisheries. It is probably a result of certain environmental conditions, preponderance of water bodies, natural and artificial, acute shortage of agricultural land and lack of other resources of animal proteins. Fish culture forms a traditional occupation of almost all households in the region, more prominent in Delta Proper and Southern Barind tract. Common culture fisheries are Rohu, Catla, Mrigala and Kalbasu. Riverine fishes are important in 24 Parganas, Murshidabad and Midnapur accounting for more than 90 per cent of riverine catch. Diamond Harbour, Kolaghat, Itindaghat and Port Canning are the main collecting and transport centres. Most of them are consumed locally. Calcutta is most important commercial market for 'Hilsa.' Sundarban areas are the favoured sites for estuarine fishery. The main area being the Hooghly-Matlah estuary. Due to lack of equipment and adequate craft and berthing facilities only 32 km of the 65 km coastline between Junput and Digha is used comprising catches of prawns, quality fish and low grade fish. Fish catch of the area does not meet local consumption so fish has to be imported from other regions including Bangladesh. Fish movement is provided by railways which finds it difficult to cope with increasing demand.

As a region rich in materials, coal, water manpower, minerals and other raw materials the main industrial concentrations have developed in the Hooghly side, Asansol Durgapur belts which account for nearly all of the total industrial employment in the region. Textile industries are located in 24-Parganas, and Nadia and smaller belts are in Bankura and Murshidabad. Calcutta's industrial belt employs about 80 per cent of the total workers in delta region in jute textiles. Rubber products are manufactured in Calcutta, 24 Parganas and in Howrah and metal industries are also concentrated in the region. Iron and steel industries employ highest percentage of metal industry workers. Aluminium and steel industries also exist alongwith wood work, leather, transport equipment, food processings, etc., concentrated in different parts of the region. North Bengal Plain accounts mostly for beverages, tobacoo, food processing and wood based industries. The Hooghly industrial belt is a zone of diverse industries but Asansol-Durgapur belt is famous for metal and chemical based industries. The industrial landscape consists of regional disparity due to uneven distribution of industrial resource base and under utilization of economic infrastructure.

The region has a fairly good transport network catering not only to international movement of goods but also inter and intra-regional traffic *i.e.*, jute, tea, coal, cereals and other industrial products. The riverside centres have experienced deterioration due to partition disrupting the river transport, more stable surface transports and oscillatory channel characteristics. The region has highest density of rail roads in the country and Calcutta-Howrah is the main convergence of the railways. Improvements need to be made by bridging up interrupted links and modernisation of tracks. There are five national highways

(Nos. 2, 6, 31, 34, 35) including Grand Trunk Road (NH-2). Dum Dum puts the region on the air map of the country being a national and international airport. It also provides air connections to Meghalaya, Tripura, Manipur, Nagaland and Assam. Problems in the region are acute. Being a region endowed with rich alluvial soil, raw materials, perennial streams thus attracting a large number of people for livelihood create population menance. Problems are physical, natural and man made.

Natural problems relate to floods from Ganga and other riverine systems in the region amounting to colossal loss of property, cattle and humans. Recurrence of floods is a regularity causing heavy damage. This recurrence can be attributed to increasing human agglomerates leading to heavy deforestation in upper stretches and silting in estuarine zone.

Industrialization and economic development in Hooghly region has led to complex problems created by man himself. Pollution caused by various industries of jute, paper and pulp, textiles, tannery, thermal power station, oil, petroleum, steel, chlorine, rubber, chemical and insectides, discharge their wastes into the river. Domestic and municipal wastes also add to the pollution load. Waste-water also affects biotic life by altering food chain and suitability of river bed for spawning and disturbs the fisheries in the region.

Chapter 2

Profiles in History of Ganga Civilization and its Cities

The rise of the Ganges civilization can be traced down from the records of Ancient Indian literature to the early fifth and sixth centuries B.C. records of the early Buddhist period. The Ganges valley remained a large jungle till small settlements proliferated into village communities landowners and peasants. The growth of population, production and distribution of goods and articles accelerated the growth of trade and commerce.

Due to a number of navigable waterways river commerce flourished greatly. In Jain texts Kampilya, Banaras and Champa representing the upper, middle and lower Ganges were counted among the ten ancient capital cities of India. By the 4th century B.C. Pataliputra began to dominate the entire Ganges basin. Around the sixth to early third century B.C. metal currency began to appear in the Ganges valley and Buddhist literature gives detailed account of traffic along the Ganga in the Jatakas. Building of houses along the river banks, cutting wood for ships and boats was a common practice during the times. The Mauryas brought about the unification of the Ganges valley with the region from Hardwar to Tamralipti falling under the rule of Ashoka. Pataliputra, the capital of the Mauryan State was a great centre for commerce and goods and the city according to Megasthenes extended nine miles along the Ganges surrounded by high wooden ramparts fitted with gates and military placements. The Arthashastra describes vigorous economic activity calling for a superintendent of ships. Thus Ganga at that time was the centre of Indian civilization. Thus maritime tradition in northern India continued in spite of political disruption in the region. Throughout the early centuries of the Christian era internal and external ocean trade continued. Towards the end of the first century A.D. the warriors from Central Asia, the Kushanas entered India and took control over part of the Ganges valley—Mathura, Pataliputra, Banaras were all in the scope of Kushan dominions. They were overthrown by Murandas in the mid-third century and this tribe has its capital either at Pataliputra or Vaishali.

Between 200 B.C. to A.D. 320 the political map of the Indo-Gangetic Plain was in period of flux and instability but commerce on the river flourished and with the rise of Guptas in 320 A.D. the valley became once more peaceful. During this time according to Fahein, the Chinese traveller, 405-411 the cities of Pataliputra, Banaras, Pundra region in Bengal flourished and traders and merchants appear throughout in literature of the times.

The Gupta empire disintegrated by the middle of the sixth century and parts of the valley were held by emperor Harsha at Kannauj on the upper Ganges where cities like Prayag, Banaras and Tamralipti thrived. Hieun Tsiang talks about these and Hardwar.

After Harsha, the Ganga plain could only be reassembled after about five hundred years. Throughout the medieval period upto the nineteenth century the river served as the main waterway of long distance trade and transportation. This Indo-Gangetic Plain was the political epicentre of great dynasties of Ancient and Medieval India.

A string of cities from Texila and Pushkalavati through Indraprastha, Mathura, Kausambi, Prayag and Pataliputra down to Tamralipti were part of the ancient Aryan Kingdoms found before 500 B.C. Pataliputra was known for Ashoka the great and Vikramaditya, Kannauj gained prominence with rise of Harsha's empire. The Sultanate empire, Khiljis, Lodhis and Mughals all added to the galore of earlier cities. Allahabad,

Mirzapur, Jaunpur and Murshidabad were new centres on the Gangetic urban scene. While the Hindus continued to workship her, foreigners used it as a major channel of communication. The Danes, Dutch, French and British sailed up the Ganga and Hooghly with towns like Calcutta, Chandranagore, Serampur becoming new landmarks. They set up their offices on the banks of Ganga and later from traders became rulers. After Calcutta came into being around 1690, Lord Clive brought around 200 boats to the city carrying royal exchequer of Murshidabad after battle of Plassey. The Ganga, thus, became a busy river and so many ships started plying that a ship building and repairing factory was set up in Calcutta.

The Britishers used the Ganga not only for trade and commerce but also for leisure and a source of entertainment. They strolled along the banks of Ganga for a breath of fresh air. The Britishers beautified the river banks and many Europeans built garden houses along the Ganga. They would also cruise along the Ganga to overcome the heat and have music and parties during the cruise.

The Britishers were the first to realise and emphasize the important of a clean Ganga. John Strachy, President of the Sanitary Commission of Bengal Presidency in 1864 wrote that around 5,000 bodies were found floating in a river whose water was used for drinking and cooking by the people of Calcutta. Showing concern over the innumerable bodies floating the English under the East India Company took the initiative of cleaning the river by setting up a Calcutta Municipality. However, the resolution adopted, opposing the practice of dropping dead bodies into the river was under severe protest by orthodox Hindus. Even the effort to meet the expense of burning bodies by public donations and a charge of Rs. 2/- by relatives for cremation was also met with great opposition. After the regime of Queen Victoria the Governor of Bengal Sir Cecil Beadon proposed shifting of two burning ghats in Calcutta from the banks of Ganga. But this too was vehemently opposed as interference in Hindu religion. However, they did succeed in making arrangements for dumping the city's wastes and carcasses at Dhapa, adjacent to Calcutta, instead of throwing these into the Ganga. Thus, while political fortunes fluctuated and dynasties changed, the Ganga continued to flow with its town becoming centres of trade, commerce, craft and culture. Expansion of river remained no longer a source of water but channel of communication and receiver of waste of the city.

Thus, Ganga has not only a place in Hindu mythology and literature but reflects the culture and heritage of India from ancient times.

GANGA, CRADLE OF CITIES

A swift moving stream from Gowmukh, the Ganga grows into a wide river just before it touches the plains at Rishikesh. It begins a 2000 km long journey henceforth and the process of pollution is induced through the heart of India—Indo Gangetic Plain before meeting the Bay of Bengal.

Before the Kumbha Mela in 1986, untreated waste and sewage flowed directly into the river from the numerous Ashrams, hotels and buildings along the river at Lakshman Jhula and Swargashram. It was then decided to lay a deep sewer. Waste water was diverted for treatment and the Rishikesh Sewage System was revamped and its pumps at Mayakund were taken up. Existing oxidation ponds at Lakkarghat recommissioned. A sewer was also laid at Muni-Ki-Reti and additional septic tanks were provided at Lakshman Jhula.

HARDWAR or the Gate of Vishnu etmologically means Hari a name of Vishnu and dwara, the Sanskrit word for door or gate, is one of the most sacred places for the people. The town is always in chaos with millions of people moving around even in the coldest months. The pilgrims gather here mainly from May through September during their long journeys to Badrinath, Kedarnath and Gangotri. Ganga water considered the holiest is filled in pots by men from all parts of the country and taken home. Even emperor Akbar according to Abul Fazal's records, used to drink the Ganga water both at home and during his travels. The life of Hardwar revolves mainly around its ghats. Millions of people bathe here in this holy river on special and holy occasions. During the time of Kumbh Mela in 1986, 70 lakh pilgrims were to bathe in pollution free waters. The problem had to be stemmed as 11 million litres of waste water flowed into Har Ki Pauri, Subhas Ghat and other sacred locations of the Ghats each day. Alongwith it the defunct pump houses of

Mayapur and Bhimgoda added to this problem. There were also problems posed by three industrial units in the city—The Indian Drugs and Pharmaceuticals Limited (IDPL), The Bharat Heavy Electricals Limited (BHEL) and Hindustan Antibiotics. The sewage system covered only 20 per cent of the city and the rest flowed into open drains falling into the Ganga. The pollution in Hardwar was so great that the BOD—a measure of organic pollution was in excess of sanctioned official levels. There were no fish in the river at Hardwar. Since Kumbh Mela was held in the city it was decided that action be taken to intercept and divert the waste water for the time being. Thus, eight out of ten drains flowing into the Ganga were stemmed. Hardwar's sewage system renovated and massive repairs on four sewage pumping stations were conducted and in some cases replaced. Construction of a sewage treatment plant and laying of outfall gravity sewers was also taken up under the Action Plan. After the Mela, the sewage was taken to a sewage farm where after treatment the sewage was used as manure while rest to generate biogas to help run the sewage treatment plant itself. A 17 lakh afforestation programme began on the occasion of the Kumbh Mela as it was felt that unless soil erosion was curbed the pollution of the river could not be controlled. Pilgrims were encouraged to plant trees and participate in the clean Ganga Project.

Organised pollution actually begins from Hardwar, one of the seven most polluted cities on the banks of the Ganga. The others are Kannauj, Kanpur, Allahabad, Varanasi, Patna and Calcutta. The industries between Rishikesh and Hardwar do not have effective effluent treatment plants. Fish have disappeared and thousands of people bathe on religious festivals like Kumbh Mela, Baisakhi and solar or lunar eclipse. Even in Rishikesh the samples collected by the U.P. Pollution Control Board indicate that water quality is not satisfactory while there is no significant entry of waste water from industries, quantity of small waste matter outlets from bathing ghats, temples and habitation cannot be ignored.

KANPUR is one of India's earliest industrial towns with varied industrial activities ranging from textile, engineering and leather works. In the early times Ganga was the only source of transportation as well as the sewerage. Kanpur is the second largest city along the Ganga but the city and adjoining areas are partly sewered. Many sewers are choked and inoperative. Sludge and sewage is carried into 13 storm water 'nullahs' for final disposal into the river. The Jajmau pumping plant was worn out. Around 70 tanneries in and around Jajmau mixed with domestic and highly toxic tannery wastes dumped their wastes into the river. The result was a severe degradation of water quality within the limits of Kanpur city. Kanpur is the most polluted town in U.P. with a large number of big and small industries and about 17.5 million residents, water itself is a scarce commodity in the city and tube wells are sinking. With increasing urbanisation and commercial establishments quantum of industrial and domestic wastes has also increased. Studies have revealed that industrial wastes from tanneries, textiles, industries and untreated domestic and human waste have contributed to increased pollution at a high level in Kanpur. Effluents of tannery affect that quality of water at "an alarming proportion". The BOD was lowest at 2.86 mg per litre at Bithoor ghat and highest at Benia ghat 30.33 mg per litre (Bannerjee, 1989).

Under the GAP the pumping station at Jajmau has been renovated and its capacity expanded. A major 130 MLD treatment plant is built and an integrated sanitation scheme for Jajmau tannery area to treat wastes, improve drainage and garbage disposal has been initiated. Tanning industries have been asked to set up effluent treatment plants before discharging their wastes in the river directly or through sewers. The U.P. Pollution Control Board's office at Kanpur has taken stringent measures for implementation of these objectives. Many tanneries not following these rules have been closed down. Penalties have been imposed to make them aware of their duties towards the community. Unfortunately the river has been reduced to a mere trickle since it has shifted its course and started to flow eight kilometres away in the district of Unnao.

ALLAHABAD or Prayag is the confluence of the rivers Ganga, Yamuna and the mythical Saraswati. The Sangam city is of great religious significance. The Rigveda indicated the importance of this city in following words:

Sitasite-Sarite-Yatra-Sangate
Tatralupta so Divmutpatanti
Ye Vai Visrajanti Dhira
Stejanasomritatvam Bhajante

At Allahabad where the white and black rivers (Ganga and Yamuna) meet, the journey to that place is believed to take you to heaven, and a person who dies at that place is immortalized.

This city does not face the problem of industrial wastes or effluents but has mainly domestic waste. Apart from this, thousands of pilgrims flank the banks on religious occasions. It is also the place for Kumbh Mela, held from time to time. The Gaughat pumping station at Allahabad has been renovated and a sewage treatment plant at Naini is constructed. Efforts for extensive renovation of the sewer networks have also been taken up in the city.

VARANASI a city known for keeping alive India's culture and heritage is one of the oldest inhabited cities. It is a leading centre of religion, philosophy, music, dance, arts, crafts and higher education and knowledge. In the city of Varanasi the river and the city are indistinguishable. The two cannot be studied separately and one exists because of the other. The river in its infrastructural facilities had no treatment plants. Eight big drains, many small ones continued to pour the city's sewage into the river. Most of its underground drains are around 200 to 300 years old. This ageing sewrage and drainage system has virtually collapsed with growing stresses of population and growth on the city. Majority of Indians still regard Varanasi as a pilgrim city. Varanasi everyday receives estimated about 70,000 pilgrims. People bathe in the river, take a round of temples and carry the water home in copper containers. Around 60,000 bodies are cremated on two ghats and 9,000 dead cattle thrown in annually in Varanasi. Pyres continuously burn day and night at ghats. In consonance with religious sentiments many half burnt bodies can be seen floating and bodies of infants and Sadhus and bodies of snake bites are not cremated but consigned to the water. At the Manikarnika and Harishchandra Ghats around 30,000 pyres are lit every year and its ashes are dumped into the river. Tourists and pilgrims also aggravate the sewage disposal and water problems in the city. Tradition and religion has choked the sacred river in Varanasi accounting for over 23 per cent of U.P. pollution load in the river. The organic debris of animal and human corpses get stuck in the bend and provide food and carnivorous birds. This bend of the Ganga flows northwards at Varanasi. Stench and decay are part of these ghats. Apart from the city being a place of pilgrimage the city has grown with the rise of industries, locomotive works and enormous interstate traffic. Certain industries situated upstream of Varanasi pour waste from coal yards, paper mills, glass works and insecticide firms. They have changed the infrastructure of the city but at times attempts to clean the river by administration were resisted by ignorant and superstitious fundamentalists nursing vested interest in the *status quo*. Varanasi is a city used intesively at the river front and pollution tends to be higher near the banks. In the lean season there is very heavy pollution due to stagnation of flow.

The GAP in Varanasi city development package has allocated Rs. 43 crores to tackle the pollution in an integrated manner. The Integrated City Development Programme for Varanasi seeks to address the problems of river and city and it made the old treatment plant at Banaras Hindu University functional and expanded its capacity. Two more diesel Locomotive Works at Konia and Dinapur were constructed. The sewage pumping stations at ghats have been renovated with sewer line. Construction of Sulabh Sauchalayas, and renovation of kunds, construction of a few important roads, improvement of lanes near ghats geared up. Special arrangements were made to deal with the problems of throwing dead bodies and animal carcasses. Five hundred tonnes of garbage everyday is collected and disposed. Regular monitoring of Ganga water has been taken up to improve the conditions of this historical city.

PATNA has emerged as another great and historical city of Pataliputra and is the capital of Bihar today. With the population of about a million people the main source of pollution is the city's sewage. Pataliputra the capital of the former Mauryan State was a flourishing centre for commerce and trade. Vigorous economic activity from its banks was the nerve centre of growth and development. Today around fifty million litres per day of sewage flows into the river. One of the largest sewage plants built at Sardpur—Patna lay defunct for more than a decade.

Under the GAP they were recommissioned in June 1986. The plant at Beur was also restarted. With funds from the GAP, the Bihar State Water and Sewage Board took up renovation and construction of pump-houses and main sewers in different parts of the city. Developing the river front has also been taken up by the administration.

CALCUTTA a metropolitan city and the capital of West Bengal is a city where growth and pollution are synonymous to each other. Apart from Calcutta and Howrah there are 37 towns and nearly 150 factories using the Ganga (Hooghly) as their water resource. While the urban, domestic and industrial waste waters generated in the Ganga Basin in West Bengal carry about 445 tonnes of BOD per day, in Calcutta metropolitan district, effluents generated from different municipal and industrial sources are directed into various channels. All the effluents are not allowed to fall directly into the Ganga (Dutta, 1986). Around 42 per cent of effluents are actually drained into the Ganga.

The schemes under the GAP are essentially sewerage jobs. The CMDA and other agencies of the West Bengal Government have devised a number of schemes to deal with the problem. Calcutta is a heavy pollution point due to large number of industries on this stretch of the river bank discharging effluents without any prior treatment. A report of the Central Inland Fisheries Research Institute revealed that fish caught in the Ganga were highly contaminated with trace elements of zinc, cadmium, mercury, thrown in by industries. Owing to pollution about 80 per cent of people in Calcutta suffer from amoebic dysentry and other infectious diseases like cholera, diarrhoea and dysentry. The Hilsa fish is becoming extinct and the crisis of the Calcutta Haldia Port Complex is deepening.

In Calcutta the prime supplier of water is Ganga. The most important task taken up by CMDC and other West Bengal Government Agencies has been renovation of existing sewers and treatment plants at Titagarh, Serampore, Bhatpara and Chandanagar to prevent overflow of wastes and effluents of acceptable standards of pollution. Construction of interceptors to divert flow of sewage to these plants for treatment was taken up. Low Cost Sanitation Schemes have been floated and the number of electric crematoria is increasing. River front development, soil erosion schemes, stopping of filling pukurs (ponds due to shortage of land) are also part of the cleaning programme.

Since the principal aim of the GAP in the first phase was to immediately reduce the pollution load on the Ganga, the areas covered under first phase included those to be taken up by Calcutta Municipal Corporation. A series of steps in different sectors through a variety of agencies or a multisectoral approach of the CMDA was expected to solve the problems of this great metropolitan city.

No other river in the world evokes such awe and reverence in the world as GANGA, also referred to with love as 'Mother Ganges' and yet man's devotion from times immemorial could not save the river from pollution. Operation Ganga, therefore, is a gigantic and difficult task. The great river is part of our civilization and culture. Except for the mountainous region the river is navigable in most of its length. Agriculture and ancient settlements flourished along its banks giving rise to the Ganges civilization. Rising from the Himalayas, entering the plains, influencing the lives of millions of people and merging into the sea from Gangotri to Ganga Sagar it passes through the shores of the most populous States of India—Uttar Pradesh, Bihar and West Bengal.

The Ganga Basin reflects certain features. Springing from hilly forested terrain the water flow is in the form of numerous surface rapids, but as it moves down the plains it slows down depositing minerals making land fertile. But in its final leg it transforms into a delta. Very slowly and sluggishly it moves till it mingles with the sea.

With increase in population in the region above the Upper Ganga Plain there is no obvious pollution. The concentration is on deforestation, especially, in catchment areas of the Bhagirathi, Yamuna and Shillanganger. This has resulted in floods in the river downstream.

Passing from Hardwar, Kanpur, Allahabad and Varanasi the river increases its pollutant input load. City sewage is a potential threat to our water resources. Hundreds of factories in most States pump out industrial wastes and effluents without treating them. U.P. alone contributes to about 55 per cent of the total pollution of the river. After crossing Meerut, Bulandshahr, Badaun, Moradabad and Etah districts the river receives domestic and industrial effluents upto Knnauj where the East-Kali a major tributary—meets the river. The tributary adds its pollutant load of Western U.P. industrial units along with wastes of dyeing industry and saree printing units localised at Farrukhabad and affects water quality but sufficient self-purification capacity of the river maintains its water quality. Water upstream Kannauj is not so polluted as there are no significant industries and less domestic waste water flows in the river.

The principal right bank tributaries of the Ganga in the State of Uttar Pradesh is Yamuna, the river flowing past Delhi and gravely polluted. Yamuna river joins the Ganga at Allabahad. Tonnes descends from Vindhyachal mountains in Madhya Pradesh and joins it soon there-after. The left bank tributaries in the State are Ram Ganga, Gomati and Ghaghra.

In the State of Bihar the main tributaries from the Himalayan region in the North are Gandak, Burhi Gandak, Ghugri and Kosi and its Southern tributary is the Son.

In West Bengal the last Indian State, ranking next to U.P. in terms of pollution load, the river becomes sluggish and is described as 'dying'. Tidal forests and swamp land exists towards the seaward side of the Delta. The forests of Sunderbans are protected by India and Bangladesh. The river Mahananda joins the Ganga from the North. The Western tributary of the delta is the Hooghly on its East bank stands Calcutta. In Bangladesh Ganga is joined by Brahmaputra and then is known as Padma. The waters of Padma joined by Meghna river at Chandpur flows into the Bay of Bengal. The length of the major tributaries of the Ganga in miles are Yamuna (860), Ram Ganga (428), Ghaghara (600), Son (487), Gandak (263), Burhi Gandak (378), Kosi (450), Mahananda (180), Damodar (368) and the Brahmaputra (1800). All join the river Ganga and pour their pollutant input load adding to the already polluted river water. An effort to keep the tributaries also free from pollution can only help to retain the Ganga clean. An integrated plan of action needs to be announced giving due attention to this aspect.

Admittedly, Ganga is a symbol of national unity and mother of Indian civilization. If it was polluted because of its ruthless exploitation, efforts towards cleaning the river under an organised programme action and follow up will help relieve the cities of the problems they are facing. Supply of healthy potable water free of bacteriological germs, toxic substances but with required minerals and oxygen has to be achieved through a joint effort of governmental, public and non-governmental organizations in unison. An integrated thinking has to be developed for sustained healthier water and healthier life.

Chapter 3

Ganga—A Prismatic Glimpse

GANGA, THE LEGEND

The descent of Ganga to the Earth has been attributed to Saint Bhagirath, descendant of Sagar who performed penance to induce Ganga to come down from heaven and purify the ashes of the sons of Sagar cursed by Kapil Muni and attain Moksha (Ramayana 1:42). On his failure Brahma advised Bhagirath to request Siva who alone could bear the burden of the falling of Ganga from the Himalayas to the plains. He prayed and accordingly Siva ascended the Himalayas and called upon Ganga to come down. Siva held the Ganga in his matted locks and saved the world from its impact. Ganga according to the legend was initially hesitant to come down as she feared that people would wash their sins in the river.

However, the Goddess Ganga did come down and Bhagirath mounted his chariot to lead the Ganga across the plains. Before merging in the sea Ganga purified the mortal remains of King Sagar's 60,000 sons, liberating their souls once and forever. Thus, Ganga for the Hindus is the river of faith, devotion, worship, health and purifier of sins. Despite the increased pollution Ganga still enjoys a sentiment of religious fervour and holiness for real believers even today. People bathe in the belief that all their sins are washed away. Lord Krishna says in the Gita "I am the Ganges among rivers." Ganga is considered most sacredly and pollution of its water hurts Hindu sentiments as was demonstrated in the Banaras riots of 1809 when Brahmins did not cook while sailing on Ganga, wash their clothes or throw saliva calling her with reverence as Gangamai (Mother Ganga). Held as a goddess her banks are thronged with millions of people from all parts of the world just for a dip in the waters.

Besides bathing there are certain religious myths and beliefs attributed to Ganga as under:

(i) Bathing at both lunar and solar eclipses removes all sins. As a result great bathing fairs—Kumbha melas are held.

(ii) Hindus believe that Ganga water has life-giving properties.

(iii) Ganges is the form of Vishnu and she came out of the Supreme Being.

(iv) Cremation of the dead at the banks of Ganga is a domestic ritual among Hindus. If death occurs at a distance ashes are consigned to the waters of Ganga. With the progress of civilization came up cities and towns along the river and all the dead were brought in the hope of salvation and burnt at the ghats.

(v) In Gaya the first rice cake (Pind) is offered to the dead ancestors to make them happy.

(vi) The Ganga water is carried to all parts of India to be used on occasions to ward of evil spirits, mixed in water tanks, put in mouth at the time of dying, sprinkled in marriages, shradhs and other Hindu auspicious occasions. The demand for the waters was so much that Ganga water at one time became a commercial item sold in the 19th century by a Calcutta businessman Baishnab Charan Seth who sold Ganga water in sealed earthen pots and even noblemen and kings used to buy 'Ganga Jal' from him.

(vii) Sick are taken to die at its banks to attain Moksha and entry to Heaven.

(viii) Saints and sages perform 'Tapas' and 'Anushthan'. They concentrate and meditate to realise spiritual vibrations, withdraw mind from physical world to regions of immortality.

(ix) Some Hindus practised the ritual of 'Antaranjali'—submerging the dying, waist downwards in the river in the belief that dying, in this state would be peaceful and the soul would live in peace.

But all dead are not burnt, some are just touched with a stick and submerged in the river. Poor people also follow the practice of dumping bodies as they cannot afford it. Bodies of lepers, animals are also immersed. Even today bodies of saints and sages are immersed into the river, a ritual known as 'JAL SAMADHI.'

(x) Water has curative powers. Rich in minerals it can cure almost all kinds of diseases with its remarkable germicidal qualities. Even scientists have admitted the efficacy of the Ganges water.

Dr. F.C Harrison of McGill University, Canada writes: "A peculiar fact which has never been satisfactorily explained is the quick death, in three or five hours, of the cholera vibrio in the waters of the Ganges. When one remembers sewage by numerous corpses of natives, often cholera casualties, and by the bathing of thousands of natives, it seems remarkable that the belief of the Hindus, that the water of this river is pure and connot be defiled and that they can safely drink it and bathe in it, should be confirmed by means of modern bacteriological research."

A well-known French physician, Dr. D. Heralle, made similar investigations into the mystery of the Ganges. He observed some of the floating corpses of men, dead of dysentery and cholera, and was surprised to find "that only a few feet below the bodies where one would expect to find millions of these dysentery and cholera germs, there were no germs at all." He then grew germs from patients having the disease, and to these cultures added water from the Ganges.

When he incubated the mixture for a period, much to his surprise, the germs were completely destroyed.

A British physician, Dr. C.E. Nelson, F.R.C.S., tells us of another striking fact. He says that "Ships leaving Calcutta for England take their water from the Hooghly river which is one of the mouths of the filthy Ganges and this Ganges water will remain fresh all the way to England. On the other hand, ships leaving England for India find that the water they take on in London will not stay fresh till they reach Bombay, the nearest Indian port, which is a week closer to England than Calcutta. They must replenish their water-supply at port Said, Suez or at Aden on the Red Sea." It is no wonder that the Indian people should hold that the Ganges is very sacred and possessed of mysterious powers. Unfortunately this self-regenerative power of river is eroded due to increase in pollutants drained into river.

(xi) The Srimad Bhagwat Gita (8/21/4) refers to the holy Ganga by saying that after making a journey through space it reaches the earth and in process purifies the three *lokas*—the three major divisions of the Cosmos.

GANGA IN HISTROY

Historically, people of all faiths, various professions have drunk the waters of Ganga. Not only Hindus but as Abul Fazl mentions in Ain-A-Akbari even emperor Akbar preferred the water of Ganga and a separate department was constituted to ensure, fresh and regular supply to the king. Akbar's successors Jahangir, Shah Jahan and Aurangzeb continued that tradition.

Great scriptures like the Vedas, the Upanishads and works of Kalidas and Kalhan were all composed in the Indo-Gangetic area. Jawaharlal Nehru called the Ganga a noble river. On his wife Kamala's death he poured the ashes in the bosom of the river and wrote in "Discovery of India"; "How many of our fore-bears she has carried thus to the sea, how many of those who will follow us will take that last journey in the embrace of her water."

While the influence of Ganga on the religious and cultural life of India is immense Ganga is not just a legend. It also forms the life line to India's people. Constituting more than 37 per cent of the population of the country the Ganga basin is also extensively cultivated with a variety of crops and other economic activities. Its waters drain through eight States: Haryana, Himachal, Punjab, Bihar, Uttar Pradesh, Rajasthan, West Bengal, Madhya Pradesh and Union Territory of Delhi. Apart from several major pilgrim centres like Allahabad, Rishikesh, Hardwar, Gangotri and Varanasi many important industrial towns are on the banks of the Ganga *e.g.*, Kanpur, Calcutta, Patna and Allahabad with large population where Ganga serves a source of water supply.

PHILOSOPHY OF GANGA

Ganga evokes a sentiment of mind—a reason for love beyond. Swami Lokeswarananda (1986) writes in his detailed treatment about the philosophical attachment of Indian people with Ganga as a legend. Ganga is not loved by Hindus alone but Non-Hindus of varied sects and religions nurse a sentiment of religious fervour and sentiment. Ganga symbolises love of Hindustanis—a composite culture of India. All this is just because she is Ganga—a river a par excellence. It is indeed difficult to explicit in words the charm of Ganga. Ganga worship has never withered through the ages despite scientific temperament attained pushing religion back into debunk. Even enchanting youth of the river has never haggerred with age. Foreign scientists are foxed to know that even scientists take dip in the Ganga on auspicious days. A British scientist was bewildered with this kind of prevailing superstition despite scientific progress not only elsewhere but even in India. But Ganga sentiment is beyond scientific truths or conservative myths. Interestingly, a non-Hindu Russian scientist plunged into Ganga when he was appraised of the holiness of the river. It is difficult to predict how the British scientist would have reacted seeing the love of Russian scientist for Ganga beyond reason.

Ganga evokes a picture of holiness, reverence, peace, pleasure, joy and beauty, and reflects the symbol of something higher than any other river. If Himalaya could be symbolised like a

'Yogin" in meditation Ganga gushes from its heart as culmination of meditational elixir. Ganga is as such a blessing of God bringing good to the righteous people and curse to the wrong doers. Ganga provides physical and spiritual nourishment and is the life and soul of India. According to Sri Ram Krishna the renowned saint, the river is believed to form a bridge between heaven and earth and her water flows as divinity. The industrial pollution could pollute the water but its sanctifying powers still remain intact for Ganga believers. The cradle of pilgrimage cities Hardwar, Prayag, Kashi, Gaya, Rishikesh, Uttarkashi, Gangotri have always attracted rich and poor alike from nooks and corners of the country. Kashi is rated as the queen of eternal cities because of its ambience of ancient and modern attributes.

Ganga etymologically means which leads to the Ulimate Reality—the Brahman. She is the sanctifier and liberator of soul from worldly sensual experiences. It dispels ignorance and helps discovery of the eternal truth that Brahman and mortal human soul is one and the same. This realization is the truth of supreme knowledge of liberation. This liberation comes when she arouses the urge to struggle for self-knowledge. The first mention of Ganga comes in Rigveda. Krishna in the Gita identifies himself as Ganga, and Valmiki is perhaps the first sage to sentimentalise 'Mother, you girdle the earth and thereby lend it grace and beauty. You are the pathway, bedecked with banners of victory, to heaven. My only prayer to you is that I may live on your bank, drink your water, watch your waves, repeat your name in my heart forever and die with my eyes fixed on you.' Shankaracharya considers Ganga as the chief of all Gods and Goddesses and is 'the redeemer of the fallen'. Vivekananda carried the river water to the West with the faith that whenever in trouble a drop or two of Ganga water would soon be calm again. Saint and sages apart, Hindus are devoted to her in a genuine and sincere way. Nehru by no means a religious person also wished that his ashes may be consigned to the Ganga. Jawahar Lal Nehru considered Ganga's story as one of Indian civilization and culture. He remarked, "She has been a symbol of India's age-long culture and civilization, changing, ever flowing and yet ever the same Ganga." Both India and Ganga seem to be marching towards a common goal, the goal of truth. Truth is another name for infinity which the Ganga achieves when she merges in the sea. God and Truth are synonymous to a Hindu. Ganga is not only his mother who shields and rears him but also teaches and guides to make him a human being. Swami Vivekananda, said 'Gita and Ganga constitute the essence of Hinduism; one its theory and the other its practice.'

GANGA IN LITERATURE

In course of his writings over several decades Rabindra Nath Tagore made many references to the river Ganga. Tagore symbolised the Himalayas as a sage in one of his poems and refers to Ganga as the rosary of river which is the most sacred river of the country to millions of people from all walks of life and different religions. Apart from being sacred he considers Ganga as a beautiful companion. Ganga served to bind centuries of Indian culture. In 'Jiban Smriti' and 'Rabindra Rachanabal' he talks about the times he spent on the river, sailing and considers it as the sacred thread of Indian Yajna, a thread connecting memories of years of wisdom, religion and meditation. The flow of the river in itself is an introduction to India.

Ganga is the only river in the world that has a history rich in legends and myths, and also forms an integral part of the terrestrial, cultural and spiritual life of a nation. Ganga formed the lifeline of the ancient Aryan civilization about 5,000 years ago providing irrigation, fertility of soil, source of water, communication and transport, scenic beauty, and at the same time carrying the rubbish and human wastes, including half burnt bodies in her sweeping flow beyond pollution. Ganga was the glory to all successors, kingdoms and builders of new civilizations. Thus, Ganga became a symbol of fertility and a mother goddess that nursed, reared and sustained life. This invested symbolism has been held till today. Though unaware and ignorant of her utility her spiritual power remains still intact.

Such a river, therefore, struck the minds of writers, painters, poets, sculptors and their creativity led to reference of Ganga in literature. In Sanskrit literature, in the epics and the Puranas she has been portrayed in her most magnificent form. The Ganga has 108 names in Sanskrit and is said to have flowed in three streams, to be Trisotra or Tripathaga.

In the Bhagavat Ganga has flown out of a pore made by the nails of Vishnu's left foot. But whatever be her mythic origin she is seen as a young, beautiful, and sparkling maiden, vivacious, wayward and sometimes unpredictable. Padma Purana mentions her as a beautiful and lovely eyed maiden who became one of the wives of Mahavishnu. Pranatoshni Tanta describes her as of sixteen years always and surrounded by Brahma and others.

The Mahabharata describes Ganga's education of Mahabhishak who came to heaven after a great deal of penance to worship Brahma and fell to the charm of Ganga. It also relates to Brahma's curse that they would be born again on the earth and live as husband and wife. It also narrates Shantanu's (Mahabhishak's earth name) surrender to Ganga. Maithili folk literature also talks about Ganga. Story of Ganga seems at one plane of exile and on the other as a purifier of sins and liberator of souls in distress.

'Ganga Lahiri' is another poem written by Jagannath Pandit, a court poet of King Shah Jahan who fell in love with a Muslim princess, married her and was excommunicated by the community. The legend has it that while he sang the glory of Ganga, the Ganga opened her arms and took the lovers in her bosom. This poem has been translated into many Indian languages. Bhartendu the famous Hindu poet wrote Ganga-Varnan as part of his theme in 'Harsha Charita'. 'Ganga Taranga' was also another sequence written by Padmakar in his poem in Brijbhasha. Nand Kishore Mishra another poet of the age wrote 'Gangabharan' talking about the glory of Ganga as a mother. Janardan Das Ratnakar also wrote many poems like 'Gangavtara.' 'Kali Kashi' another epic Ratnakar could not finish, but Ganga a collection of devotional songs is on record. Surdas, Tulsidas, and other modern Indian poets have referred to Ganga for its scenic beauty, religous fervour or symbolic reflection on life.

Many Hindi novelists like Keshava Prasad Misra in 'Ganga Jal', Laxminarayan Lal Prem 'Apavitra Nadi,' Shiva Prasad Singh 'Gali Age Mudti Hai,' and 'Alag Alag Vaitarni', Nagarjun 'Varun Ke Bete' refer to Ganga directly or indirectly in their works. Jagannath's Ganga Lahiri was also translated by Marathi medieval poets. Vaman and Moropant wrote poems like *Ganga Prarthana, Ganga Vakili, Gangastava,* etc. In the 19th century, Bapu Saheb Kurundwadkar wrote a long poem called *Gangavarnan*. Other medieval Marathi poets also wrote on the Ganga like Tukaram, Jonarbai, Mukteshwar and Narahari. Musicals like *Jaya Ganga, Jaya Bhagirathi* have become quite popular. *Ambi Bhagirathache Putra* of G.N. Dandekar has made reference of identical nature of the holy river. Varied languages of different regions even deprived of Ganga's flow described its national acceptance and worship of people of India living in areas other than Indo-Gangetic Plain. Similies and metaphors acredit rainbow attributes of the river. This magnificence of description is not only a literary marvel but a fitting tribute to the holy river. Uma Shankar Joshi's poems *Gangotri*, Yone Naguchi's, The call of the Ganges and T.S. Eliot's wasteland are just a few out of a crowded galaxy of poets and writers who have chosen to write on the Ganga as an expression of their sublimation, love and reverence towards the river. Subramanyam Bharati of Tamilnadu, Vallathore of Kerala, Bindre of Karnataka, Maya Dharman Singh of Orissa, Biharilal Chakravarty of Bengal were drawn towards the beauty and holiness of Ganga and have expressed their sentiments in no uncertain words. Their works are rated in different Indian languages as a literary piece. *Mahaprasthaner Pathey* of Prabodh Kumar Sanyal and *Amrit Kumbhar Sandhane* of Samresh Basu are excellent expressions of Ganga worship. But no amount of literature is enough to present totality of the river as it seeks to evoke in the organic life of human beings devoted to Ganga Seva. The river seems to evoke ambience of the kind like Hardy's Egdon Health or Synge's Sea. Ganga symbolises a cumulative picture of history, geography and culture.

Ganga enjoys the stature of a deity which could never be installed in a temple in various parts of the country and her very existence is a temple for Ganga worshippers. A cradle of pilgrimage cities are beaded along the expanse of the river from Gangotri to Ganga Sagar. Sir J.C. Bose an eminent scientist burst into emotional prose and Shanti Swaroop Bhatnagar into Urdu poetry. But rhapsody on Ganga is from G.V. Desani in his Hatter who was a transnational. He expressed his sublimity of Ganga in various words "Blessed by her stars and the little crescent moon, caressed by her waves, the deathless spirit of the beautiful Ganga too sleeps." The silver ripples of the river appear to come out of womb of the night in the form of gushing stars. This is the mother Ganga also "betimes, like an exquisite woman, winning, over-powering, awe-inspiring". Ganga's trickle makes a music which fascinates through comely flower decked, dew laden, fair, dainty, elegant, gay and a lovely superb maid.

A river inspiring such ecstacy, diction and insight is both a heritage and a hope. It was Bhagiratha who brought her down to earth and guided its course. It seems only another Bhagiratha can save her from being rudiculed as a 'Polluted Nalla'.

FLORA AND FAUNA OF GANGA BASIN

Varied and fascinating accounts of vegetation, animals and birds, along the banks of the Ganga have been given by travellers. In the densely forested areas instances of tigers, bisons, buffaloes, rhinoceroses roaming freely are given in the 19th century.

Travellers indicated that in the Rajmahal Hills, the boundary line between Bengal and Bihar, many varieties of deer, hog-deer, wild buffaloes, patridges floukins, jungle-fowl, chukores and in certain open spots quails in great numbers were witnessed. Birds are found in great numbers flocking, perching and flying in every direction on the sand of banks. Crocodiles known as Makara are the vehicle of the anthropomorphic Ganga in Indian art. Alligators are a common sight in the shallows and the banks of the river but their indiscriminate slaughter on large scale for commercial gains has lead to the near extinction of the species in modern times leading to the announcement by the Government of India to be a protected species.

The river scenery in many parts presents such a picture to the travellers in which specimens of the gaour, resembling buffaloes but with a more finer coat, were known to feed themselves into the valley. It found way to the fair at Hardwar but attempts to tame it were unsuccessful. The animal life found beyond Gangotri has been described by Captain Hodgson the first European traveller to go beyond Gangotri. He described, bears in higher mountains, Boorul—an animal between deer and goat, Pheir an animal of similar kind also exist. Abundance of Monaul Pheasants and Chakors, are in existence near village where snow is found in the larger part of the year. In lower mountains black patridges, tigers, bears and leopards have been found.

The Sunderbans region was inhabited by tigers, leopard and rhinoceros. The birds of the aquatic kind include Indian Jabiru (a great heron, the Argali, Violet heron the Monichjore (small

Caboga white heron) Lohaugunj heron; the Indian crane, Stork, Ibis called at Calcutta, Junghil, the flamiengo, the gulls and terns, grey headed goose, pink headed duck, black bellid Anhinga and white pelican. While many trees are found in the course of their journey up and down the Ganga, the Banyan tree absolutely enchanted the travellers.

However, these varied accounts of travellers give us a glimpse of enriched region with nature's bounties in full freedom. But today with the growing demands of pupulation and more and more land under cultivation much of the original natural vegetation and a large number of animals have disappeared.

LIFE-LINE OF THE NORTH

The Ganga is among the first 10 mighty rivers in the world. A total population of 45 million people dwells in its basin. Average water run off amounts about thousand million cubic metres, traversing through three major States, Uttar Pradesh, Bihar and West Bengal of the country. Befittingly, therefore, the fate of the river is equated with the destiny of the millions of people living down the Indo-Gangetic Plain. Ganga's intriguing fluctuations are watched with anxiety, pollution spells panic and sharing this water resource generates conflict among neighbouring States. Naturally, the question arises, why ? Perhaps it signifies the pivotal importance of the river.

While the major contributions of Ganga to India are examined the foremost importance super imposes itself a source of irrigation. The Ganga basin has the largest irrigation facility, fed with network of canals playing a dominant role. Four major irrigation projects operate in Uttar Pradesh: (i) The Upper Ganga Canal System, (ii) Lower Ganga Canal System, (iii) The three pumped canal schemes of Dalmau, Bhoupali and Zamania and the Ram Ganga project to augment capacity of the above two canals.

Further augmentation of the flow of the water in Upper and Lower Ganga Canal System, Tehri Dam and Kotli Bhel Dam are constructed.

(iv) A new canal is introduced in U.P. known as Madhya Ganga Canal System.

Ganga apart from the mainstay of irrigation in U.P. is also a supplier of fresh water to industries on all riparian states. The hydro power generation potential of the Ganga cannot be denied. Four major thermal plants in the country depend on Ganga water—Panki in Kanpur, Barauni in Bihar, Farakka in Murshidabad and Bandel in Hooghly. Uttar Pradesh is an industrial State with cities like Kanpur well known for cotton, jute and leather goods, Varanasi's silk and light engineering, Allahabad light engineering and Hardwar for BHEL and antibiotics. Most of UP's sugar factories are situated along the banks of 1500 kms run in the State.

In comparative perspective Bihar is neither a substantial user of Ganga water for irrigation nor for industries. Only some light engineering industries exist in Patna and Bhagalpur along the 450 km stretch.

West Bengal's industries depend largely on Ganga water. Calcutta receives the main supply of water for its entire industrial hub from Ganga. Jute industries are concentrated on the banks of the river with light and medium engineering factories.

Ganga also serves as a source of water for domestic and municipal requirements. Ganga plain has 29 Class-I, 71 Class-II and Class-III towns and cities, and many smaller towns of IV, V and VI categories. Requirement for water in these human settlements rises day by day with population increase and urbanisation. It is assumed that the demand for domestic water from Ganga would go upto 62,000 million cubic metres by the end of the century.

The navigational aspect of Ganga has always had a promise. The Calcutta port has been constantly under threat due to silting and minimum yearly flow of 40,000 cusecs keeps the river navigable but the threat of fish and wild life persists.

Naturally, pollution in Ganga has a multiprong impact. The concern, anxiety and action to cleanse Ganga of pollution is no less than providing oxygen to the sagging life of this water resource. Experts advocate implementation of: (1) Augmentation of flow in the river (2) Pollution control (3) Controlled use of water territorially and sectorally (4) Adoption of long term pollution strategy to keep the water healthy. But the task is stupendous and is not merely a technical issue but has social, economic and political implications.

Chapter 4

The Quality of Ganga Water

Water is the elixir of life and basic for existence of ecosystem. Despite its abundance this vast resource has come to exist as a scarce commodity. Urbanization brings about changes in the ecological balance and breeds pollution. Pollutant input load into lentic and lotic waters has assumed alarming proportions. While aquatic systems have the ability to assimilate certain amounts of waste and maintain normal functions, the continued pollutant spills tend to degrade water quality.

A large number of industries in towns and cities along river banks are mainly responsible for pollution of India's greatest river the GANGA. Disposal of untreated effluents and sewage wastes pollute the river water. The contamination of the Ganga is bipronged—organic and inorganic existing in the form of colloidal and dissolved matters, and biological organisms. River water is used for drinking purposes with minimum purification leading to many water borne diseases like hepatitis in Calcutta, Kanpur, etc., mainly due to polluted water. The pollution level of a river in time and space can be measured by the BOD (biological oxygen demand) of the river water. The purity of water is gauged in terms of dissolved oxygen used by fish and aquatic plants.

It is nature's own processes that maintains the delicate oxygen water equation normally. Water is essential for life as it supplies oxygen for survival. Natural waters are usually unpolluted and saturated with dissolved oxygen. As a universal solvent water is used practically for every human enterprise from drinking, washing, bathing, cooking, religious rituals to commercial aspects. While mechanism of nature helps to combat assaults on river water quality, the influx of sewage, industrial wastes, and agricultural washouts, create inorganic and organic substances of varied nature which fertilize water and promote growth of micro-organisms. Sometimes they act as toxic substances detrimental to growth of flora and fauna, thus, destroying its regenerative capacity. Rivers today are no better than drains and sewers because of unplanned water management.

Two kinds of organisms make their appearance in the river waters when garbage and wastes are dumped. (i) Aerobes—which consume oxygen and populate the upper layers of oxygenated water. (ii) Anaerobes—which do not need oxygen and, therefore, remain deep within the water. The aerobes use oxygen from water as they consume organic waste and at the same time multiply. The oxygen demand becomes greater as the amount of waste increases. The inorganic waste is consumed by algae and gets converted into useful organic material. Algae release oxygen through photosynthesis. Apart from algae there are certain other organisms like protozoa, fungi, planktons, and bacteria, forming part of the micro-biological profile of the river. The climax of pollution ends up in a process called 'eutrophication'. Eutrophication is a process of large biological productivity with a reduced number of individual species. Unicellular organisms in water like protozoa help destroy many disease bacteria and germs in the waste. Oxygen for their survival is supplied through water. Algae oxygenate water. Such natural bio-chemical processes operate within the water to maintain the oxygen. Water eco-balance begins to disrupt when quality of waste added becomes huge and oxygen level depletes due to bacterial and fungal growth. Depletion of oxygen in upper layers of water results in movements of aerobes upwards and waste is consumed by them. In the process they liberate hydrogen from organic waste which combines with sulphur of the waste to produce hydrogen sulphide producing a bad odour. Turbidity in water prevents sunlight from reaching marine life leading to their death. Toxicity pervades the river and deterioration of the Ganga has outpaced its self-defence limits. Capacity of 'self-purification' existent in every river is dependent on volume of water and the release of wastes into it. Industrial effluents account for 20 per cent of the total pollution but the remaining 80 per cent comes from municipal waste waters. Accompanying this pollutant load is discharge from agricultural farms including fertilizer, insecticides and pesticides run off from land.

The main sources of pollution as mentioned earlier according to the GGA are urban liquid wastes (Sewage and sullage), industrial liquid waste, cattle bathing and throwing of dead bodies in the river. Added to it are surface run off from cultivated land applied with chemical fertilizers and manures as well surface run off from lands dumped with urban solid wastes and industrial solid wastes. In India pesticide contamination in the country amounts to 300 gm per hectare compared to 1.79 gm in the West. This kind of contamination poses great health problems. This situation is aggravated when countries like India use agro-chemicals identified by the WHO as toxic and hazardous, and banned in Western countries. Pesticide poisoning cases are commonly reported in hospitals. Even fruits, dairy products and oil seeds have also found to be contaminated causing many diseases. Regulating use of insecticides and pesticides is the only alternative apart from proper pest management. The Insecticides Act, in quality control and enforcement can also help in lowering the levels of toxicity. Since millions of tonnes of food is lost to pests every year the importance of pesticides in agriculture cannot be ignored and its total disuse impossible. Pesticides remain a need for crop prospects, so efforts must be made to minimize their toxic side-effects.

It is a known fact that influx of pollutants into rivers is enormous and its spills through indiscriminate utilization of river water likely to cause enormous damage to the agricultural lands due to salinity and alkanity of soil and water logging caused because of seepage from canal systems. This problem can be tackled by providing proper drainage, necessary linking of canals and improved water management techniques.

The conflict between development and environment is inevitable. The gains of development substantially outweigh the adverse environmental impacts. In developing countries the economic development becomes necessary and essential.

The untreated sewage and industrial effluents dumped into the river at numerous places, residues of pesticides and insecticides used in farms aggravate the situation in a cumulative manner, especially, during the lean season. The sedimentation problem in canals is another aspect needing attention. The main cause of sedimentation was the increased activity of man and the encroachment on the environment in the form of cutting and burning of bushes and forests, overgrazing of grasslands and defective methods of cultivation (Banerjee, 1989). Siltation is an equally serious problem faced by the Ganga. It is estimated that 3 tons of silt is carried away per square kilometre of the Ganga's catchment areas. The rich top soil is carried away from hills to the sea as a result of deforestation. Siltation causes floods and stagnation of flow in the lean season.

Ganga is a major contributor to most of the water-borne diseases in the country. The water is used for drinking, bathing, washing and as a dump yard of effluents and domestic wastes. The waters develop and transfer bacteria of cholera, gastro-enteritis, jaundice, diaorrhoea and other enteric diseases as a permanent feature of the river. Drinking of this water results in epidemic diseases, taking millions of lives in the country. Certain pathogenic micro-organisms more resistant to disinfection than bacteria are also found in the river. The causes of pollution are endless but not accidental.

The quality of water is measured in terms of three main yardsticks : (1) Total solids (2) dissolved solids (3) suspended solids. The parameters for measuring the purity of the river are (i) *Dissolved Oxygen Content in Water*: The dissolved oxygen concentration depends not only on the relative dilutions but on rate of oxidation of organic material and recreation. The more the oxygen, the better the quality of water, (ii) *Biochemical Oxygen Demand and Chemical Oxygen Demand*: The demand of oxygen from water to break down of organic and inorganic wastes and sewage is known as biochemical oxygen demand. The more oxygen taken out from water the less becomes the content of dissolved oxygen, thus, increasing the pollution in the river with a high BOD and COD load. In the Ganga around 40 million gallons of sewage water is discharged daily. The analysis of this water shows a much more higher and alarming rate of BOD concentration (270 mg per litre) than the prescribed rate of 33.4

mg per litre. The amount of dissolved oxygen needed by bacteria to decompose the wastes determines the quality of wastewater. The problem arises when demand for dissolved oxygen exceeds available supply. This in turn leads to huge quantities of organic pollutants, organic debris and other non-degradable pollution resultant of leather tanning, sugar refining, metal packing and other industrial activities. The transformation of a healthy river into a virulent sewer becomes complete with the voluminous discharge of waste and filth. Thus, BOD is a measure of oxygen required to sustain organisms like aerobes, protozoa, etc. and COD is the measure of extent of oxygen utilized to convert organic material into inorganic mater, (iii) *The Coliforms in the Water*: The coliforms are bacteria, *E. coli* (*Colon bacillus*) measured in terms of the 'most probable number' (MPN). The coliform group indicates the presence of contaminating organisms. They do not cause diseases but faecal coliforms and pathogens are common disease spreading bacteria. The pathogenic organisms can cause cholera, jaundice, typhoid and other intestinal diseases. Other parameters determining the water quality are the pH Value (a measure of hydrogen ion concentration acidity). Nitrogen, sulphates, chlorides, radioactivity and properties of temperature, turbidity, conductivity and hardness are other parameters of water quality.

Due to the varied pollutional load of industries and quantity of discharge, organic, inorganic, metallic and non-metallic substances are found in the river. Metallic loads include lead, mercury, cadmium, arsenic, selenium, etc., while non-metallic loads are phenol, cyanide, pesticides and petro-chemicals poisonous for mammals and aquatic life. Organic load is higher in industries of thermal power, paper pulp, petro-chemical and distillery effluents. Toxic metal ions tie up enzymes required for microbial growth and thereby interfere in the self-purification of the river, and non-metallic toxic wastes yield bacterial growth.

Reports on the Ganga water claim that the water quality of Ganga has been found to be in a state for drinking only after conventional treatment and disinfection. Ganga water quality ranked between third and fourth grade scale in terms of water quality.

The Central Pollution Control Board has fixed standards for different uses of water, expressed as A, B, C and D classes of water. The Ganga Action Plan objective is to bring river quality to Class 'B'. Regular monitoring of the ruler quality is essential not to assess problems faced but also to measure the success and failure of the Action Plan. There is also the need to enlist the support of the people and the government for successful implementation of the cleaning and water pollution abatement programme.

Millions of litres of drain sewage is dumped into the Ganga every day without treatment. The Ganga is subjected to toxic, metallic and chemical wastes that pose grave public health hazards for the people. All the rivers have a self-purification capacity and the capacity of Ganga to keep its waters pure is well known and even mentioned in the scriptures. But loads of domestic sewage and sullage, industrial effluents pollute the river to the extent that it does not have any self-regenerative capacity.

Vegetables grown in sewage farms are also harmful. A factor to be considered is the difference between developed and developing countries for treatment wastes before use. A primary treatment alone can be sufficient in the advanced countries but in poor countries there can be terrible break of diseases as slum dwellers and poor people even use the polluted waters for bathing and washing. In case such water is discharged after primary treatment it can cause many diseases. In the advanced countries the practice is to release these treatment effluents into a dry water course leading to a river—a process not possible in poor and developing countries. In the process bacteria in the water are killed by the oxygenation of the flowing water. In India simpler system and non-conventional sewage treatments are needed as simpler systems incorporating secondary treatment facilities eliminate harmful bacteria in the sewage water.

The lack of maintenance of pumping plants is due to lack of funds for operation and maintenance of the pumping plants. Many plants are out of order due to financial factors. In many cases power bills are not paid for several years and the power agencies are compelled to disconnect the power connections.

Certain by products of sewage treatment like biogas, NPK content in manurial matter, fish, algae, etc. can be commercially used and utilized to cover the cost of these treatment plants and thus, save the Ganga and people subsisting on its water from

Table 4.1: Pollution Control Standards

Characteristics	Sewage IS 4764-73	Industrial effluents		Inland surface water IS 2296-74
		Inland surface water IS 2490-74	Public sewers IS 3306-74	
BOD (5 day 20°C), mg/l	20	30.0	500.0	3.0
COD, mg/l	—	250.0	—	—
pH	—	5.5-9.0	5.5-9.0	6.0-9.0
Total suspended solids, mg/l	30	100.0	600.0	—
Temperature °C	—	40.0	45.0	—
Oil & grease, mg/l	—	10.0	100.0	0.1
Phenolic compounds, mg/l	—	1.0	5.0	0.005
Cyanides, mg/l	—	0.2	2.0	0.01
Sulphides, mg/l	—	2.0	—	—
Fluorides, mg/l	—	2.0	—	1.5
Total residual chlorine, mg/l	—	1.0	—	—
Insecticides, mg/l	—	Zero	—	Zero
Arsenic, mg/l	—	0.2	—	0.2
Zinc, mg/l	—	5.0	15.0	—
Chloride, mg/l	—	—	600.0	600.0
Sulphates, mg/l	—	—	—	1000.0
Ammoniacal Nitrogen, mg/l	—	50.0	50.0	—
Nitrates, mg/l	—	—	—	50.0
Radioactive materials:				
alpha-emitters, micro curie/ml	—	10^{-7}	—	10^{-9}
beta-emitters, micro curie/ml	—	10^{-6}	—	10^{-8}
Dissolved Oxygen, mg/l	—		—	40% of saturation value, or 3 mg/l whichever is higher
Coliform organism (monthly average) —MPN per 100 ml	—	—	—	under 5.000

Source: Deodhar and Nath, 'Water Pollution and its Control in India', Indian Science Congress, 1981.

serious health hazards. Instead of self-financed project of maintenance of clean waters, additional money should be allocated for healthier Ganga. The economic cost of such projects should be measured in terms of social cost and not in narrow Cost: benefit ratio.

Table 4.2: Water Pollution from Selected Industries

Industry	Waste water generated (million) m^3/year	Pollutants				Remarks
		BOD (mg/1)	COD(mg/1)	Suspended soilds (mg/1)	Others (mg/1)	
Cotton textiles	1,530	200-300	400-800	350-550	Chlorides 500-1,400	Highly alkaline; discoloured and hot effluent.
Nitrogenous fertilizer	52	—	—	-	Total nitrogen 1,250 Fluoride 15 Arsenic 1.6 Phosphates 70	Nitrogen in ammonia is deadly to fish; nitrates are hazardous to livestock and humans; effluents are highly alkaline and slightly brownish.
Paper	450	160	725	410	—	High or low pH depending on processs used.
Pesticides	—	upto 700	3,000-10,000	—	Organic nitrate 0-500 Suphate 3,000-20,000	pH can vary from acid (2) to alkaline (10); effluent is highly toxic to aquatic life.
Oil refinery	15	160	320	350	Phenol 20 Sulphides 9 Phosphates 49	
Integrated steel plants	40	75-100	200-2,000	—	Phenol 20-60	Acid effluent; high in suspended solids.
Sugar	225	300-2,000	600-4,000	220-800	Total nitrogen 10-40	

Source : Deodhar and Nath, 'Water Pollution and its Control in India,' Indian Science Congress, 1981.

Chapter 5

Biological Profile of Ganga Eco-system

Rene Dubos (1975) rightly stated on Quote "Now and then in my dreams, I see a place where one might initiate and publicize programme to give environmental movement a positive and constructive philosophy that would compliment the present defensive attitude of environmental conservation and protection." Nevertheless, this dream of Dubos is not singular, the global environmental community is aspiring to develop a technology and a system that could control and co-ordinate use and reuse of flowing river water for drinking, domestic, industrial and agricultural purposes. But there are severe handicaps in transforming this dream into reality. The problems causing impediment in development of an unpolluted, safe and hygienic water are many and beyond control of technological knowhow. The defensive exploration and environmental research movement emphasizes creative aspects of riverine water management. Hillman (1961) rightly pointed out that "While adequate coverage of all work with a particular group of plants must perforce touch most fields of botanical research, it is impossible to consider each of the problems in its general context, to do so would be to write an encyclopedia." However, in the broad spectrum of riverine biology and its involvement in self-purification, water quality and disease spread is strength of this treatise.

The ecology of rivers and pollution in India have been reviewed elsewhere (Singh *et al.*, 1982; Patil, 1983; Bilgrami, 1986; Ajmal and Raziuddin, 1988; Sinha, 1988; Sinha, 1989; Sharma *et al.*, 1990; Shrotriya and Dubey, 1990; Trivedy *et al.*, 1990; Trivedy, 1990). There are excellent works on riverine biology (Saxena *et al.*, 1966; Parmasivam and Sreenivasan, 1981; Venkateswarlu and Sampat Kumar, 1982; Somashekar, 1984, Trivedy and Goel, 1984; Dubey *et al.*, 1986; Kudesia and Verma, 1986; Reddy and Vekateswarlu, 1987 and Malviya *et al.*, 1990). The literature suggests that Ganga is peculiar of all other rivers in the country for its biological attributes.

The topic on Ganga ecosystem *vis-a-vis* biological gamut has been investigated earlier by Tripathi, 1991 (Shuklaganj); Gupta, 1991 (Nanamau and Bithoor); and Kaur, 1991 (Kanpur). Biological profile of river Ganga portrays its significance. Organisms including algae and coliforms are known to interact in multiple ways with human beings and cattle propagating several fatal diseases. The quantitative estimation of coliforms, and qualitative and quantitative assay of algae is specialized area emphasizing in-depth treatment.

Perhaps there exists a direct co-relation between chemical constituents of organic and inorganic nature coupled with physical attributes of the water body and microbe profile. The biology of a riverine system is subject to changes in physico-chemical characteristics and complementary impacts of microbes. Involvement of microbes in self-purification of water pollution and public health hazards is documented in the literature (Shukla, 1989; Nigam, 1986; Gupta, 1990; Tripathi, 1990). The problem of monitoring, abatement, pollution, control of health hazards and resultant biology of river ganga *inter alia* physico-chemical attributes in unison constitutes environmental priorities for exploitation, use and reuse of river water. It becomes incumbent, therefore, that microbiological profile need be recorded with special reference to organisms growing luxuriantly.

Literature supports isolation of soil fungi from different parts of the world (Gaddard, 1911; Jensen, 1931; Ma, 1933; Sabet, 1935; Meredith,1938; Ellis, 1940; Akhtar *et al.*, 1987). Recently considerable attention has been paid to record soil fungi in India (Chaudhuri and Sachar, 1934; Subramanian, 1952; Saksena, 1953; Agrawal 1955; Srivastava, 1955; Singh, 1968; Mukerji *et al.*, 1969; Mishra 1986). Mycoflora of Ganga water of Bihar (Bilgrami and

Dutta Munsi, 1979) and Rishikesh (Shanker and Prasad, 1983) has also been studied. Despite pioneering works of Cooke (1954 a,b,c, 1970; 1977), and Cooke and Matsuura (1966; 1969) on fungal comments of fresh water ecosystem, little is known about fungi of fresh water bodies. There are few reports on fungi of Ganga waters (Bilgrami and Dutta Munsi, 1979; Shanker and Prasad, 1983). However, there is no record of the mycoflora of Ganga.

Dermatophytes are essentially related to diseases of cutinized and keratinized animal and human tissues. Keratinolytic organisms possess proteolytic enzymes capable of degrading keratinized substrate. Keratinophilic organisms explicit competitive ability for keratin in natural condition. During the past two decades, diversity in keratinophilic forms has been investigated in India (Dey and Kakoti, 1953; Randhawa *et al.*, 1959; Padhye, and Thirumalachar, 1962; Randhawa and Sandhu, 1963; Randhawa and Sandhu 1965; Garg, 1966; Padhye *et al.*, 1966; Padhye, 1968; Tewari, 1969; Roy *et al.*, 1972; Kushwaha, 1976; Kushwaha and Agrawal, 1976; Jain and Agrawal, 1977; Verma, *et al.*, 1982; Deshmukh and Agrawal, 1983; Randhawa and Khan, 1983; Nigam and Kushwaha, 1984; Jain *et al.*, 1985; Kushwaha *et al.*, 1985; Jain 1986; Awasthi and Shukla, 1988) and different parts of the world (Daniels, 1954; Durie and Frey, 1955; Ajello *et al.*, 1965; Carmichael, 1962; Ajello and Alpert, 1972; Caretta and Piontelli, 1975; Guarro *et al.*, 1981; Michel Brun and Monier, 1981; Abdel *et al.*, 1982; Deshmukh and Agrawal, 1982; Mercantini *et al.*, 1983; Calvo *et al.*, 1984).

Decomposable organic matter of plants and animals, and washouts from agricultural land render water unfit for drinking and bathing. Such circumstances promote the growth of algae along the river-bed as well as in the water. The thriving capacity of algae in such conditions are subject to the adaptability for change in quality of soil and periodicity, largely depending on the environmental conditions produced by the quality of the ecosystem. However, role of algae in primary production of riverine ecosystem is not major.

There is existence of literature on algal infestations in natural water-bodies, sewage and water-works systems. Algae play significant role in water supplies. Excessive growth of algae causes problems in water filtration and purification. Throughout the world difficulties due to growth of algae in water-works systems have been witnessed in United States, United Kingdom, France, Switzerland and India. Common troubles posed in water filtration are: (1) Algae not only bring about clogging of filters but also enrich them with organic content and thereby support luxuriant growth of other forms. (2) Algae produce odours, tastes and their prolific growth modifies pH and turbidity of water. (3) Algae corrode metal tanks and form pits in their walls, disintegrate concrete blocks, channels and tanks. (4) Algae harbour large number of bacteria of pathogenic nature. (5) Some of the algae are toxic and make the water unsuitable for drinking and other commercial uses. But besides such problems they also form a biological film and oxygenate water, and perform Yeo-men's service in biology of waters.

Algal sources of pollution are known (Ingram and Palmer, 1952; Palmer, 1956; 1959; 1980; Kulikova, 1960; Rai and Kumar, 1980; Siddiqui, 1980; Parmasivam and Sreenivasan, 1981; Tripathi and Singh, 1983; Pandey and Tripathi, 1984; Tripathy, 1985; Wa, 1986). Algae also act as indicators of pollution (Rawsom, 1956; Palmer, 1957; Cholnoky, 1958; Fogg, 1960; Brook, 1965; Patrick, 1965; Round, 1965; Vaidya, 1967; Woodwell, 1967; Ware *et al.*, 1968; Whitton, 1970; 1975; Moore, 1974; Keeney *et al.*, 1976; Das *et al.*, 1978; Hosmani and Bharati, 1980; Prasad and Singh, 1982; Kant, 1983). Significance of such pollution indicators in unravelling facts for monitoring water quality has wide application in potable waters. There are certain algal species widely used as biological indicators of water quality (Patrick, 1973; James and Evison, 1979; Whitton, 1979; Venkateswarlu 1981; Mohanty, 1983; Kannan and Krishnamurthy, 1985). Algae are also involved in self-purification of waters (Shrivastava and Kaushik, 1966; Bhargava, 1981; Nigam, 1986; Shukla, 1989).

Terrestrial, aquatic and aero-algal allergins have been reported from North America (Lewis and Elvin Lewis, 1976). Aero-allergins are known to cause Allergic Rhinitis, Bronchial Asthma, Hypersensitivity and Pneumonitis. Algal infestation in rivers acts as stock for algal allergins.

The dynamic features of aquatic and terrestrial habitats have supported diverse algal flora. Algal flora of various localities presents interesting qualitative and quantitative variance in time and space. Literature supports excellent work on algal flora from various parts of the world. There are preliminary survey reports on algal flora of Allen Forest lake, (Ahmad, 1963 and 1972) and algae of some temporary and certain rain waters reservoirs (Bitta, 1965). Nair (1965) studied algae of potable waters. Algal flora of Panki rice fields has been recorded elsewhere (Shukla, 1982).

Theoretically, lotic waters are a continuous fermentation system of the plug-flow type in which part of biomass is in free planktonic suspension and the rest of the part is fixed bottom benthos-biomass. The dominant properties of the system are, nevertheless, clearly recognizable (a) there is no feedback of information in the system, that reacts at any location and have no influence on future events upstream, (b) the system may be considered as a single batch reaction and the actual state of affairs in subsequent river represents stages of one and the same situation, observed at the uppermost reach as a function of the residence flow time from one point of observation to the next, (c) every flowing water reach of a suitable length may be considered as an individual batch in which the water quality is constant and the biocenosis is in a state of 'equilibrium' with the ecological conditions (d) in large lotic waters true phytoplankton represent the major part of the biomass in the system. Continuous inoculation of a reach from an upstream location is essential for maintaining the level of concentration of organisms in the water mass.

Polluting matter subjected to self-purification may include compounds completely inert against organisms or products influencing the metabolism (as substrates, inhibitors and toxicants) of the phyto or zoocenosis in a stream. Bulk of literature in the field of self-purification deals with pollutants like those utilizable substrates and the biocenological effects produced by 'saprobic' conditions. In contrast to clean flowing waters, however, the aspects of the trophic structure and the energy relationships in situations of organic pollution have not yet found an adequate exploitation. Biological self-purification representing an activity of community metabolism may include ecosystem development. It must be admitted, that essential facts for quantitative considerations are unknown adequately.

Response of certain algae, to major pollutants of Ganga is unknown with special reference to calcium chloride, zinc sulphate, lead acetate, nickel acetate and cobalt acetate. Algae are related to food cycle in nature and form basic link in energetics of ecosystems.

Few algae have been reported earlier from flowing waters (Shrivastava and Kaushik, 1966; Nigam, 1986). Phytoplankton diversity and density influences water quality in multiple ways *vis-a-vis* colour, turbidity and odour. Planktons are also linked with fish for their food value. In view to depict totality of algal infestation in time and space, it is prudent that both algae collected from nature and those observed in culture need be studied. Response of algae to major pollutants of Ganga appears of considerable importance to unravel basic mysteries of Ganga eco-system.

While describing various allergenic pollen floras (Shivpuri, 1963, 1964; Shivpuri and Singh, 1971; Vishnu Mittre and Khandelwal, 1973; Chanda *et al.*, 1972, 1976, 1980; Desh Pande and Chetaley, 1976) pollen morphology of grains needs to be recorded in detail with the objective to make use of this information in taxonomy. It is exceedingly important to describe regional pollen floras in time and space and account changes to counter balance their impact on allied sciences.

There are few reports concerned with description of complete pollen calendar giving inforamtion about period of pollen emission, their qualitative and quantitative abundance and succession. Natural resources of pollen could be used in allergenic disease control.

Allergenic pollen calendar of Ganga basin is a key to undertake remedial measures in time and space. Herein, this treatise dwells in providing a conceptual synthesis of aeropalynological implications of Ganga basin flora *vis-a-vis* public health hazards and physiological interweb.

Presence of decomposable organic matter of plants and animals, washouts from agricultural land and industries promotes the growth of macrophytes along the river bed as well as in the water. Limnological survey and input of human activities on river Ganga is on record (Bilgrami and Dutta, 1979). Aquatic weeds of South-East Asia have been described by Varshney and Rozovka, 1976. The thriving capacity of such macrophytes are subject to the adaptability for change in quality of soil and periodicity, largely depending upon the environmental conditions produced by the quality of the ecosystem. However, role of macrophytes in primary production of river in ecosystem is not major.

Assessment of pollutional impact on macrophytes was undertaken in river Ganga (Tripathi, 1991). The richness of fauna of any fresh water formation primarily depends upon the oxygen and carbondioxide percentage in water. A number of physical and chemical parameters are responsible for the distribution and diversity of fishes in particular. Due to proliferating industries and urbanisation the water quality deteriorated, consequently, affecting the fish diversity and distribution. Fish diversity and density of Cooum river has been studied Kulti Estuary (David, 1959; Abraham, 1962), and Daha one and the same situation observed at the upper most reach as function of the residence flow time from one point of observation to the next; (c) every flowing water reach of suitable length may be considered as an individual batch in which the water quality is constant and biomass is in a state of equilibrium with the ecological conditions; (d) in large lotic waters true potamoplanktons represent the major parts of the biomass in the system, continuous-inoculation of a reach from an upstream location is essential for maintaining the level of organisms concentration in the water mass.

Polluting matter subjected to self-purification may include compounds completely inert against organisms or product which are influencing the metabolism (as substrate inhibitors, toxicants etc.) of phyto or zoocoenosis in a stream. The majority of work on self-purification deals with pollutants, utilizable substrates and the biocoenological effect produced by 'saprobic' conditions. In contrast to clean flowing waters, however, the aspects of trophic structure and energy relationships under organic pollution have not yet found an adequate emphasis. Biological self-purification representing an activity of community metabolism may be included in concept of ecosystem development. It must be admitted, however, that essential data for quantitative consideration are lacking to a large extent.

Ecological studies of algal communities of Ganga river at Varanasi have been described elsewhere (Rai, 1978). In view of significance of algae in pollution, disease spread, self-purification and as an important constituent of food-web linked with fish productivity, in-depth study of algae appears necessary. The growth and infestation of algae would remain incompletely known unless studies are made taking into consideration algal infestation in natural waters and occurrence observed in sand and water cultures. There are two distinct categories of algae growing in lotic waters visualized benthic and planktonic. The algal flora also shows variance in various layers and zones of water. It, therefore, becomes incumbent that a representative study should take into account collection of algae from various zones and layers of the river (David and Ray, 1966). A calendar of fish density and diversity need be prepared. Urban domestic and industrial discharge in our country is drained into the environments as waste without due consideration of the deleterious effects caused to the environment (Chandra, Singh, Srivastava and Mehrotra, 1981). Many instances of the fish mortality due to indiscriminate industrial waste discharges have been reported in the literature (Carpenter, 1924, 1925; Jhingran, 1970; Hora, 1942; Hora and Nair, 1944; Bimachar and David, 1946; Ellis, Westfall and Ellis, 1946; Ganapati and Alikunhi, 1950; David, 1956; Banerjee, Motwani, and Karamchandani, 1956; Klein, 1957; Banerjee and Motwani, 1960; Qasim and Siddique, 1960; De and Mukherjee, 1971; Hussain, 1976). Many investigators like Sinha, (1988); Sinha *et al.*, (1989); Gupta and Pandey (1980); Mahadevan and Krishnaswamy (1983); Sinha and Karim (1984) studied the physico-chemical aspects of water courses to estimate the quantum and nature of pollution.

In the past, pollution in the lotic waters caused particularly from industrial sources occupied priority of consideration around organic matter, nutrients, and suspended solids. Existence of enough literature on pollution by metals has accumulated only in the recent past to develop an understanding of the distribution and fate of trace metals following their discharge into aquatic environment. Japanese had the most bitter experience with the toxicity of mercury and cadmium in outbreak of neurologic disease. Minimata disease was spread due to mercury discharge by plastic industry into aquatic environment (Fujiki, 1972) and itai-itai disease was result of cadmium poisoning (Kobayashi, 1971). Incidence of non-parasitic elephantiasis has also been reported in the immediate vicinity of Lake Abdeya, Ethiophia because of heavy metal pollution in the lake (Klein, 1977; Gibbs,

1973), Henrikson and Wright (1978), Harding and Whitton(1978), Shephard, Mc Intosh and Atchison (1980) discussed the distribution and fate of metals in the aquatic system. In India, Pandey and Das (1980) reported metal pollution in Nainital lake, Kedesia and Sharma (1981) discussed metal pollution in Kalinadi due to industrial discharge. Somashekhar and Ramaswamy (1982) noticed alarming signs of metal contamination in Cauvery river. Masironi (1970), Adaychi and Douabul (1985) are the few notable workers to pin point co-relation between the various pollutants.

However, response of certain algae to measure pollutants of Ganga is little known. Effect of calcium, chromium, cobalt, copper, and lead to algae has been investigated. Algae are related to food cycle in nature and form basic link in energetics of ecosystems. They serve as an important constituent of fish-food and play significant role in fisheries (Fish, 1951; Ross, 1955; Alikunhi, 1958; Moitra, 1956 and Singh, 1961). Attempts have also been made to investigate effects of feeding fish on algae. Preference of fish for a particular alga or algae constituting fish-food could be ascertained through examination of gut content.

Macrophytic infestation although comparatively of less consequences in biology of water, still its massive growth is likely to cause problems in the lean periods of river when the water quantity reduces. Macrophytes appear likely the source of subsidiary protein food and could be harnessed as fertilizer suppliment to promote crop yields. In view of such multiple uses of micro-organisms and macrophytes the studies were targeted to study algal and macrophytic profile of the river. Nevertheless, prospects and aspects of pollution and its control, quality of water, public health hazards and disease spread through coliforms, fungi and algae is major. These vital perspectives is intermeshing fabric of ensuing chapter. Herein, biological profile of Ganga *vis-a-vis* coliforms, fungi, algae, macrophytes, allergenic pollen and fish is described to depict significant components of flora and fauna.

COLIFORM

Coliform growth in various calendar months shows a systematic pattern. Herein, for the sake of convenience coliform density in mid-stream water and bank water has been recorded.

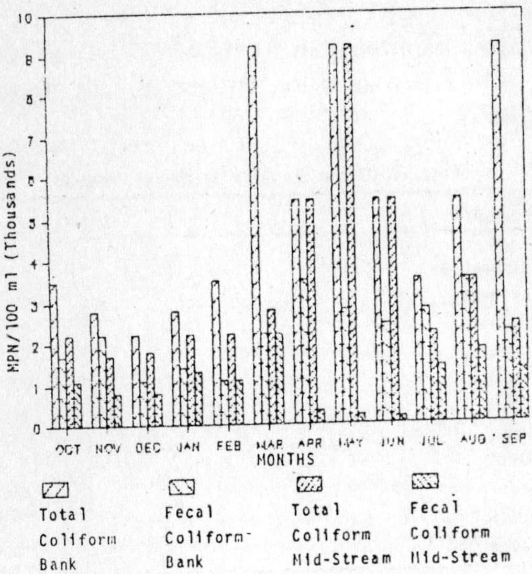

Fig. 1
Total and Fecal Coliform Bacteria in Bank and Mid-stream Waters

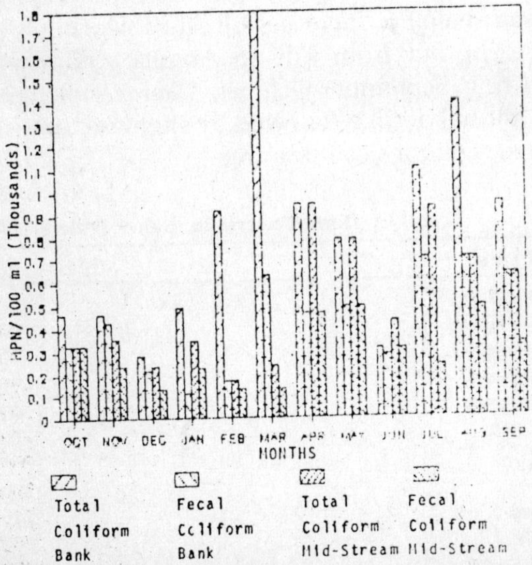

Fig. 2
Total and Fecal Coliform Bacteria in Bank and Mid-stream Waters

Total Coliform

Total coliforms expressed in terms of MPN per 100ml shows that there is an increase from January to May thenafter it declines gradually.

Total Coliform Bacteria in (Bank) Waters

Months	MPN/100ml
October	3500
November	2800
December	2200
January	2800
February	3500
March	9200
April	5400
May	9200
June	5400
July	3500
August	5400
September	9200

Fecal Coliform

Data shows that fecal coliform level declines during January, February and thenafter increases till April. During May there is a sharp decline but from July to August coliforms show an increase during September-October. There is a decline again. There is a gradual decline followed by slight increase in November and then again a declining trend.

Fecal Coliform Bacteria in (Bank) Waters

Months	MPN/100ml
October	1700
November	2200
December	1100
January	1400
February	1100
March	2200
April	3500
May	2800
June	2400
July	2800
August	3500
September	2200

MID-STREAM

Total Coliform

Observations show that during January, February total coliform density was identical. Coliform number thenafter increased upto May and then marked a decline in July. During August there was again an increase followed by a declining trend.

Total Coliform Bacteria in Flowing Waters (Mid–Stream)

Months	MPN/100ml
October	2200
November	1700
December	1800
January	2200
February	2200
March	2800
April	5400
May	9200
June	5400
July	2200
August	3500
September	2400

Fecal Coliform

A slight decline from January to February followed by high MPN count in March is seen. From April to June again there is a decline followed by considerable increase in fecal-coliforms during July, August period. From September to December there is a gradual decline.

Fecal Coliform Bacteria in Flowing Waters (Mid–Stream)

Months	MPN/100ml
October	1100
November	790
December	790
January	1300
February	1100
March	2200
April	330
May	230
June	170
July	1400
August	1800
September	1300

FUNGI

SYSTEMATIC ENUMERATION OF FUNGI

Phycomycetes Mucorales Mucoraceae *Mucor* Micheli

Mucor hiemalis **Wehmer:** Gilman Joseph C., A Manual of Soil Fungi, 38, Pl. 1, D,E, 1957.

Sporangiophores usually unbranched, erect, then prostrate by wilting. Turf about 1 cm high close and fine, cottony, white, rarely grayish-yellow. Sporangia spherical, gray or brownish-yellow. Columella free, spherical or oval, colourless. Spores usually unequal, the majority elongate or ellipsoid, smooth, hyalin with thin membrane.

Dimensions: Sporangia 51.2-52.4µ in diameter; columella free, spherical 28-48µ in diameter or oval columella 26×22 to 36×29.2µ; spores 3.2-9.2×2-5.6µ.

Distribution: Found in sand and water culture.

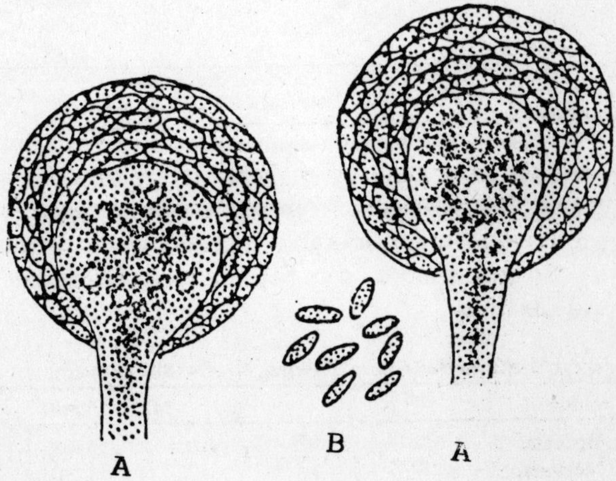

Fig. 3: *Mucor hiemalis* (A) Sporangiophore (B) Sporangiospores

Mucor racemosus **Fresenius:** Gilman, *op. cit.*, 31, 1957.

Sporangiophores erect, close, forming a yellow-brown turf of very variable height, branched irregularly in groups. All the branches terminated by sporangia which are very unequal. Sporangia small, globose, unequal, erect, or at times incurved,

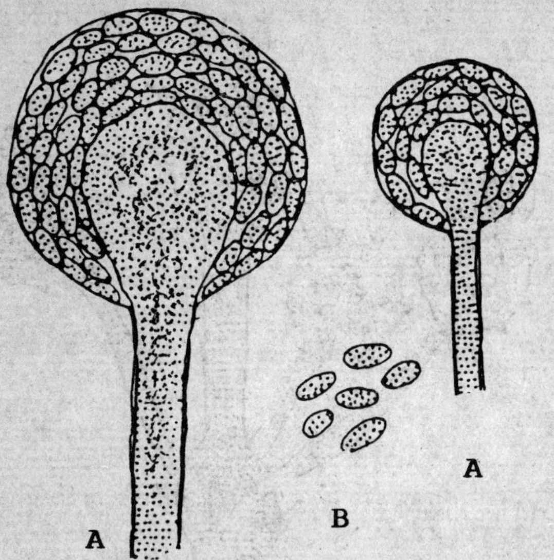

Fig. 4: *Mucor racemosus* (A) Sporangiophore (B) Sporangiospores

pale yellow or yellow brown, hyalin. Walls of sporangium not diffluent, but fragile, persistent, incrusted, leaving a collarette. Columella free, globose, ovoid, or broadly cruciform, campanulate. Spores rarly globose, more often elliptic, smooth, yellow in mass. Zygospores globose brown with conic warts yellow or red brown, suspensors straighter than the zygospore and not swollen.

Dimensions: Sporangia 20-68μ in diameter; columella 18-60μ long×7.6-32μ wide and 8-42μ in greatest diameter; spores 5.6-10×4-7.6μ.

Distribution: Found in sand and water culture.

CHOANOPHORANCEAE
CUNNINGHAMELLA Matruchot

***Cunninghamella bertholettiae* Stadel:** Gilman, *op.cit.* 66, 1957.

Turf gray, filaments firm and interwoven. Conidiophores erect, irregularly-cymosely branched, nonseptate. Terminal

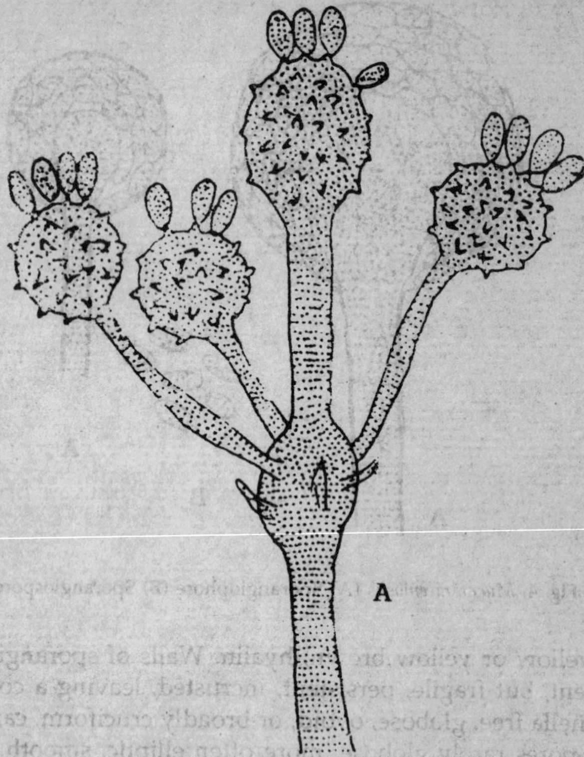

Fig. 5: *Cunninghamella bertholettiae* (A) Sporangiophore

vesicles ovate. Lateral branches variable, alternately arranged in groups, numerous. Lateral vesicles round in shape. Terminal conidia ovate, smooth; lateral conidia similar to terminal but slightly smaller.

Dimensions: Terminal vesicles 24-24.8µ wide×32-33.2µ long; lateral vesicles 22.4-23.2µ in diameter; terminal conidia 4.4-4.8µ wide×8.8 - 10µ long.

Distribution: Found in sand and water culture.

ASCOMYCETES SPHAERIALES CHAETOMIACEAE
CHAETOMIUM Kunze and Schmidt

Chaetomium sps. Kunze and Schmidt: Gilman, *op.cit.*, 175, Pl. VI, C,E, 1957.

Perithecia superficial, thin-membranous, with an apical tuft of bristles and usually with an ostiole. Asci clavate, evanescent. Spores simple, hyalin to dark brown, more or less compressed.

Dimesnsions: Ascospores 8-9.2μ long×5.6-8μ wide.

Distribution: Found in sand culture.

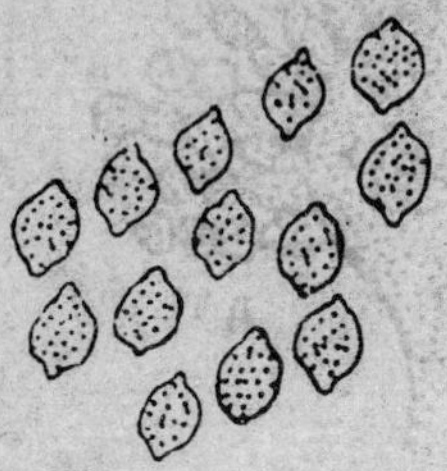

Fig. 6: *Chaetomium sps.*
A—Ascospores

FUNGI IMPERFECTI SPHAEROPSIDALES SPHAERIOIDACEAE *PHOMA* (Fries) Desmazieres

***Phoma* sps. (Fries) Desmazieres.** Gilman, *op.cit.*, 198, Fig. 64, 1957.

Pycnidia globose or slightly lens-shaped with a small papilla at the apex, membranous to leathery or almost carbonous, black. Spores small, egg-shaped, spindle-shaped, cylindric or almost spherical, one celled, hyalin, usually with 2 oil drops.

Dimensions: Spores 4.4-7.2μ long×3-4μ wide.

Distribution: Found in sand culture.

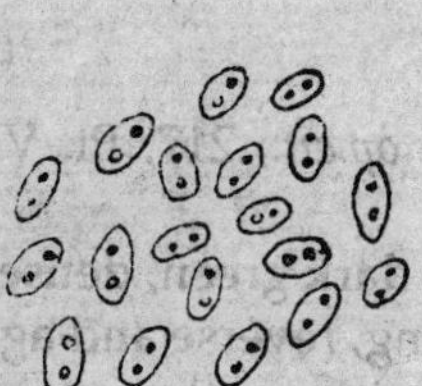

Fig. 7: *Phoma sps.*
A- Conidia

MONILIALES MONILIACEAE
ASPERGILLUS (Micheli) Corda

***Aspergillus chevalieri* (Mangin) Thom and Church.** Gilman, *op. cit.*, 219, 1957.

Colonies on czapeck's agar restricted, plane, closely felted, bluish gray in centre to orange at margin. Conidial heads abundant, pale blue-green, appearing radiate from divergent conidial chains. Stalks enlarging to somewhat globose vesicular apex.

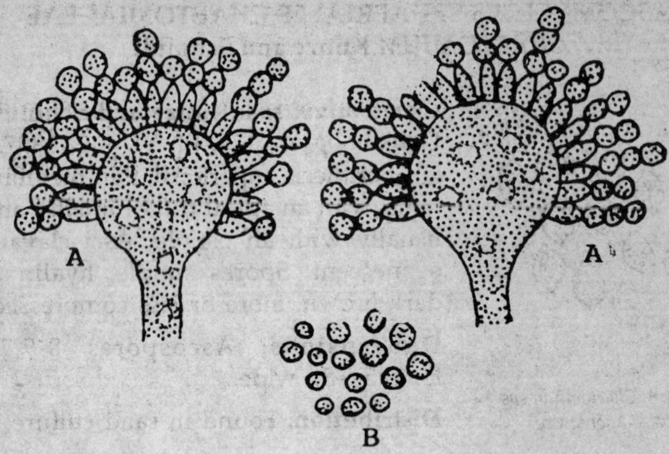

Fig. 8: *Aspergillus chevalieri* (A) Conidial head (B) Conidia

Phialides in a single series which are closely packed. Conidia subglobose and spinulose.

Dimensions: Vesicle 25.6-34μ in diameter; phialides 4.8-7.5×3-3.6μ; conidia 4.4-5.6μ in diameter.

Distribution: Found in sand culture.

***Aspergillus clavatus* Desmazieres.** Gilman, *op.cit.*, 217, Pl. VII, E, 1957.

Colonies on czapek's agar gray-green to dark green, densely matted, heavy, rapidly growing and spreading, reverse and agar more or less brownish. Conidiophores with walls smooth colourless, gradually enlarged into a clavate vesicle. Phialides in a single series, densely covering the fertile area, bearing long chains which frequently adhere more or less into masses. Conidia elliptical, green smooth.

Dimensions: Conidiophores 14.8-19.6μ in diameter; phialides 7.2-9.6×2-2.8μ; conidia 2.4-2.8X3.6-4.4μ.

Distribution: Found in sand and water culture.

***Aspergillus flavus* Link.** Gilman, *op.cit.*, 225, Pl. VIII A, 1957.

Colonies on czapek's agar widely spreading, with floccosity limited to scanty growth of a few aerial hyphae in older areas. Conidial areas ranging in colour from sea-foam yellow, lime

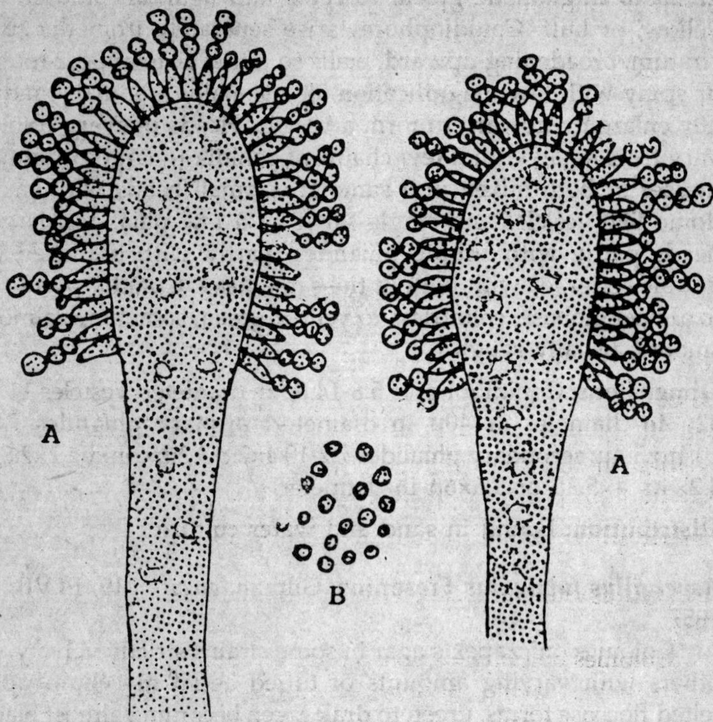

Fig. 9: *Aspergillus clavatus* (A) Conidial head (B) Conidia

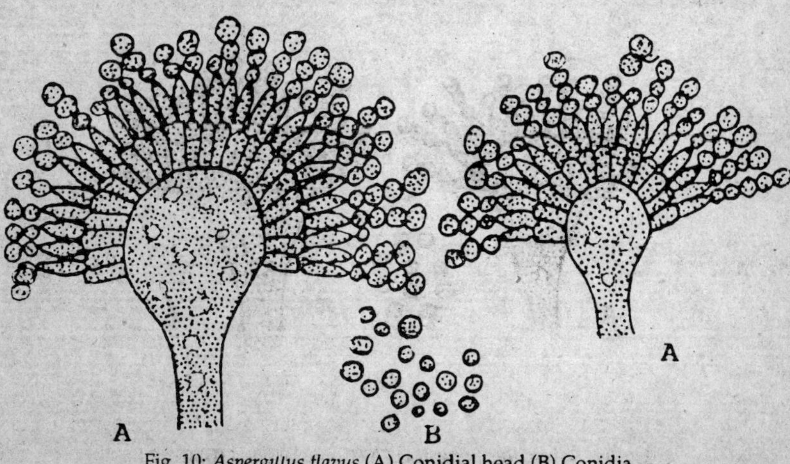

Fig. 10: *Aspergillus flavus* (A) Conidial head (B) Conidia

green to mignonette-green. Reverse and agar uncoloured or yellow, or buff. Conidiophores arise seperately from the substratum, broadening upward, walls so pitted as to appear rough or spiny with low magnification, occasionally granular, gradually enlarging upward to form a vesicle. Heads in every colony vary from small with a few chains of conidia to large columnar masses or both mixed in a same area; small heads with small dome like vesicles and single series of a few phialides; larger heads partly with simple phialides, partly with branched or double series, or with both in the same head. Conidia pyriform to almost globose, colourless to yellow-green, sometimes almost smooth, usually rough.

Dimensions: Conidiophores 5.6-14.8μ in diameter; vesicles 11.2-32μ in diameter or 40μ in diameter; primary phialides 7.2-10.4μ×3-4μ; secondary phialides 7.2-10.4×2.8-3.2μ; conidia 2×2.8μ, 3.2×4μ, 4×5.2μ or 5.6×6μ in diameter.

Distribution: Found in sand and water culture.

Aspergillus fumigatus **Fresenius.** Gilman, *op.cit.*, 219, Pl.VII, F, 1957.

Colonies on czapek's agar in some strains strictly velvety, in others with varying amounts of tufted aerial mycelium upto felted floccose forms, green to drak green becoming almost black in age, spreading. Reverse and substratum, colourless to yellow. Conidiophores short, usually densely crowded, arising directly

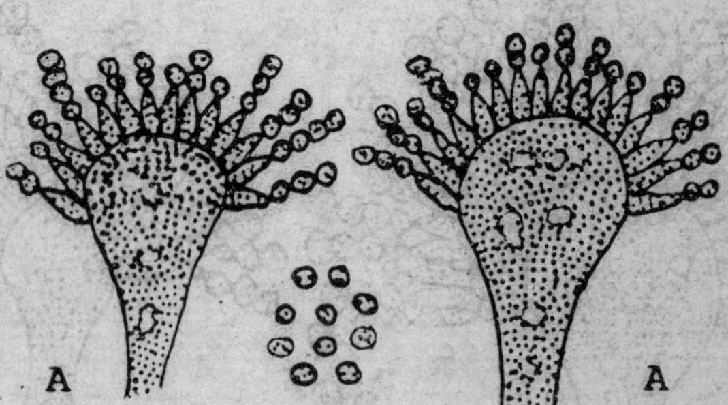

Fig. 11: *Aspergillus fumigatus* (A) Conidial head (B) Conidia

from submerged hyphae or non-septate, gradually enlarged, upward with apical flask shaped vesicles, fertile usually only on the upper half, bearing phialides in one series, crowded and closely packed. Conidia dark green in mass and globose.

Dimensions: Conidiophores 2-8.4μ in diameter; vesicles 21.2-30μ in diameter; phialides 6-8×2-4μ; conidia 2.4-4μ in diameter.

Distribution: Found in sand and water culture.

***Aspergillus luchuensis* Inui.** Gilman, *op.cit.*, 228, 1957.

This form shows a single series of phialides with conidia

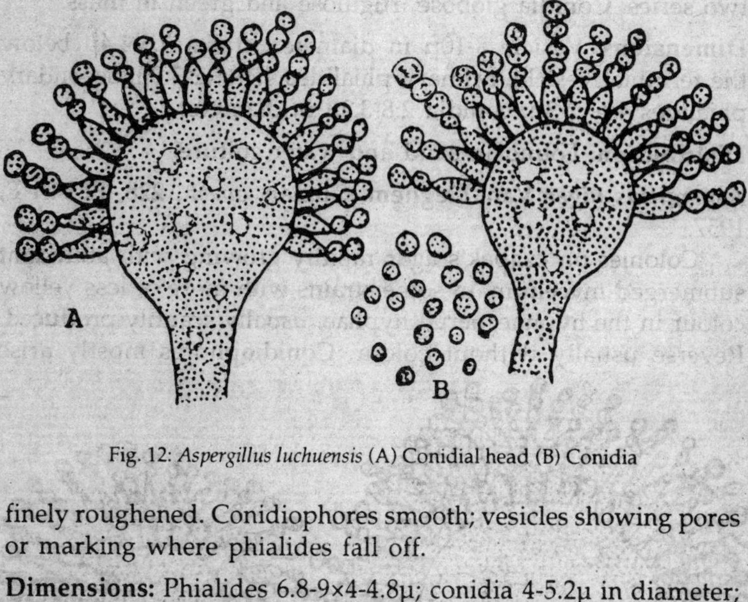

Fig. 12: *Aspergillus luchuensis* (A) Conidial head (B) Conidia

finely roughened. Conidiophores smooth; vesicles showing pores or marking where phialides fall off.

Dimensions: Phialides 6.8-9×4-4.8μ; conidia 4-5.2μ in diameter; conidiophores 9.2-15.2μ in diameter; vesicles 30.4-40μ in diameter.

Distribution: Found in sand culture.

***Aspergillus nidulans* (Eidam) Winter.** Gilman, *op.cit.*, 220, 1957.

Colonies on czapek's agar plane, spreading broadly, dark cress-green from abundant conidial heads during the first two weeks. Reverse of colony in shades of purplish-red, becoming dark in age. Conidial heads short columnar. Stalks sinuous, with smooth walls, in shades of cinnamon-brown. Phialides in

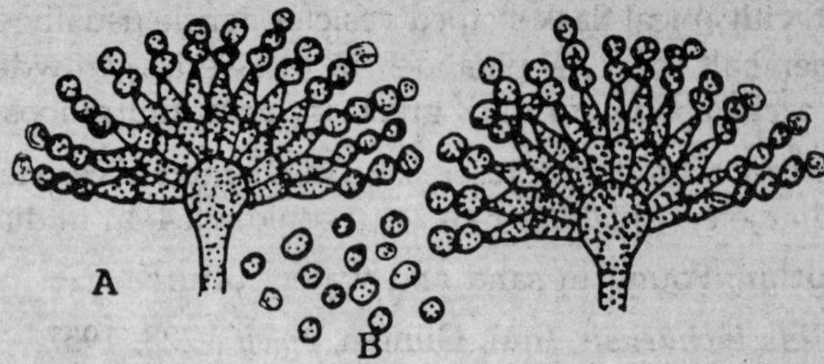

Fig. 13: *Aspergillus nidulans* (A) Conidial head (B) Conidia

two series. Conidia globose, rugulose and green in mass.

Dimensions: Vesicle 8-10μ in diameter; stalks 3.6-4.4μ below the terminal vesicle; primary phialides 4.4-6×2-3.2μ; secondary phialides 4-6×2-3μ; conidia 2.8-3.2μ in diameter.

Distribution: Found in sand and water culture.

***Aspergillus niger* Van Tieghem.** Gilman, *op.cit.*, 228, Pl.VIII E, 1957.

Colonies on czapek's agar rapidly growing with abundant submerged mycelium, in some strains with more or less yellow colour in the hyphae, aerial hyphae, usually scantily produced. Reverse usually without colour. Conidiophores mostly arise

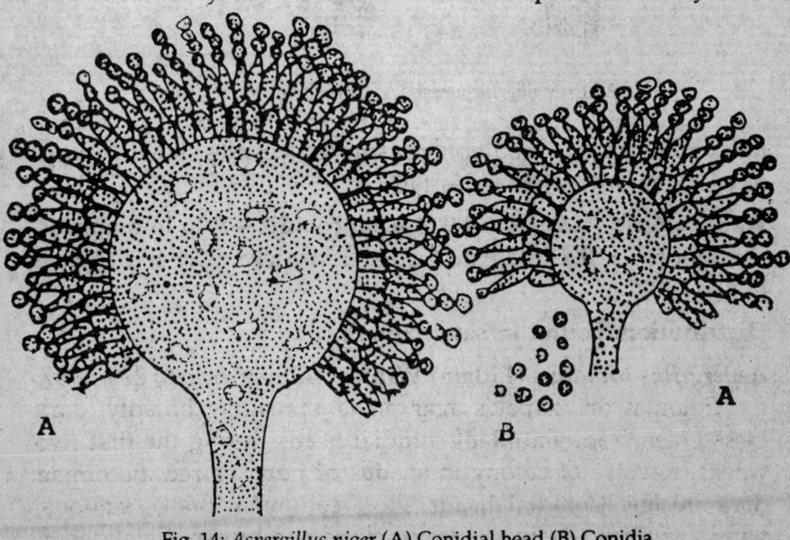

Fig. 14: *Aspergillus niger* (A) Conidial head (B) Conidia

directly from the substratum, smooth, septate or non-septate. Conidial heads fuscous, blackish-brown, purple brown, in every shade to carbonous black, varying from small almost columnar masses of a few conidial chains to the more common globose or radiate heads; phialides typically in two series, thickly covering the vesicle, primary varying greatly in length. Conidia globose, at first smooth, but later spinulose with colouring substance.

Dimensions: Vesicles commonly 21.6-52µ in diameter, upto 100µ in diameter; conidiophores 6-10.8µ in diameter; secondary phialides 6-10×2-4µ; conidia 2.4µ in diameter.

Distribution: Found in sand and water culture.

Aspergillus ustus **(Bainier) Thom and Church.** Gilman, *op.cit.*, Pl. VIII C, 1957.

Colonies more or less felted, floccose, with fine hyphae from white through shades of gray, olive-gray, yellow, yellow-brown toward fuscous, with often a greenish cast, but no true green colour, in old cultures purplish vinaceous at times. Reverse through shades of yellow, orange and brown. Few branches of aerial hyphae septate, sinuous, with rather thin, smooth, usually partly coloured some shade of brown; heads hemispherical to almost columnar; phialides colourless, semi-radiate, loosely arranged into 2 series. Conidia globose, spinulose, with chains forming fairly compact columns in old cultures.

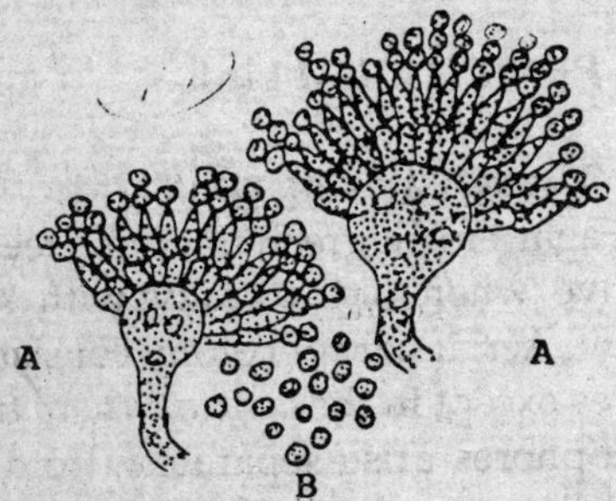

Fig. 15: *Aspergillus ustus* (A) Conidial head (B) Conidia

Dimensions: Vesicles 10-20.4µ in diameter; primary phialides 5.2-8×3-3.6µ, secondary 6-9.2×2-3.2µ; conidia 2.8-4µ in diameter.

Distribution: In sand culture.

BOTRYTIS Micheli

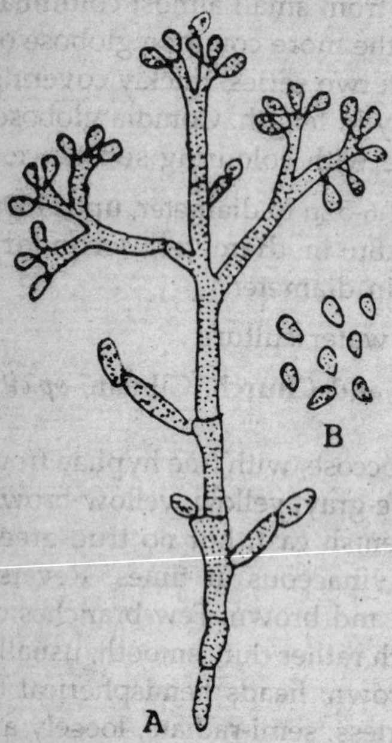

Fig. 16: *Botrytis terrestris*
(A) Conidiophore (B) Conidia

***Botrytis terrestris* Jensen.** Gilman, *op.cit.*, 299, 1957.

Colonies at first white, later gray; sterile hyphae creeping, hyalin, branched, septate. Conidiophores erect, ascending, septate, branched; primary and secondary branches verticillate, dichotomous, or alternate. Conidia produced on the ends of the branches, forming a more or less compact triangular cluster, obovate, somewhat angled, uniform, hyalin to light gray. Clusters of conidia separate very easily.

Dimensions: Sterile hyphae 1.6-3.2μ in diameter; conidiophores 2-4μ in diameter; conidia 3.2-4.8×2.4-3.2μ.

Distribution: In sand and water culture.

PENICILLIUM Link

***Penicillium citrinum* Thom.** Gilman, *op.cit.*, 251, Pl. IX C, 1957.

Colonies on bean agar bluish green to clear green, becoming olive to brownish-olive when old, usually with sterile white margin; reverse yellow. Aerial part of colony consists of densely standing conidiophores except in the centre, where tufts of aerial hyphae arise. Conidiophores arise separately from submerged hyphae or from mycelium on the surface. Conidial fructifications in two stages; each metulae producing of compact verticil of phialides. Conidial chains in columns, a separate column arising from each verticil of phialides, so that the fructification may appear double, triple or more complex. Conidia globose, green, slightly granular.

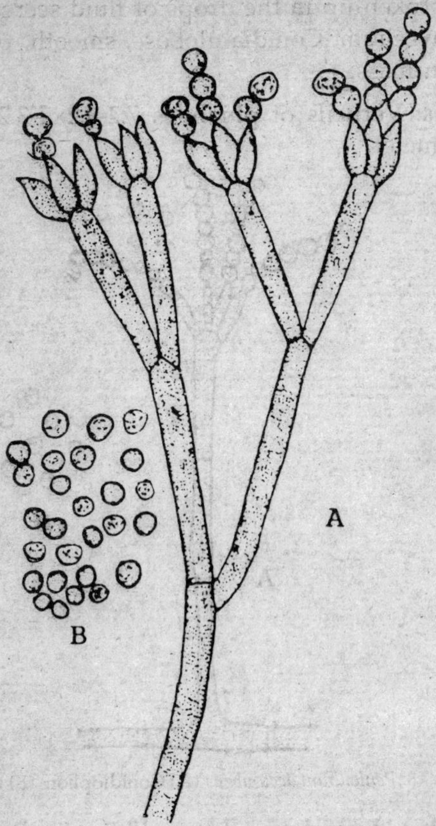

Fig. 17: *Penicillium citrinum* (A) Conidiophore (B) Conidia

Dimensions: Primary metulae 15.9-30.0×2.4-3.1μ, verticil of phialides 5.9-7.1×1.8-3.0μ; conidia 1.4-3.0 or 3.4μ in diameter.

Distribution: In sand and water culture.

***Penicillium decumbens* Thom.** Gilman, *op.cit.*, 245, 1957.

Colonies on potato agar white to gray, gray-green, ultimately yellowish brown. Reverse colourless. Surface growth consisting of trailing stolon-like hyphae sparsely developed and so close to the substratum as to appear only as fertile hyphae, bearing the conidiophores as short branches; in old cultures with dense tufts of sterile secondary mycelium scattered over the surface. Conidial fructifications consisting of single verticil of phialides, bearing conidial chains first in loose columns but soon enveloped and broken up in the drops of fluid secreted abundantly from the mycelium. Conidia globose, smooth, pale green, then brownish in mass.

Dimensions: Verticils of phialides 7.2-9.2×2-3.2μ; conidia 2.8-3.6μ in diameter.

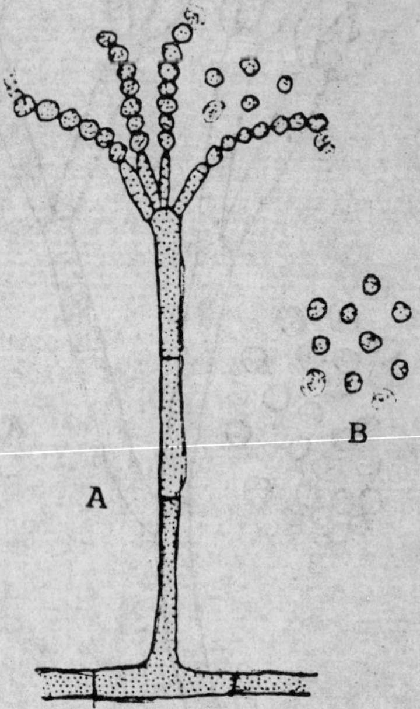

Fig. 18: *Penicillium decumbens* (A) Conidiophore (B) Conidia

Distribution : In sand and water culture.

TRICHODERMA (PERSOON) HARZ

***Trichoderma koningii* oudemans.** Gilman. *op.cit.*, 213, 1957.

Colonies on czapek's agar spreading, floccose, white at first, becoming light green in four to five days; may show various shades of light green, but never becomes deep green; reverse colourless. Vegetative hyphae septate, hyalin. Conidiophores arise as branches of aerial mycelium, alternate or opposite, di or trichotomously branched. Conidia oblong to elliptical, smooth, hyalin.

Dimensions: Conidiophore upto 24.8μ in height×2.8-3.2μ in diameter; conidia 2.8-4.8×2 2.8μ.

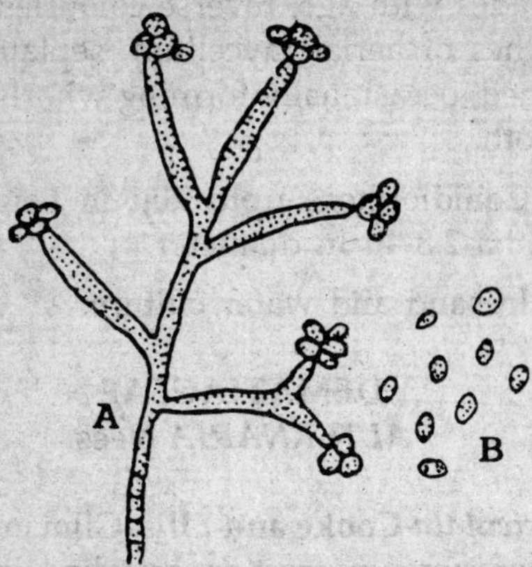

Fig. 19: *Trichoderma koningi* (A) Conidiophore (B) Conidia

Distribution: In sand and water culture.

***Trichoderma lignorum* (Tode) Harz**. Gilman, *op.cit.*, 213, Pl.VII, B, 1957.

Colonies on czapek's agar broadly spreading, hyalin, fruiting areas appear as tufts, white at first, and becoming various

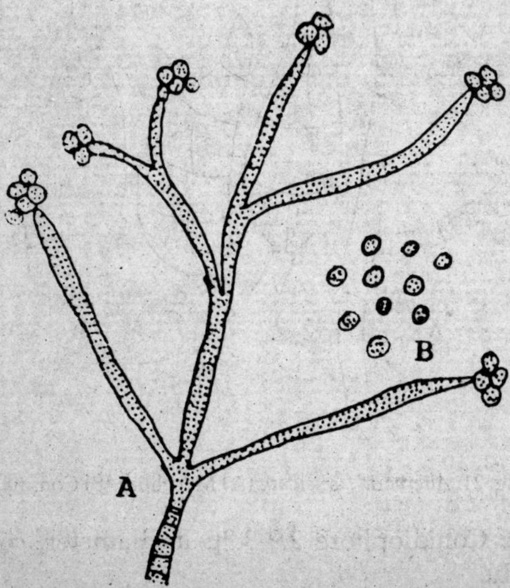

Fig. 20: *Trichoderma lignorum* (A) Conidiophores (B) Conidia

deep green shades with age; reverse colourless. Conidiophores arise as branches of aerial mycelium, septate, di or trichotomously branched, occasionally forming whorls. Conida globose to ovate, smooth.

Dimensions: Conidiophore upto 70μ in height × 2.8-3.2μ in diameter; conidia 2.8-4μ in diameter.

Distribution: In sand and water culture.

DEMATIACEAE
ALTERNARIA Nees

Alternaria fasciculata **Cooke and Ellis.** Gilman. *op.cit.*, 347, 1957.

Conidiophores brown, erect or ascending, irregularly curved, solitary or caespitose, septate, diameter uniform; conidia dark brown, oblong ovate, minutely apiculate, endochrome transversely two to seven times septate with usually several longitudinal septa, the apical cell short or elongate into a straight hyalin beak.

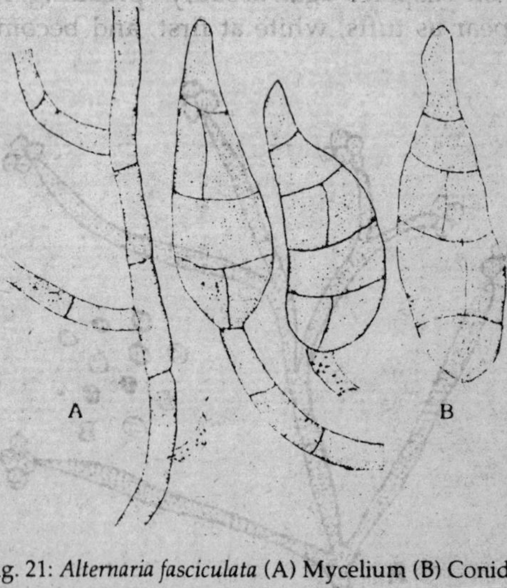

Fig. 21: *Alternaria fasciculata* (A) Mycelium (B) Conidia

Dimensions: Conidiophore 2.9-3.9μ in diameter; conidia 33.4-89.9×8.9-17.5μ.

Distribution: In sand and water culture.

CLADOSPORIUM LINK

***Cladosporium herbarum* (Persoon) Link.** Gilman, *op.cit.*, 334, Pl. XI D, 1957.

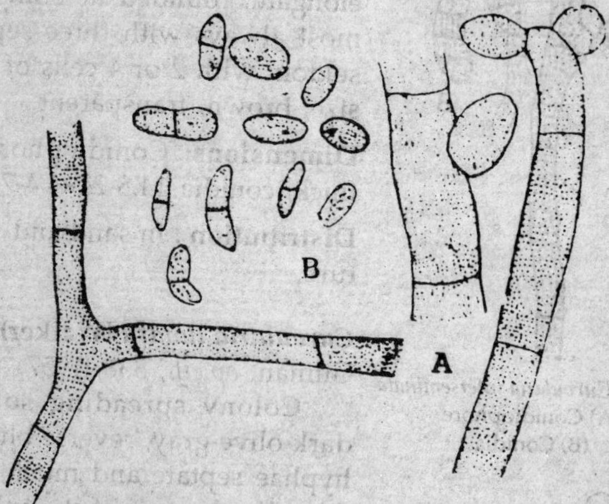

Fig. 22: *Cladosporium herbarum* (A) Conidiophore (B) Conidia

Turf matted, yellow-green, later black-green; conidiophores erect, little branched, septate, brown or olive-green. Conidia terminal, by extension of the tip falsely lateral, on short knee-like swellings, single or at times in chains, of various shapes, elongate, oval, and then usually one celled, or cylindrical-ellipsoid and then with one-to four-septa, smoky-brown or olive-green, slightly constricted at the septa, with a finely granulate or spiny wall of very different diameter and length.

Dimensions: Conidiophores 4.8-10µ in diameter; conidia one to 4 celled smoky brown.

Distribution: In sand and water culture.

CURVULARIA Boedijn

***Curvularia interseminata* (Berkeley and Ravenel).** Gilman, *op.cit.*, 337, 1957.

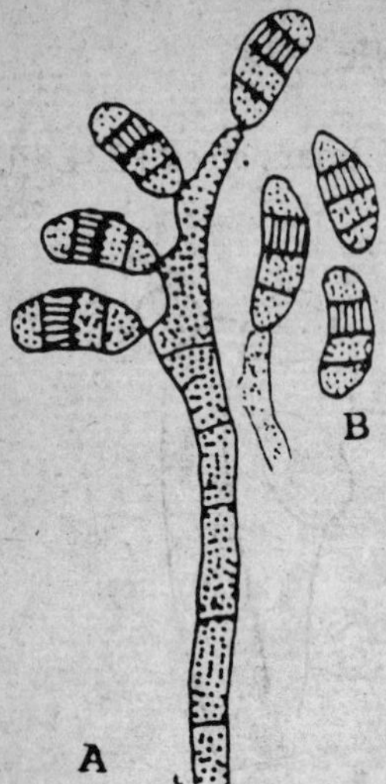

Fig. 23: *Curvularia interseminata*
(A) Conidiophore
(B) Conidia

Colonies broadly spreading, brown-black, velvety. Conidiophores grouped thickly together, erect, unbranched or branched, multiseptate, bent or with geniculations, slender elongate, rounded at both ends, almost always with three septa, more seldom with 2 or 4 cells of the same size, brown, transparent.

Dimensions: Conidiophores 3.6-4μ thick; conidia 14.8-24×4.4-7.2μ.

Distribution : In sand and water culture.

***Curvularia lunata* (Walker) Boedijin.** Gilman. *op.cit.*, 338, 1957.

Colony spreading subfloccose, dark olive-gray, reverse bluish black; hyphae septate and much branched. Conidiophores erect, unbranched. Spores borne more or less in a whorl at tip of conidiophore, three-septate, curved, brown.

Dimensions: Conidiophores more than 100μ long×2.8-4μ in diameter; spores 16.4-28×7.2-9.6μ.

Distribution: In sand and water culture.

Fig. 24: *Curvularia lunata*
(A) Conidiophore
(B) Conidia

TUBERCULARIACEAE
FUSARIUM Link

***Fusarium chlamydosporum* Woolenweber and Reinking.** Gilman, *op. cit.*, 359, 1957.

Fruiting layer, floccose, hyalin or pink, plectenchymatic below at times forming somewhat warty sclerotial knots, variously coloured, hyalin, carmine-purple-red, sulphur-yellow, or dark brown. The dark colour is caused by the occurrence in the

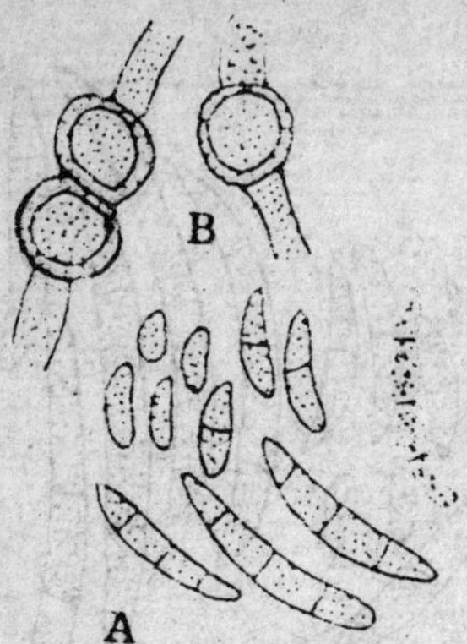

Fig. 25: *Fusarium chlamydosporum* (A) Microconidia & Macroconidia (B) Chlamydospores

mycelium of numerous, globose to pear shaped, smooth to rough or spiny, intercalary or terminal, single, paired, catenulate or muriform chlamydospores. Conidia small, spindle-ellipsoid, not globose, lemon-shaped, usually one celled seldom septate, three septate spores sickle-shaped, weakly pedicellate.

Dimensions: Chlamydospores 10-16μ in diameter; conidia 0-septate 4-10.8X2.8-4μ, 1-septate 11.6-16×2.8-4μ, 3-septate 26.4-32×2.8-4μ.

Distribution: In sand and water culture.

***Fusarium moniliforme* Sheldon.** Gilman, *op.cit.* 361, 1957.

Microconidia produced in chains and remaining connected or held in false heads, later becoming scattered over the bright yellow to rosy white aerial mycelium as a transparent shining powder, one-or two-celled, spindle- egg-shaped. Macroconidia delicate awl-shaped slightly crescent-shaped or almost straight, tapering at both ends, often constricted at the tip, bright in mass, three - to five-, seldom six- to seven-septate.

Dimensions: Conidia 0-septate 4-18×1.6-4μ, 1-septate 9.6-30.4×2.4-5.2μ, 3-septate 20-60×2.4-4.8μ, 5-septate 38-70×2-4.8μ, 7-septate 60-90×2-4.8μ.

Distribution: In sand and water culture.

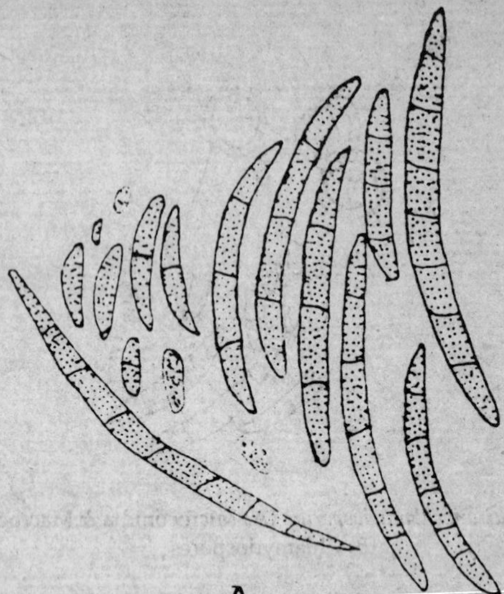

Fig. 26: *Fusarium moniliforme* (A) Microconidia and Macroconidia

***Fusarium oxysporum* Schlechtendahl.** Gilman, *op.cit.*, 367, 1957.

Stroma brownish-white to violet, plectenchymatic, smooth, extended or coloured green to blue-black by erumpent, sclerotial hard bodies, and more or less wrinkled, under moist conditions, usually covered by fascicled, medium high aerial myce-

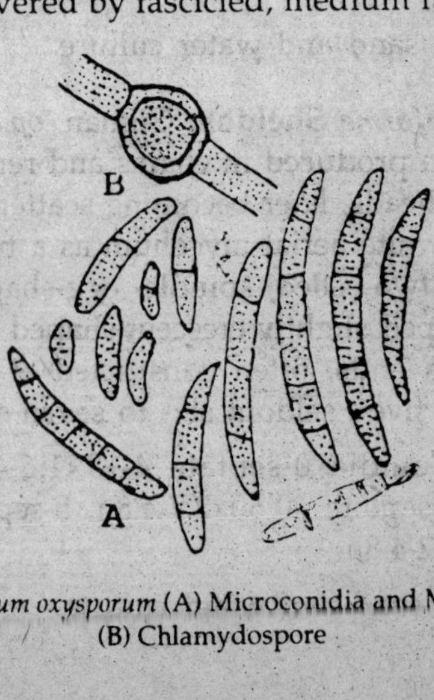

Fig. 27: *Fusarium oxysporum* (A) Microconidia and Macroconidia (B) Chlamydospore

lium, later forming sporodochia, more seldom pionnotes with three-septate spindle-sickle-shaped conidia, curved or almost straight. Smaller conidia, one-or two-celled, oval to reniform, are numerous in the aerial mycelium but are lacking in the typical fruiting layers of macroconidia. Chlamydospore terminal and intercalary, globose, smooth or wrinkled, one-celled seldom two-celled, in hyphae.

Dimensions: Conidia 0-septate 5.2-16×2-4μ, 1-septate 10-26.4×2-4.4 μ, 3-septate 19.2-46.4×2.4-6μ, 5-septate 30-60 × 3.6-4.8μ; chlamydospores 10-15μ in diameter.

Distribution : In sand and water culture.

***Fusarium poae* (Peck) Wollenweber.** Gilman, *op.cit.*, 358, 1957.

Fruiting layer cobwebby or felty-wooly, white to rose with various paired or whorled branches, of richly branched hyphae and conidiophores. Conidia usually one-celled (at most 4 per cent, one- to four-septate) in large numbers on the conidiophores. In addition to the typical globose, lemon-shaped, or

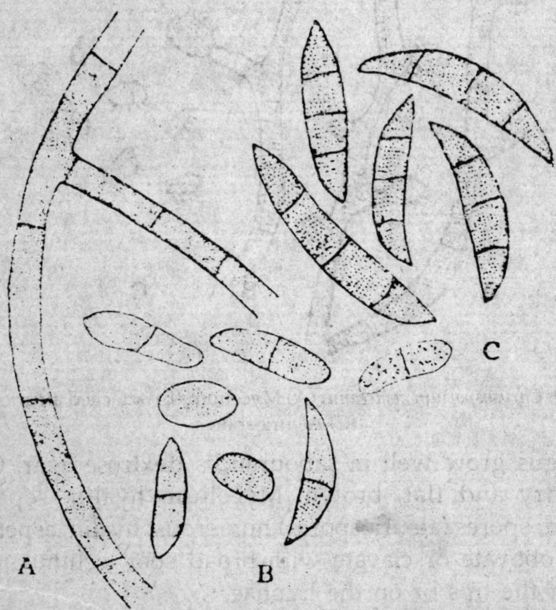

Fig. 28: *Fusarium poae* (A) Mycelium
(B) Microconidia and Macroconidia

pear-shaped one- or two-celled microconidia, long spindle-shaped, ellipsoid to sickle-shaped conidia occur.

Dimensions: Oval to lemon-shaped conidia; non-septate 6.5-8.9×3.9-6.0μ, sickle-shaped conidia three septate 15.9-34.9×3.4-5.0μ.

Distribution: In sand and water culture.

KERATINOPHILIC FUNGI AND DERMATOPHYTES

Chrysosporium evolceanui **(Randhawa and Sandhu) Garg.** Shrivastava J.N., Studies on Soil Inhabiting Dermatophytes and Keratinophilic Fungi, Ph. D. Thesis, Kanpur University 49, Fig. 3, 1985.

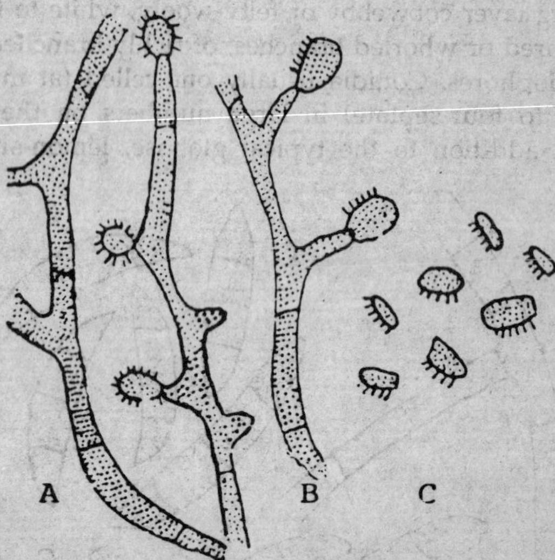

Fig. 29: *Chrysosporium evolceanui* (A) Mycelium (B) Attached aleuriospores (C) Aleuriospores

Fungus grow well in sabouraud's dextrose agar. Colonies white, dry and flat, brown. Mycelium hyalin, septate and branched; spores (aleuriospores) numerous, hyalin, aspetate, thin walled, obovate or clavate with broad soar, echinulate, borne singly at the tips or on the hyphae.

Dimensions: Mycelium 2.4-4.2μ thick; aleuriospores 4-10×2-4.4μ.

Distribution: Found in sand culture.

Chrysosporium indicum **(Randhawa and Sandhu) Garg.** Shrivastave, *op.cit.*, 55, Fig.5, 1985.

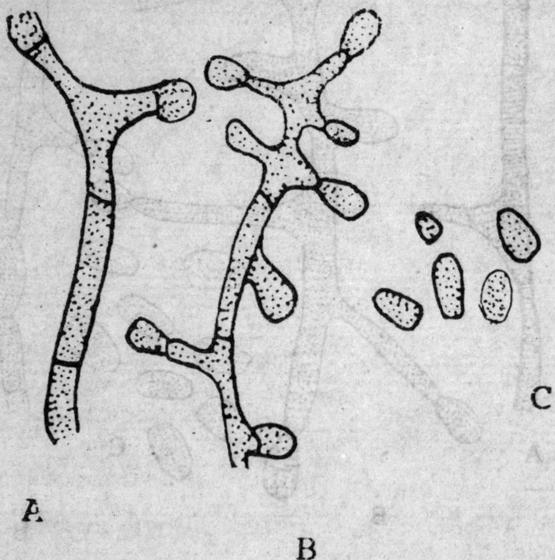

Fig. 30: *Chrysosporium indicum* (A) Mycelium (B) Attached aleuriospores (C) Aleuriospores

Fungus grow on sabouraud's dextrose agar medium. Colonies white, floccose become powdery on maturity. Mycelium hyalin, septate and branched. Terminal and lateral conidia sessile or on the short protrusions or slightly swollen lateral branches, solitary, occasionally subtended by another conidium, subhyalin, smooth or slightly echinulate, thin walled, obovoid, often cymbiform, one celled with fairly wide basal scar, borne singly on the mycelium.

Dimensions: Mycelium 2-4μ in width, aleuriospores 4-8x2-4μ.

Distribution: Found in sand culture.

Chrysosporium tropicum **Carmichael.** Shrivastava *op.cit.*, 58, Fig.6, 1985.

Fungus grow well in Sabouraud's dextrose agar. Colonies white felty, powdery dense at the centre, less than 1mm high margin, not well defined, fimbricate, reverse pale creamy to yellow. Vegetative hyphae septate, branched, hyalin and smooth.

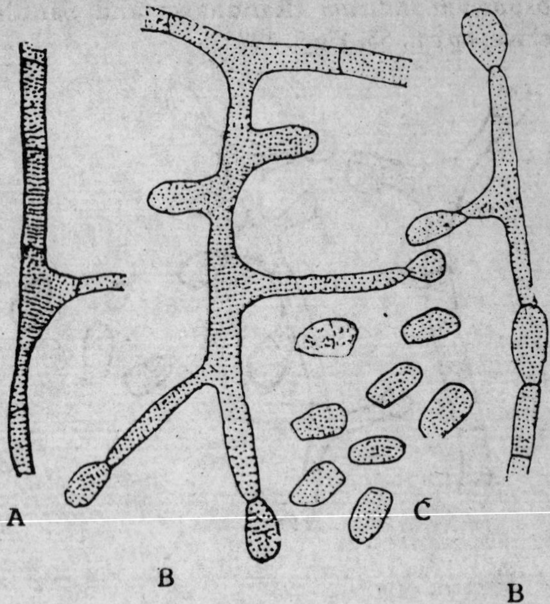

Fig. 31: *Chrysosporium tropicum* (A) Mycelium (B) Attached aleuriospores (C) Aleuriospores

Aleuriospores borne at the tips of the hyphae directly or laterally on short or long branches which sometimes become ampulliform. Aleuriospores single celled, hyalin, thin walled, smooth, subglobose to clavate with broad basal scar. Intercalary conidia rare, subhyalin, smooth and slightly thick walled, cylindrical to barrel shaped chlamydospores, smooth and thick walled, globose with narrow basal scar.

Dimensions: Vegetative hyphae 0.5-4μ wide, aleuriospores 4-8×2-4μ.

Distribution : Found in sand culture.

Microsporum fulvum **Uriburu.** Shrivastava. *op.cit.*, 66, Fig.8, 1985.

Fungus growing well on sabouraud's dextrose agar medium. Colonies floccose, white light buff, powdery in old cultures. Mycelium septate; microconidia few single celled, clavate; conidiophores long and delicate; macroconidia numerous, ellipsoid, rough walled, septate, septa thick and borne on lateral side

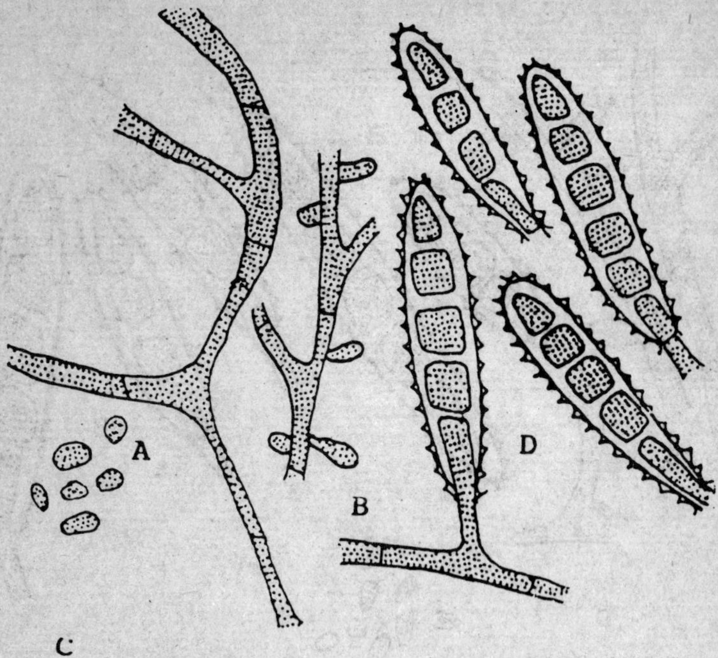

Fig. 32: *Microsporum fulvum* (A) Mycelium (B) Attached microconidia (C) Microconidia (D) Macroconidia

branches.

Dimensions: Microconidia 4-6×2-4μ, macroconidia measuring 20-42×6-10μ.

Distribution: Found in sand culture.

Microsporum gypseum **(Bodin) Guiart and Grigorakis.** Shrivastava, *op. cit.*, 68, Fig.9, 1985.

Colonies grow well on sabouraud's dextrose agar medium. Colonies powdery, buff or light brown in colour, sometimes develop white aerial mycelium, which later becomes powdery and light brown in the centre, reverse reddish brown or orange. In young cultures few, small single celled microconidia present. Macroconidia rough, thick walled, ellipsoid or fusiform with a narrow rounded tip and broad faceted and collard base. Some of the conidia are appendiculate. Racquet hyphae, nodular bodies and chlamydospores present.

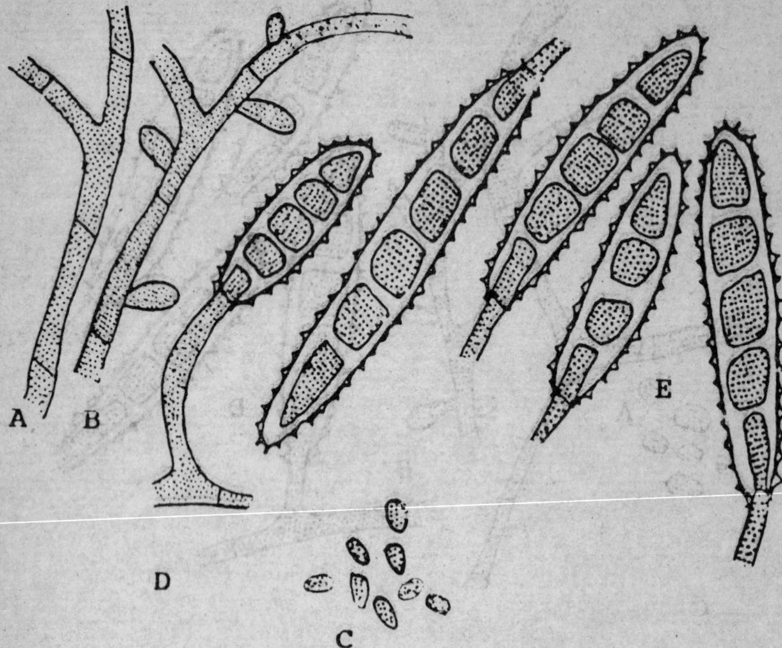

Fig. 33: *Microsporum gypseum* (A) Mycelium (B) Attached microconidia (C) Microconidia (D) Attached macroconidia (E) Macroconidia

Dimensions: Microconidia measuring 4-8×2-4μ, macroconidia 22-62×8-8.4μ.

Distribution : Found in sand culture.

Trichophyton simii **Stockdale, Mackenzie and Austwick.** Shrivastava, *op. cit.*, 72, Fig.11, 1985.

Fungus grow well in sabouraud's dextrose agar medium. Colonies white or pale buff; velvety or finely granular with a central umbo and a diffuse margin. The vinaceous pigment observed on the reverse of colonies. Mycelium septate; microconidia single celled, abundant; macroconidia smooth, thin walled, cylindrical to fusiform, multiseptate, borne on lateral and side branches, rapidly degenerating with formation of intercalary conidia; chlamydospores clavate fusiform.

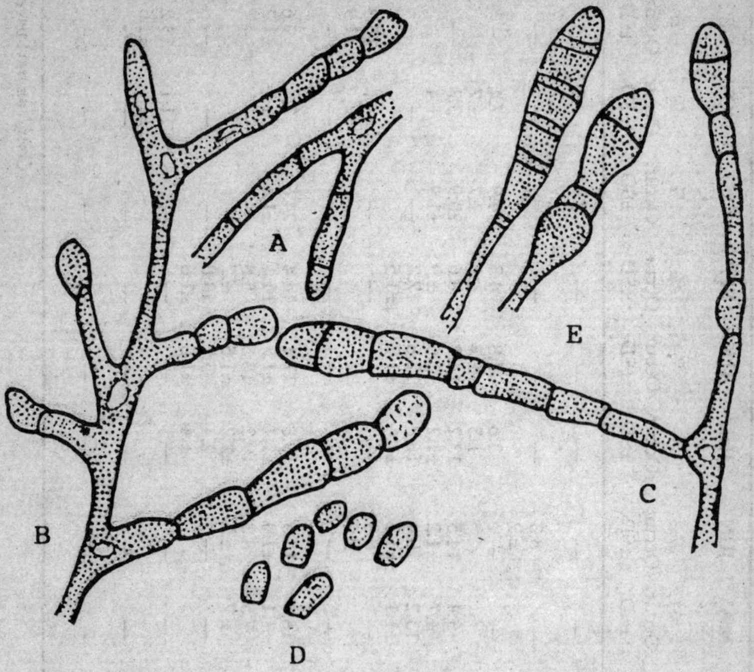

Fig. 34: Trichophyton simii (A) Mycelium (B) Attached conidia (C) Intercalary conidia (D) Microconidia (E) Macroconidia and chlamydospore

Dimensions: Macroconidia measuring 28-60×6-8μ; microconidia 4-8×2-4.4μ.

Distribution: Found in sand culture.

Fungi From Sand Culture

Twenty six species of fungi belonging to twelve genera were isolated.

January : During January 11 species of 8 genera were isolated with following percentage—*Alternaria fasciculata* 14.2±1.5, *A. nidulans* 7.9±1.2, *A. niger* 3.1±0.4, *Cladosporium herbarum* 4.7±0.5, *Cunninghamella bertholettiae* 15.8±1.1, *Fusarium chlamydosporum* 3.1±0.4, *F. moniliforme* 6.3±0.9, *F. oxysporum* 6.3±0.7, *Mucor hiemalis* 17.4±2.1, *Penicillium citrinum* 15.8±1.5 and *Trichoderma koningii* 4.7±0.6.

Bottom Sand Fingi of River Ganga

S.No.	ORGANISM	JAN % Occur. ±SD	FEB % Occur. ±SD	MAR % Occur. ±SD	APR % Occur. ±SD	MAY % Occur. ±SD	JUN % Occur. ±SD	JUL % Occur. ±SD	AUG % Occur. ±SD	SEP % Occur. ±SD	OCT % Occur. ±SD	NOV % Occur. ±SD	DEC % Occur. ±SD
1	*Alternaria fasciculata*	14.2 ±1.5	18.3 ±1.5	13.3 ±.9	—	—	—	—	—	—	3.7 ±.7	9.2 ±.7	5.7 ±.7
2.	*Aspergillus chevalieri*	—	—	15.5 ±1.2	—	—	8.3 ±.9	—	—	—	—	—	—
3.	*Asoergillus clavatus*	—	—	—	—	—	—	3.6 ±.5	2.8 ±.4	4.3 ±.5	4.9 ±.7	7.4 ±.7	—
4.	*Aspergillus flavus*	—	—	4.4 ±.4	15.6 ±1.0	9.4 ±.7	14.5 ±1.2	13.2 ±1.2	14.2 ±1.6	8.6 ±.9	20.9 ±1.9	3.7 ±.4	—
5.	*Aspergillus fumigatus*	—	—	—	13.7 ±1.2	13.2 ±1.1	10.4 ±1.2	16.8 ±1.6	12.8 ±1.7	17.3 ±1.5	—	—	—
6.	*Aspergillus luchuensis*	—	10.2 ±.9	11.1 ±.9	—	—	8.3 ±.9	—	—	—	—	—	15.3 ±1.9
7.	*Aspergillus nidulans*	7.9 ±1.2	6.1 ±.7	—	—	—	—	—	—	—	—	—	—
8.	*Aspergillus niger*	3.1 ±.4	8.1 ±.9	20.0 ±1.7	19.6 ±1.9	18.8 ±1.1	12.5 ±1.0	10.8 ±1.1	17.1 ±1.7	14.4 ±1.1	24.6 ±1.6	7.4 ±.7	9.6 ±1.2
9.	*Aspergillus ustus*	—	—	—	9.8 ±.9	7.5 ±.7	8.3 ±.5	19.2 ±2.2	18.5 ±1.5	5.8 ±.7	—	—	—
10.	*Botrytis terrestris*	—	—	—	—	—	—	2.4 ±.4	5.7 ±1.0	13.0 ±1.6	—	—	—
11.	*Chaetomium sps*	—	—	—	—	—	—	—	—	—	—	11.1 ±1.2	19.2 ±1.6
12.	*Cladosporium herbarum*	4.7 ±.5	4.0 ±.4	—	—	—	—	—	—	—	—	—	—
13.	*Cunninghamella bertholettiae*	15.8 ±1.1	—	—	—	—	—	—	—	—	—	—	—

Contd. on next page

S.No.	ORGANISM	JAN % Occur. ±SD	FEB % Occur. ±SD	MAR % Occur. ±SD	APR % Occur. ±SD	MAY % Occur. ±SD	JUN % Occur. ±SD	JUL % Occur. ±SD	AUG % Occur. ±SD	SEP % Occur. ±SD	OCT % Occur. ±SD	NOV % Occur. ±SD	DEC % Occur. ±SD
14.	*Curvularia interseminate*	—	—	—	—	—	—	—	—	—	—	3.7 ±.4	5.7 ±.7
15.	*Curvularia lunata*	—	6.1 ±.5	—	—	—	—	—	—	—	—	11.1 ±1.0	7.6 ±.7
16.	*Fusarium chlamydosporum*	3.1 ±.4	—	—	—	—	—	—	—	—	—	—	—
17.	*Fusarium moniliforme*	6.3 ±.9	—	—	—	—	—	—	—	—	—	3.7 ±.4	—
18.	*Fusarium oxysporum*	6.3 ±.7	16.3 ±1.5	—	—	—	—	—	—	—	—	12.9 ±1.5	9.6 ±1.2
19.	*Fusarium poae*	—	8.1 ±.9	—	—	—	8.3 ±.7	—	—	—	—	—	3.8 ±.4
20.	*Mucor hiemalis*	17.4 ±2.1	14.2 ±1.1	11.1 ±.9	11.7 ±.9	15.0 ±1.0	12.5 ±1.07	4.8 ±.7	10.0 ±1.1	10.1 ±1.3	23.4 ±1.8	14.8 ±1.2	7.6 ±.5
21.	*Mucor racemosus*	—	—	—	—	3.7 ±.4	—	3.6 ±.5	4.26 ±.5	7.2 ±.7	3.7 ±.7	5.5 ±.7	5.7 ±1.1
22.	*Penicillium citrinum*	15.8 ±1.5	8.1 ±.7	15.5 ±1.1	21.5 ±1.4	26.7 ±2.1	16.6 ±1.07	10.8 ±.9	11.4 ±1.2	15.9 ±2.3	13.5 ±1.4	9.2 ±1.1	9.6 ±.9
23.	*Penicillium decumbens*	—	—	—	—	5.6 ±.7	—	14.4 ±1.7	2.8 ±.4	2.9 ±.49	4.9 ±.7	—	—
24.	*Phoma sps*	—	—	—	7.8 ±.9	5.6 ±.5	—	—	—	—	—	—	—
25.	*Trichoderma lignorum*	—	—	8.8 ±.7	—	—	—	—	—	—	—	—	—
26.	*Trichoderma koningii*	4.7 ±.6	—	—	—	—	—	—	—	—	—	—	—

February : Seven genera and 10 species were isolated with following percentage occurrence—*Alternaria fasciculata* 18.3±1.5, *Aspergillus luchuensis* 10.2±0.9, *A. nidulans* 6.1±0.7, *A. niger* 8.1±0.9, *Cladosporium herbarum* 4.0±0.4, *Curvularia lunata* 6.1±0.5, *Fusarium oxysporum* 16.3±1.5, *F. Poae* 8.1±0.9, *Mucor hiemalis* 14.2±1.1 and *Penicillium citrinum* 8.1±0.7.

March : Five genera comprising 8 species were isolated during March with following percentage occurrence—*Alternaria fasciculata* 13.3±0.9, *Aspergillus chevalieri* 15.5±1.2, *A. flavus* 4.4±0.4, *A. luchuensis* 11.1±0.9, *A. niger* 20.0±1.7, *Mucor hiemalis* 11.1±0.9, *Penicillium citrinum* 15.5±1.1 and *Trichoderma lignorum* 8.8±0.7.

April : Seven species of 4 genera were isolated during April with following percentage occurrence—*Aspergillus flavus* 15.6±1.0, *A. fumigatus* 13.7±1.2, *A. niger* 19.6±1.9, *A. ustus* 9.8±0.9, *Mucor hiemalis* 11.7±0.9, *Penicillium citrinum* 21.5±1.4 and *Phoma sps.* 7.8±0.9

May : The fungal community comprised 4 genera spread over 9 species isolated with following percentage occurrence—*Aspergillus flavus* 9.4±0.7, *A. fumigatus* 13.2±1.1, *A. niger* 18.8±1.1, *A. ustus* 7.5±0.7, *Mucor hiemalis* 15.0±1.0, *M. racemosus* 3.7±0.4, *Penicillium citrinum* 20.7±2.1, *P. decumbens* 5.6±0.7 and *Phoma sps.* 5.6±0.5.

June : June isolates comprised 4 genera spread over 9 species with following percentage occurrence—*Aspergillus chevalieri* 8.3±0.9, *A. flavus* 14.5 ±1.2, *A. fumigatus* 10.4±1.2, *A. luchuensis* 8.3±0.9, *A. niger* 12.5±1.0, *A. ustus* 8.3±0.5, *Fusarium poae* 8.3±0.7, *Mucor hiemalis* 12.5±1.0 and *Penicillium citrinum* 16.6±1.0.

July : Four genera belonging to 10 species were isolated with following percentage occurrence—*Aspergillus clavatus* 3.6±0.5, *A. flavus* 13.2±1.2, *A. fumigatus* 16.8±1.6, *A. niger* 10.8±1.1, *A. ustus* 19.2±2.0, *Botrytis terrestris* 2.4±0.4, *Mucor hiemalis* 4.8±0.7, *M. racemosus* 3.6±0.5, *Penicillium citrinum* 10.8±0.9, *P. decumbens* 14.4±1.7.

August : Four genera spread over 10 species were isolated with following percentage occurrence—*Aspergillus clavatus* 2.8±0.4, *A. flavus* 14.2±1.6, *A. fumigatus* 12.8±1.7, *A. niger* 17.1±1.7, *A. ustus* 18.5±1.5, *Botrytis terrestris* 5.7±1.0, *Mucor hiemalis* 10.0±1.1, *M. racemosus* 4.2±0.5, *Penicillium citrinum* 14.4±1.2, *P. decumbens* 2.8±.49.

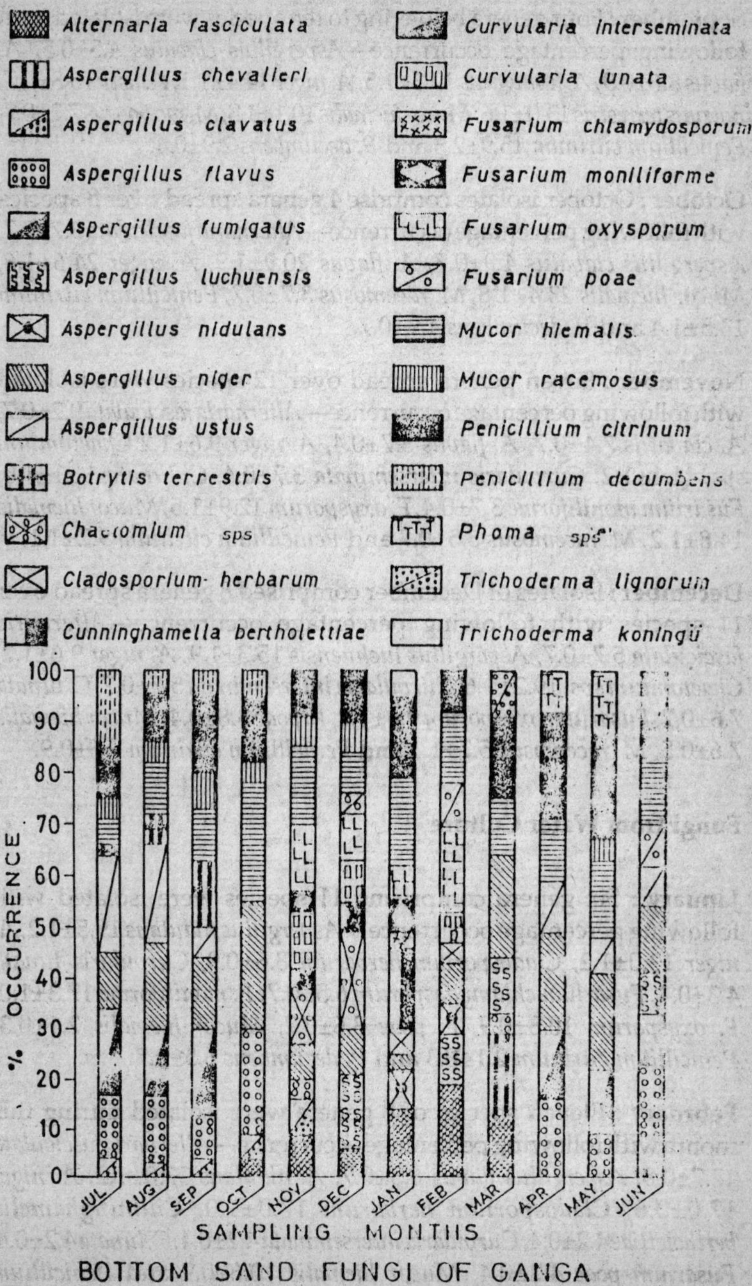

BOTTOM SAND FUNGI OF GANGA

September : Four genera belonging to ten species were isolated with following percentage occurrence—*Aspergillus clavatus* 4.3±0.5, *A. flavus* 8.6±0.9, *A. fumigatus* 17.3±1.5, *A. niger* 14.4±1.1, *A. ustus* 5.8±0.7, *Botrytis terrestris* 13.0±1.6, *Mucor hiemalis* 10.1±1.3, *M. racemosus* 7.2±0.7, *Penicillium citrinum* 15.9±2.3 and *P. decumbens* 2.9±0.4.

October : October isolates comprise 4 genera spread over 8 species with following percentage occurrence—*Alternaria fasciculata* 3.7±0.7, *Aspergillus clavatus* 4.9±0.7, *A. flavus* 20.9±1.9, *A. niger* 24.6±1.6, *Mucor hiemalis* 23.4±1.8, *M. racemosus* 3.7±0.7, *Penicillium citrinum* 13.5±1.4 and *P. decumbens* 4.9±0.7.

November : Seven genera spread over 12 species were isolated with following percentage occurrence—*Alternaria fasciculata* 9.2±0.7, *A. clavatus* 7.4±0.7, *A. flavus* 3.7±0.4, *A. niger* 9.6±1.2 *Chaetomium sps.* 11.1±1.2, *Curvularia interseminata* 3.7±0.4, *C. lunata* 11.1±1.0, *Fusarium moniliforme* 3.7±0.4, *F. oxysporum* 12.9±1.5, *Mucor hiemalis* 14.8±1.2, *M. racemosus* 5.5±0.7 and *Penicillium citrinum* 9.2±1.1.

December : Isolates of December comprised 8 genera spread over 11 species with following percentage occurrence—*Alternaria fasciculata* 5.7±0.7, *Aspergillus luchuensis* 15.3±1.9, *A. niger* 9.6±1.2, *Chaetomium sps.* 19.2±1.6, *Curvularia interseminata* 5.7±0.7, *C. lunata* 7.6±0.7, *Fusarium oxysporum* 9.6±1.2, *F. poae* 3.8±0.4, *Mucor hiemalis* 7.6±0.5, *M. racemosus* 5.7±1.1 and *Penicillium citrinum* 9.6±0.9.

Fungi from Water Culture

January : Six genera comprising 11 species were isolated with following percentage occurrence—*Aspergillus nidulans* 19.5±1.2, *A. niger* 13.0±1.2, *Cladosporium herbarum* 8.6±0.9, *Curvularia lunata* 4.3±0.4, *Fusarium chlamydosporum* 6.5±0.7, *F. moniliforme* 17.3±1.0, *F. oxysporum* 10.8±0.9, *F. poae* 8.6±0.7, *Mucor hiemalis* 2.1±0.3, *Penicillium citrinum* 2.1±0.3 and *P. decumbens* 6.5±0.7.

February : Eleven species of 8 genera were isolated during this month with following percentage occurrence—*Alternaria fasciculata* 12.7±0.8, *Aspergillus flavus* 6.3±0.7, *A. nidulans* 17.0±1.2, *A. niger* 17.0±0.6, *Cladosporium herbarum* 17.0±1.0, *Cunninghamella bertholettiae* 4.2±0.4, *Curvularia interseminata* 4.2±0.4, *C. lunata* 4.2±0.4, *Fusarium poae* 4.2±0.4, *Mucor hiemalis* 10.6±0.9 and *Penicillium citrinum* 2.1±0.3.

March : Seven genera spread over 9 species were isolated with following percentage occurrence—*Aspergillus flavus* 6.6±0.4, *A. fumigatus* 3.3±0.1, *A. niger* 23.3±1.0, *Cladosporium herbarum* 16.6±0.95, *Cunninghamella bertholettiae* 6.6±0.4, *Fusarium poae* 3.3±0.3, *Mucor hiemalis* 30.0 ±0.9, *Penicillium citrinum* 3.3±0.3 and *Trichoderma lignorum* 6.6±0.4.

April : Six genera including 9 species were isolated during April with following percentage occurrence—*Aspergillus flavus* 2.8±0.3, *A. niger* 17.1±1.0, *Cladosporium herbarum* 8.5±0.7, *Cunninghamella bertholettiae* 2.8±0.3, *Mucor hiemalis* 28.5±1.1, *M. racemosus* 2.8±0.3, *Penicillium citrinum* 2.8±0.3, *Trichoderma koningii* 8.5±0.5 and *T. lignorum* 25.7±0.9.

May : May isolates included 8 species belonging to 6 genera. Their percentage occurrence is indicated below—*Aspergillus clavatus* 13.7±0.7, *A. fumigatus* 3.4±0.3, *A. niger* 13.7±0.7, *Botrytis terrestris* 13.7±0.7, *Cladosporium herbarum* 13.7±0.7, *Mucor hiemalis* 13.7±0.7, *Penicillium citrinum* 10.3±0.7 and *P. decumbens* 17.2±0.9.

June : Five genera spread over 9 species were isolated with following percentage occurrence—*Alternaria fasciculata* 22.2±0.9, *Aspergillus flavus* 3.7±0.3, *A. fumigatus* 3.7±0.3, *A. niger* 18.5±0.9, *Cunninghamella bertholettiae* 3.7±0.7, *Mucor hiemalis* 18.5±0.9, *M. racemosus* 7.4±0.7, *Penicillium citrinum* 7.4±1.7 and *P. decumbens* 14.8±0.7.

July : Four genera spread over 8 species were isolated with following percentage occurrence—*Aspergillus clavatus* 6.2±0.4, *A. flavus* 6.2±0.4, *A. fumigatus* 6.2±0.4, *A. niger* 28.1±1.2, *Botrytis terrestris* 9.3±0.7, *Mucor hiemalis* 25.0±1.4, *M. racemosus* 12.5±0.9 and *Penicillium citrinum* 6.2±0.4.

August : Isolates of August comprised 4 genera and 9 species with following percentage occurrenece—*Alternaria fasciculata* 15.6±0.7, *Aspergillus clavatus* 6.2±0.4, *A. flavus* 6.2±0.4, *A. fumigatus* 6.2±0.4, *A. niger* 12.5±0.7, *Mucor hiemalis* 31.2±1.1, *M. racemosus* 6.2±0.4, *Penicillium citrinum* 9.3±0.7 and *P. decumbens* 6.2±0.4.

September : Six genera spread over 10 species were isolated with following percentage occurrence—*Alternaria fasciculata* 19.5±1.2, *Aspergillus clavatus* 12.1±0.9, *A. flavus* 4.8±0.4, *A. fumigatus* 2.4±0.3, *A. niger* 19.5±1.2, *Botrytis terrestris* 4.8±0.4,

Fungi of Ganga Water

S.No.	ORGANISM	JAN % Occur. ±SD	FEB % Occur. ±SD	MAR % Occur. ±SD	APR % Occur. ±SD	MAY % Occur. ±SD	JUN % Occur. ±SD	JUL % Occur. ±SD	AUG % Occur. ±SD	SEP % Occur. ±SD	OCT % Occur. ±SD	NOV % Occur. ±SD	DEC % Occur. ±SD
1.	*Alternaria fasciculata*	—	12.7 ±.8	—	—	—	22.2 ±.9	—	15.6 ±.7	19.5 ±1.2	2.2 ±.3	—	—
2.	*Aspergillus clavatus*	—	—	—	—	13.7 ±.7	—	6.2 ±.4	6.2 ±.4	12.1 ±.9	—	7.6 ±.7	—
3.	*Aspergillus flavus*	—	6.3 ±.7	6.6 ±.4	2.8 ±.3	—	3.7 ±.3	6.2 ±.4	6.2 ±.4	4.8 ±.4	4.5 ±.4	5.1 ±.4	—
4.	*Aspergillus fumigatus*	—	—	3.3 ±.1	—	3.4 ±.3	3.7 ±.3	6.2 ±.4	6.2 ±.4	2.4 ±.3	2.2 ±.3	—	—
5.	*Aspergillus nidulans*	19.5 ±1.2	17.0 ±1.2	—	—	—	—	—	—	—	9.0 ±.7	20.5 ±1.0	12.5 ±1.2
6.	*Aspergillus niger*	13.0 ±1.2	17.0 ±.6	23.3 ±1.0	17.1 ±1.0	13.7 ±.7	18.5 ±.95	28.1 ±1.2	12.5 ±.7	19.5 ±1.2	15.9 ±1.2	—	22.5 ±1.2
7.	*Botrytis terrestris*	—	—	—	—	13.7 ±.7	—	9.3 ±.7	—	4.8 ±.4	—	—	—
8.	*Cladosporium herbarum*	8.6 ±.9	17.0 ±1.0	16.6 ±.9	8.5 ±.7	13.7 ±.7	—	—	—	—	—	—	±5.0 ±.4
9.	*Cunnighamella bertholettiae*	—	4.2 ±.4	6.6 ±.4	2.8 ±.3	—	3.7 ±.3	—	—	—	—	—	—
10.	*Curvularia interseminata*	—	4.2 ±.4	—	—	—	—	—	—	—	4.5 ±.4	12.8 1.2	—
11.	*Curvularia lunata*	4.3 ±.4	4.2 ±.4	—	—	—	—	—	—	—	18.1 ±1.0	25.6 1.4	10.0 ±.7

Contd. on next page

S.No.	ORGANISM	JAN % Occur. ±SD	FEB % Occur. ±SD	MAR % Occur. ±SD	APR % Occur. ±SD	MAY % Occur. ±SD	JUN % Occur. ±SD	JUL % Occur. ±SD	AUG % Occur. ±SD	SEP % Occur. ±SD	OCT % Occur. ±SD	NOV % Occur. ±SD	DEC % Occur. ±SD
12.	*Fusarium chlamydosporum*	6.5 ±.7	-	—	—	—	—	—	-	—	6.8 ±.7	5.1 ±.4	—
13.	*Fusarium moniliforme*	17.3 ±1.0	—	—	—	—	—	—	—	19.5 ±1.2	20.4 ±1.2	10.2 ±.7	15.0 ±1.0
14.	*Fusarium oxysporum*	10.8 ±.9	—	—	—	—	—	—	—	—	—	—	10.0 ±.9
15.	*Fusarium poae*	8.6 ±.7	4.2 ±.4	3.3 ±.3	—	—	—	—	—	—	9.0 ±.9	7.6 ±.7	5.0 ±.4
16.	*Mucor hiemalis*	2.1 ±.3	10.6 ±.9	30.0 ±.9	28.5 ±1.1	13.7 ±.9	18.5 ±.9	25.0 ±1.4	31.2 ±1.1	—	2.2 ±.3	—	7.5 ±.7
17.	*Mucor racemosus*	—	—	—	2.8 ±.3	-	7.4 ±.7	12.5 ±.9	6.2 ±.4	7.3 ±.7	—	2.5 ±.3	2.5 ±.3
18.	*Penicillium citrinum*	2.1 ±.3	2.1 ±.3	3.3 ±.3	2.8 ±.3	10.3 ±.7	7.4 ±1.7	6.2 ±.4	9.3 ±.7	4.8 ±.4	4.5 ±.4	2.5 ±.3	2.5 ±.3
19.	*Penicillium decumbens*	6.5 ±.7	—	—	—	17.2 ±.9	14.8 ±.7	—	6.2 ±.4	4.8 ±.4	—	—	—
20.	*Trichoderma koningii*	—	—	—	8.5 ±.5	—	—	—	—	—	—	—	—
21.	*Trichoderma lignorum*	—	—	6.6 ±.4	25.7 ±.9	—	—	—	—	—	—	—	7.5 ±.7

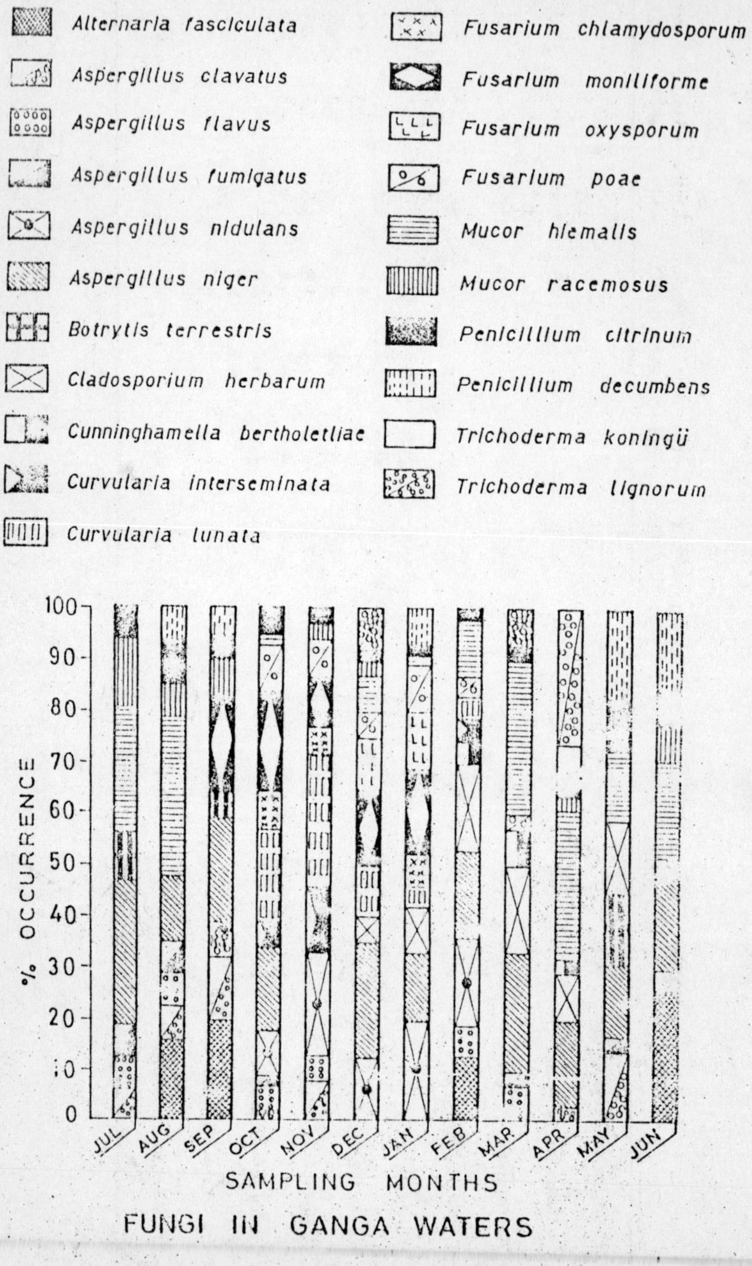

FUNGI IN GANGA WATERS

Fusarium moniliforme 19.5±1.2, *Mucor racemosus* 7.3±0.7, *Penicillium citrinum* 4.8±0.4 and *P. decumbens* 4.8±0.4.

October : Six genera spread over 12 species were isolated with following percentage occurrence—*Alternaria fasciculata* 2.2±0.3, *A. flavus* 4.5±0.4, *A. fumigatus* 2.2±0.3, *A. nidulans* 9.0±0.79, *A. niger* 15.9±1.2, *Curvularia interseminata* 4.5±0.4, *C. lunata* 18.1±1.0, *Fusarium chlamydosporum* 6.8±0.7, *F. moniliforme* 20.4±1.2, *F. poae* 9.0±0.9, *Mucor hiemalis* 2.2±0.3 and *Penicillium citrinum* 4.5±0.4.

November : November isolates comprised 6 genera spread over 10 species with following percentage occurrence—*Aspergillus clavatus* 7.6±0.7, *A. flavus* 5.1±0.4, *A. nidulans* 20.5±1.0, *Curvularia interseminata* 12.8±1.2, *C. lunata* 25.6±1.4, *Fusarium chlamydosporum* 5.1±0.4, *F. moniliforme* 10.2±0.7, *F. poae* 7.6±0.7, *Mucor racemosus* 2.5±0.3 and *Penicillium citrinum* 2.5±0.3.

December : Eleven species belonging to 7 genera were isolated during this month with following percentage occurrence—*Aspergillus nidulans* 12.5±1.2, *A. niger* 22.5±1.2, *Cladosporium herbarum* 5.0±0.4, *Curvularia lunata* 10.0±0.7, *Fusarium moniliforme* 15.0±1.0, *F. oxysporum* 10.0±0.9, *F. poae* 5.0±0.4, *Mucor hiemalis* 7.5±0.7, *M. racemosus* 2.5±0.3, *Penicillium citrinum* 2.5±0.3 and *Trichoderma lignorum* 7.5±0.7.

Kerationophilic Fungi from Sand Culture

Three genera spread over 6 species were isolated from Ganga sand.

January : Two genera and 3 species were isolated with following percentage occurrence—*Chrysosporium indicum* 50.0±0.5, *Microsporum fulvum* 37.5±0.7 and *M. gypseum* 12.5±0.3.

February : Isolates of February comprised 2 genera spread over 4 species were isolated with following percentage occurrence—*Chrysosporium indicum* 33.3±0.5, *C. tropicum* 33.3±0.5, *Microsporum fulvum* 25.0±0.5 and *M. gypseum* 8.3±0.3.

March : Four species of 3 genera were isolated with following percentage occurrence—*Chrysosporium indicum* 25.0±0.4, *C. tropicum* 50.0±0.7, *Microsporum gypseum* 12.5±0.3 and *Trichophyton simii* 12.5±0.3.

Keratinophilic Fungi and Dermatophytes of Ganga

S.No.	ORGANISM	JAN % Occur. ±SD	FEB % Occur. ±SD	MAR % Occur. ±SD	APR % Occur. ±SD	MAY % Occur. ±SD	JUN % Occur. ±SD	JUL % Occur. ±SD	AUG % Occur. ±SD	SEP % Occur. ±SD	OCT % Occur. ±SD	NOV % Occur. ±SD	DEC % Occur. ±SD
1.	*Chrysosporium evolceanui*	—	—	—	18.1 ±.4	30.7 ±.7	18.1 ±.4	25.0 ±.5	33.3 ±.5	23.0 ±.7	—	—	—
2.	*Chrysosporium indicum*	50.0 ±.5	33.0 ±.5	25.0 ±.4	18.1 ±.7	23.0 ±.7	27.2 ±.7	41.6 ±.7	25.0 ±.5	23.0 ±.7	40.0 ±.7	27.2 ±.7	35.7 ±.9
3.	*Chrysosporium tropicum*	—	33.3 ±.5	50.0 ±.7	45.4 ±.9	38.4 ±.7	36.3 ±.7	25.0 ±.5	33.3 ±.7	30.7 ±.7	40.0 ±.7	45.4 ±.9	35.7 ±.9
4.	*Microsporum fulvum*	37.5 ±.7	25.0 ±.5	—	—	—	—	—	—	15.3 ±.4	10.0 ±.3	18.1 ±.4	21.4 ±.5
5.	*Microsporum gypseum*	12.5 ±.3	8.3 ±.3	12.5 ±.3	9.0 ±.3	—	—	—	—	—	10.0 ±.3	9.0 ±.3	7.1 ±.3
6.	*Trichophyton simii*	—	—	12.5 ±.3	9.0 ±.3	7.6 ±.3	18.1 ±.4	8.3 ±.3	8.3 ±.3	7.6 ±.3	—	—	—

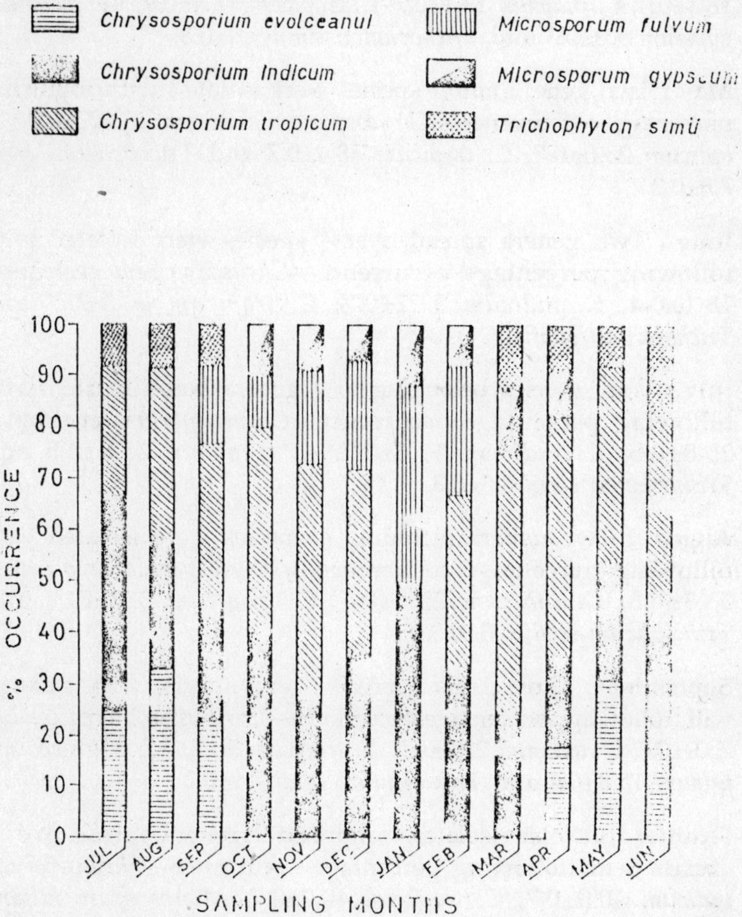

OCCURRENCE OF KERATINOPHILIC FUNGI AND DERMATOPHYTES OF GANGA BOTTOM SAND

April : Three genera spread over 5 species were isolated with following percentage occurrence—*Chrysosporium evolceanui* 18.1±0.4, *C. indicum* 18.1±0.7, *C. tropicum* 45.4±0.9, *Microsporum gypseum* 9.0±0.3 and *Trichophyton simii* 9.0±0.3.

May : Two genera and 4 species were isolated with following percentage occurrence—*Chrysosporium evolceanui* 30.7±0.7, *C. indicum* 23.0±0.7, *C. tropicum* 38.4±0.7 and *Trichophyton simii* 7.6±0.3.

June : Two genera spread over 4 species were isolated with following percentage occurrence—*Chrysosporium evolceanui* 18.1±0.4, *C. indicum* 27.2±0.7, *C. tropicum* 36.3±0.7 and *Trichophyton simii* 18.1±0.4.

July : Four species belonging to 2 genera were isolated with following percentage occurrence—*Chrysosporium evolceanui* 25.0±0.5, *C. indicum* 41.6±0.7, *C. tropicum* 25.0±0.5 and *Trichophyton simii* 8.3±0.3.

August : Two genera including 4 species were isolated with following percentage occurrence—*Chrysosporium evolceanui* 33.3±0.5, *C. indicum* 25.0±0.5, *C. tropicum* 33.3±0.7 and *Trichophyton simii* 8.3±0.3.

September : Three genera spread over 5 species were isolated with following percentage occurrence—*Chrysosporium evolceanui* 23.0±0.7, *C. indicum* 23.0±0.7, *C. tropicum* 30.7±0.7, *Microsporum fulvum* 15.3±0.4 and *Trichophyton simii* 7.6±0.3.

October : October isolates comprised 2 genera spread over 4 species with following percentage occurrence—*Chrysosporium indicum* 40.0±0.7, *C. tropicum* 40.0±0.7, *Microsporum fulvum* 10.0±0.3 and *M. gypseum* 10.0±0.3.

November : Two genera spread over 4 species were isolated with following percentage occurrence—*Chrysosporium indicum* 27.2±0.7, *C. tropicum* 45.4±0.9, *Microsporum fulvum* 18.1±0.4 and *M. gypseum* 9.0±0.3.

December : The fungal community comprised 2 genera spread over 4 species were isolated with following percentage occurrence—*Chrysosporium indicum* 35.7±0.9, *C. tropicum* 35.7±0.9, *Microsporum fulvum* 21.4±0.5 and *M. gypseum* 7.1±0.3.

ALGAE

SYSTEMATIC ENUMERATION OF ALGAE

CHLOROPHYTA, CHLOROPHYCEAE, ULOTRICHALES, ULOTRICHACEAE, *ULOTHRIX* Kuetzing

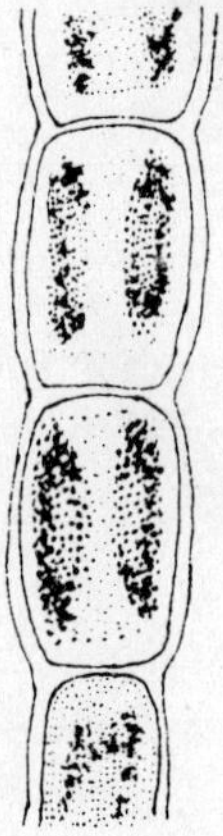

Fig. 35: *Ulothrix zonota*

***Ulothrix zonata* (Weber and Mohr) Kuetzing.** Tiffany, L.H. and M.E. Britton, The Algae of Illinois, 26, Pl. 4, Fig. 35, 1952.

Vegetative cell cylindric or swollen; cell wall thick at maturity; chromatophore usually a median band with several large pyrenoids.

Dimensions: Vegetative cells 11.2-44.8µ broad and 10.4-98.4 µ long.

Distribution: Either attached or free floating in Ganga water.

CHAETOPHORACEAE
STIGEOCLONIUM Kuetzing

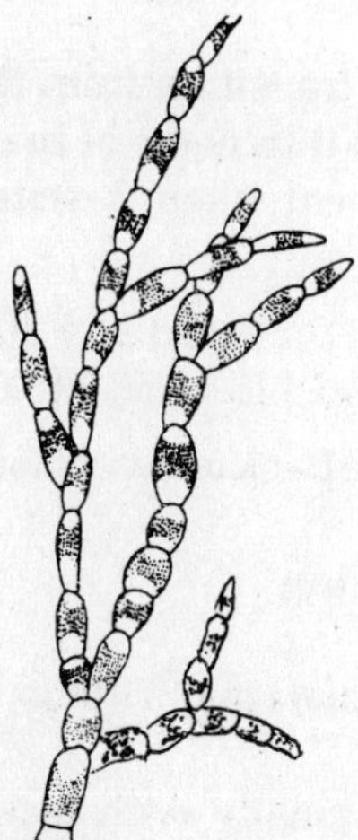

Fig. 36: *Stigeoclonium glomeratum*

***Stigeoclonium glomeratum* (Hazen) Collins.** Tiffany and Britton, *op. cit.*, 36, Pl. 9, Fig. 69, 1952.

Vegetative cell, cylindric or stightly swollen; branching usually opposite, generally densely fascicled at the top; branchlets often setiferous.

Dimensions: Vegetative cells 10.8 µ - 14.4µ broad and 19.2-90.8 µ long.

Distribution: Free floating in Ganga water.

CLADOPHORALES CLADOPHORACEAE
CLADOPHORA Kuetzing

***Cladophora glomerata* (Linnaeus) Kuetzing.** Tiffany and Britton, *op. cit.*, 45, Pl. 13, Fig. 93, 1952.

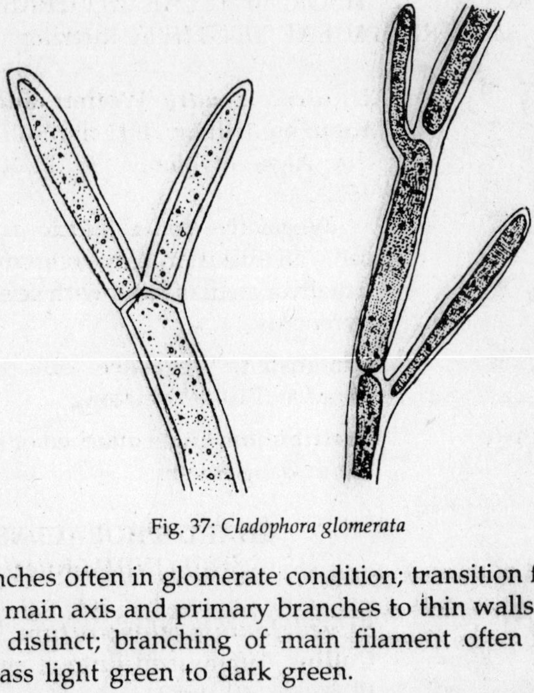

Fig. 37: *Cladophora glomerata*

Branches often in glomerate condition; transition from thick walls of main axis and primary branches to thin walls of branch clusters distinct; branching of main filament often Y-shaped; plant mass light green to dark green.

Dimensions: Cells of the main axis 42.9-151.0μ broad and 300.7-974.3μ long, of branches 34.6-60.9μ broad and 148.3-358.9 μ long.

Distribution: Found in shallow water on the bank of river Ganga.

RHIZOCLONIUM Kuetzing

***Rhizoclonium hieroglyphicum* (Agardh) Kuetzing.** Tiffany and Britton, *op.cit.*, 46, Pl. 13, Fig. 91, 1952.

Vegetative cells usually simple, occasionally with 1-celled and nearly colourless rhizoidal branches; filaments slightly or not at all constricted at the cross-walls; cell-walls usually thin, but may be thick in terrestrial forms; plants unattached, aquatic,

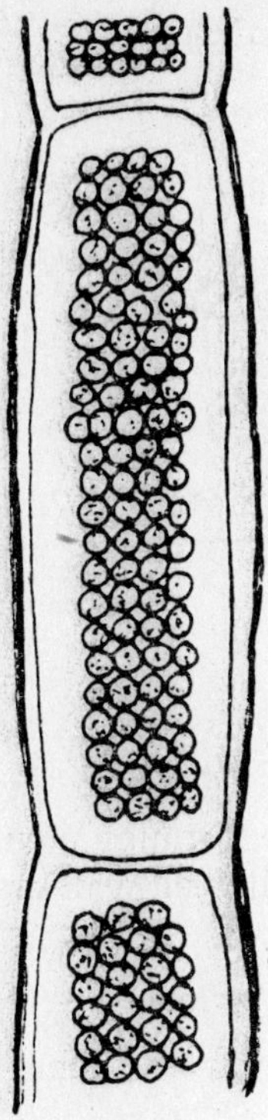

Fig. 38: *Rhizoclonium hieroglyphicum*

subaerial or terrestrial; floating forms grass-green or yellow-green, straight, thin walled and without rhizoidal branches.

Dimensions: Vegetative cells 10.4-.35.2μ broad and 9.2-113.2μ long.

Distribution: Found in stagnant water of river Ganga.

CHLOROCOCCALES
HYDRODICTYACEAE
HYDRODICTYON Roth

***Hydrodictyon reticulatum* (Linnaeus) Lagerheim.** Tiffany and Britton, *op. cit.*, 110, Pl. 29, Fig. 289, 1952.

Cells united in radiate, plate-like or saccate colonies with a definite number of cells, mostly multinucleate; chromatophores parietal, entire or reticulate when young-diffuse in older cells; phrenoids single or many in a cell. (Fig. 39)

Dimensions: Cells 5.2-249.2μ in diameter and up to 1.4 cm long.

Distribution : Free floating on the bank of river Ganga.

SCENEDESMACEAE *SCENEDESMUS* Meyen

***Scenedesmus acuminatus* (Lagerheim) Chodat.** Tiffany and Britton, *op.cit.*, 123, Pl. 35, Fig. 368, 1952.

Cells fusiform, arcuate or lunate, with pointed ends; cell-wall smooth and without spines; coenobia curved. (Fig. 40)

Dimensions: Cells 3.2-7.2μ broad and 24.8-40.8μ long.

Distribution: In sandy material of river Ganga.

***Scenedesmus arcuatus* Lemmermann.** Tiffany and Britton, *op.cit.*, 123, Pl. 35, Fig. 360, 1952.

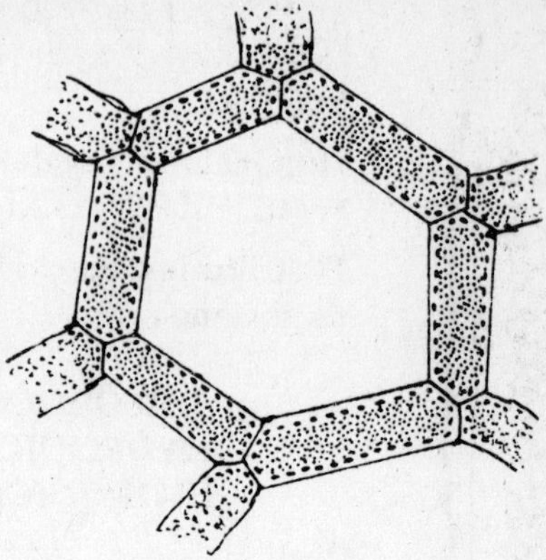

Fig. 39: *Hydrodictyon reticulatum*

Cells ovoid or angular, in a double row, forming a curved coenobium with small interstices between cells; cell-wall smooth without teeth or spines.

Dimensions: Cells 2.9-9.1μ broad and 8.9-17.2μ long.

Distribution: Found along with other algae of river Ganga.

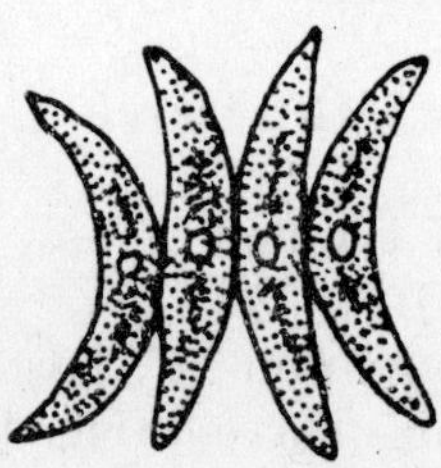

Fig. 40: *Scenedesmus acuminatus*

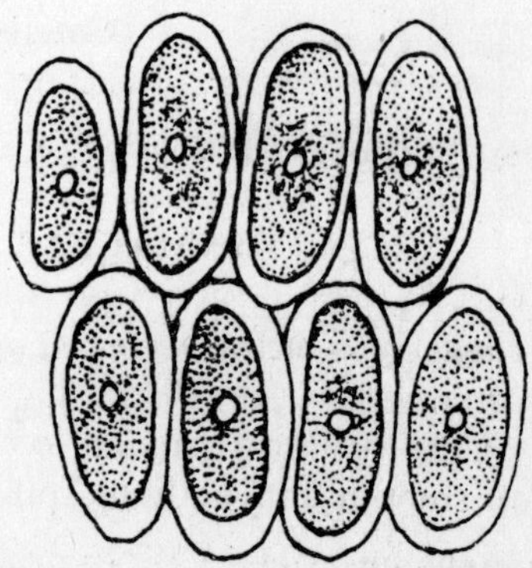

Fig. 41: *Scenedesmus arcuatus*

***Scenedesmus bijuga* (Turpin) Langerheim.** Tiffany and Britton, *op. cit.*, 123, Pl. 35, Fig. 371, 1952.

Cells oblong-ellipsoid to ovoid with broadly rounded ends, usually in a single series; cell-wall smooth, without teeth or spines.

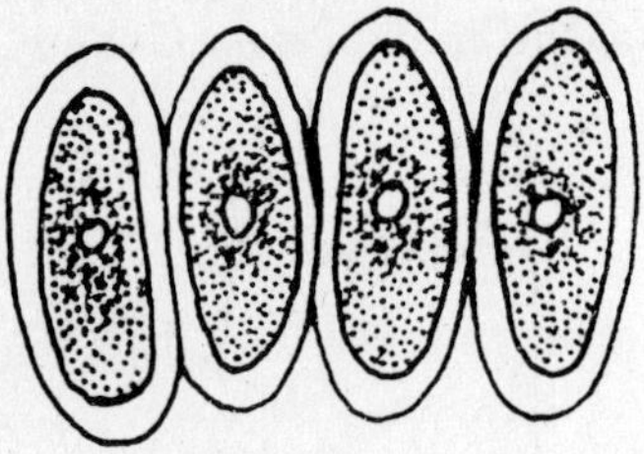

Fig. 42: *Scenedesmus*

Dimensions: Cells 4.0-7.1μ broad and 6.9-17.9μ long.

Distribution: Mixed with other algae of river Ganga.

Scenedesmus quadricauda var. ***quadrispina*** **(Chodat) G.M. Smith.** Tiffany and Britton, *op.cit.*, 122, Pl. 35, Fig. 358, 1952.

Cells broadly ovoid, about twice as long as wide; coenobia 4-celled; poles of outer cells with spines.

Dimensions: Cells 3.4-8.1μ broad and 8.2-15.0μ long, spines 2.6-10.4μ long.

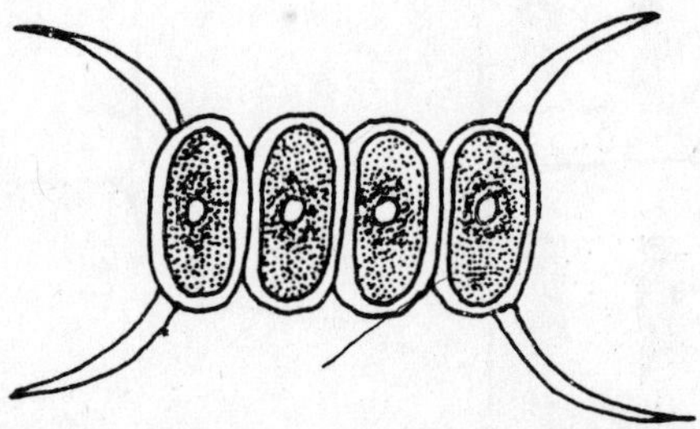

Fig. 43: *Scenedesmus quadricauda* var. *quadrispina*

Distribution: Found along with other algae of river Ganga.

ZYGNEMATALES ZYGNEMATACEAE
MOUGEOTIA C.A. Agardh

Mougeotia robusta **(De Bary) Wittrock.** Tiffany and Britton, *op. cit.*, 132, Pl. 38, Fig. 394-395, 1952.

Chromatophores with irregularly distributed phyrenoids; sporangia not dividing either gametangium; zygotes ovoid to subglobose, spore-wall brown, scorbiculate; pits scattered.

Dimensions: Vegetative cells 24.4-33.6μ broad and 74.4-256.8μ long; zygote 32.8-41.6μ broad and 45.6-57.6μ long; chromatophores 10-20 irregularly distributed phyrenoids.

Distribution: Mixed with spirogyra at the bank of river Ganga.

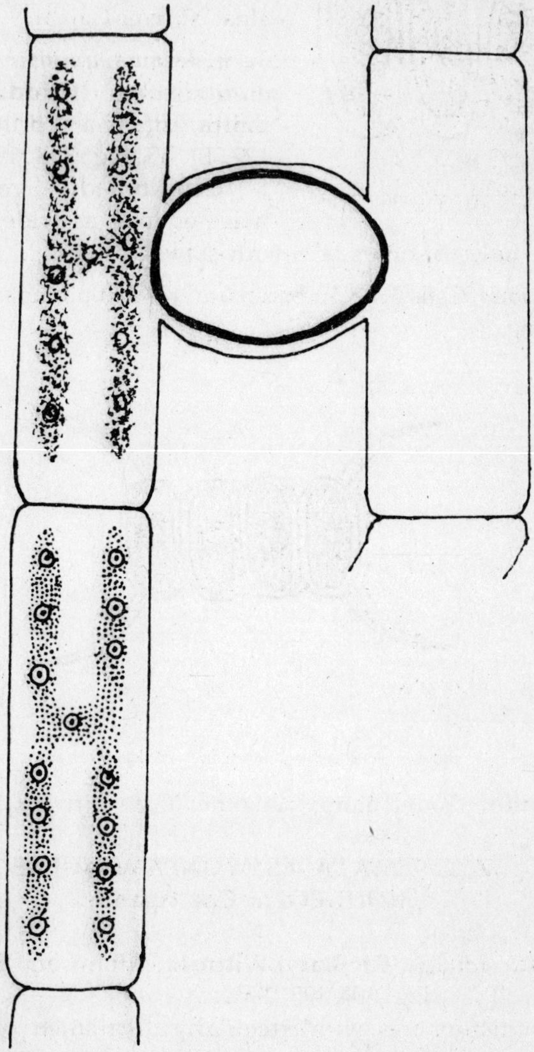

Fig. 44: *Mougeotia robusta*

ZYGNEMA C.A. Agardh

***Zygnema cruciatum* (Vaucher) C.A. Agardh.** Tiffany and Britton, *op.cit.*, 138, Pl. 41, Fig. 439, 1952.

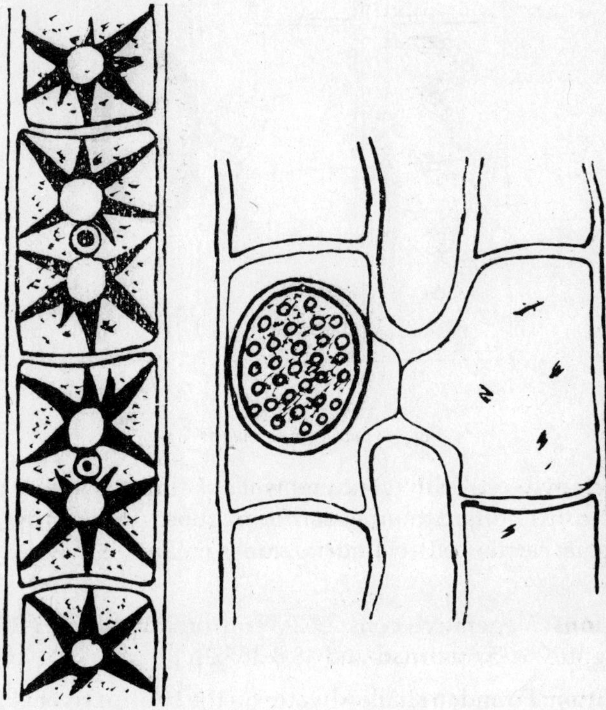

Fig. 45: *Zygnema cruciatum*

Vegetative cells usually with 2 distinctly stellate central phrenoids, connected by a cytoplasmic isthmus containing the nucleus; conjugation scalariform; zygote globose to ovoid, in one of the gametangia, receptive gametangia cylindric or enlarged, median spore-wall brown and scorbiculate.

Dimensions: Vegetative cells 28.8-37.2μ broad and 29.2-57.2μ long; zygote 29.6-38.8μ broad and 32.4-41.2μ long.

Distribution: In shallow water of river Ganga.

SPIROGYRA Link

***Spirogyra singularis* Nordstedt.** Tiffany and Britton, *op.cit.*, 146, Pl. 43, Fig. 454, 1952.

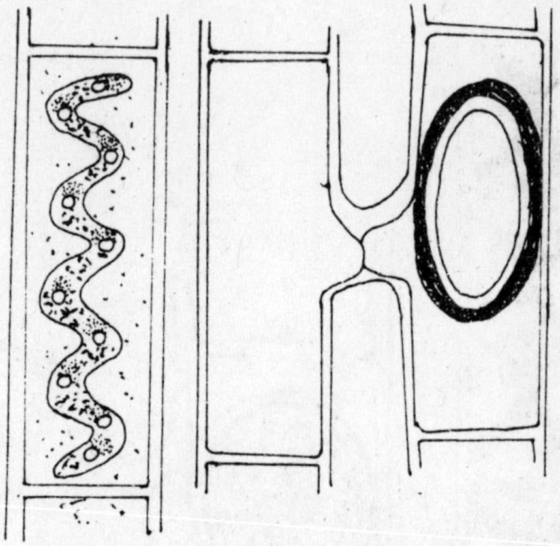

Fig. 46: *Spirogyra singularis*

Vegetative cells with plane end walls; 1 chromatophore making 3-7 turns; conjugation scalariform, tubes formed by both gametangia; fertile cells cylindric, rarely enlarged; zygote ellipsoid.

Dimensions: Vegetative cells 29.2-39.6μ broad and 59.2-235.2μ long; zygote 26-35.6μ broad and 38.8-155.2μ long.

Distribution: Found in shallow water on the bank of river Ganga.

Spirogyra sps. Vegetative cells longer than broad; 1 chromatophore making 5-7 turns. (Fig. 47)

Dimensions: Vegetative cells 120.8-134.6μ long and 4.4-33.6μ broad.

Distribution: Found in shallow water on the bank of river Ganga.

Spirogyra sps. Vegetative cells longer than broad; 2 chromatophores forming 6-7 turns. (Fig. 48)

Dimensions: Vegetative cells 36.7-42.8μ broad and 189.7-205.0μ long.

Fig. 47: *Spirogyra sps.* Fig. 48: *Spirogyra sps.* Fig. 49: *Spirogyra sps.* Fig. 50: *Spirogyra sps.*

Distribution: Mixed with other algae of river Ganga.

Spirogyra sps. Vegetative cells longer than broad; 3 chromatophores forming 6-8 turns. (Fig. 49)

Dimensions: Vegetative cells 35.1-41.3μ broad and 162.1-175.9μ long.

Distribution: Found in shallow water on the bank of river Ganga.

Spirogyra sps. Vegetative cells longer than broad; 4 chromatophores forming 8-9 turns. (Fig. 50)

Dimensions: Vegetative cells 53.5-59.6μ broad and 139.2-159.1μ long.

Distribution: Found in shallow water on the bank of river Ganga.

DESMIDIACEAE
COSMARIUM Corda

***Cosmarium granatum* Brebisson.** Tiffany and Britton, *op.cit.*, 186, Pl. 53, Fig. 565, 1952.

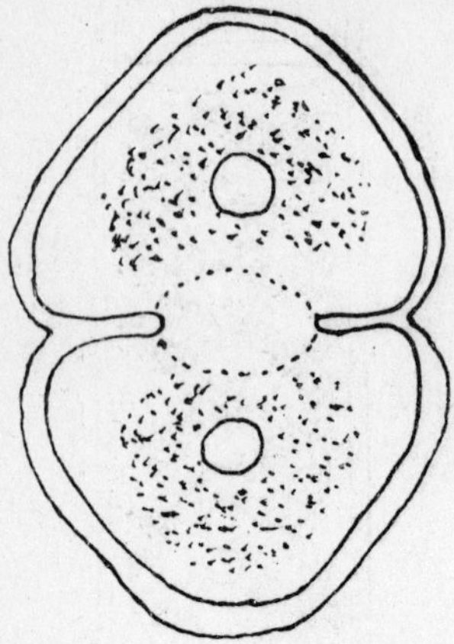

Fig. 51: *Cosmarium granatum*

Cells longer than wide, subrhomboid elliptic, deeply constricted, sinus linear, slightly dilated at the apex; semicells truncate-pyramidate, basal angles rounded, sides straight, slightly convex or rarely slightly concave, subparallel at the base and converging toward the apex, upper angles obtuse, apex narrowly truncate and straight; vertical view elliptic; lateral view of semicell elliptic-ovate; cell wall finely puncates; chromatophores axial; pyrenoid single central.

Dimensions: Cells 18.7-30.1μ wide and 26.1-52.0μ long; isthmus 6.0-9.4μ.

Distribution: Mixed with other algae of river Ganga.

CHRYSOPHYTA BACILLARIOPHYCEAE
PENNALES FRAGILARIACEAE
SYNEDRA Ehrenberg

Synedra ulna **(Nitzsch) Ehrenberg.** Tiffany and Britton, *op.cit.*, 237, Pl. 63, Fig. 713, 1952.

Cells linear in girdle view, with widened extremities, solitary; valves linear to linear-lanceolate, gradually narrowed toward the ends, with broadly rounded poles; transverse striations finely but plainly punctate; pseudoraphe narrowly linear, with central area varying, often absent. (Fig. 52)

Dimensions: Cells 4.4-9.6μ wide and 49.6-349.6μ long, transverse striations 8-11 in 10μ.

Distribution: In sandy material on the bank of river Ganga.

Synedra ulna var. **aequalis (Kuetzing) Hustedt.** Tiffany and Britton, *op.cit.*, 237, Pl. 63, Fig. 714.

Cells with straight sides in valve view, scarcely narrowed toward the rounded ends. (Fig. 53)

Dimensions: Cells 1.8-9.6μ broad and 50.8-352.4μ long; transverse striations 8-10 in 10μ.

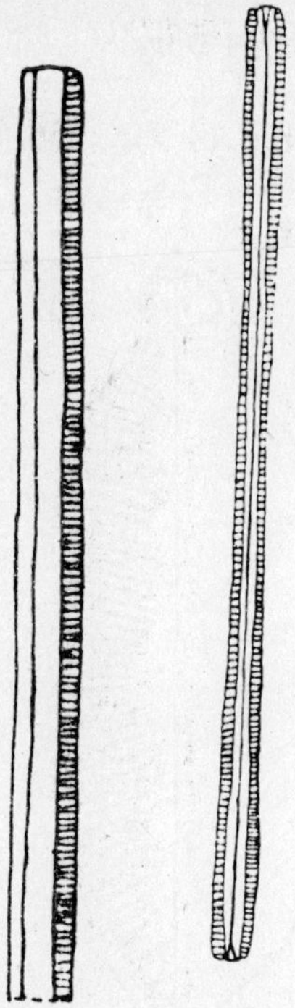

Fig. 52: *Synedra ulna*

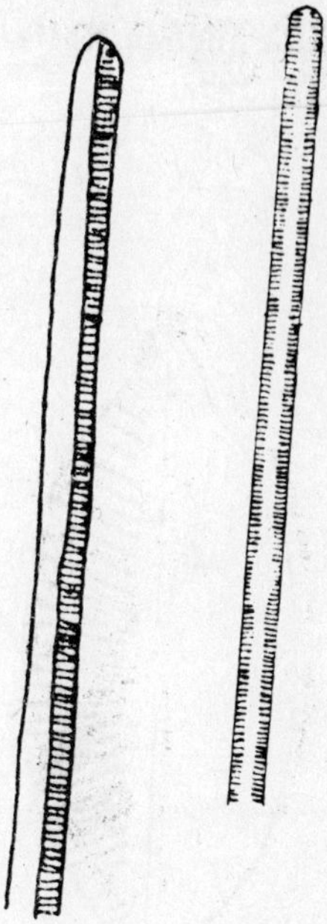

Fig. 53: *Synedra ulna* var. *aequalis*

Distribution: Found in sandy material of algae.

ACHANTHACEAE *COCCONEIS* Ehrenberg; Grunow

***Cocconeis placentula* Ehrenberg.** Tiffany and Britton, *op.cit.*, 241, Pl. 64, Fig. 734, 1952.

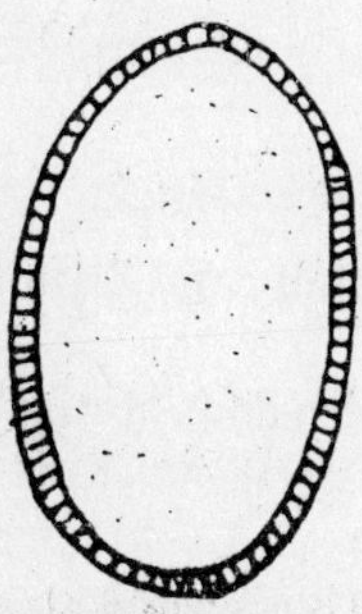

Fig. 54: *Cocconeis placentula*

Cells flat or slightly curved; valves elliptic, transverse striae sometimes radial; hypovalve with filamentous, straight raphe and intramarginal hyaline ring, central area round, small; epivalve with narrow linear pseudoraphe.

Dimensions: Cells 7.2-38.4μ broad and 10.4-72.4μ long.

Distribution: In sandy material on the bank of river Ganga.

NAVICULACEAE *NAVICULA* Bory

Navicula anglica **Ralfs.** Tiffany and Britton, *op. cit.*, 256, Pl. 67, Fig. 764, 1952.

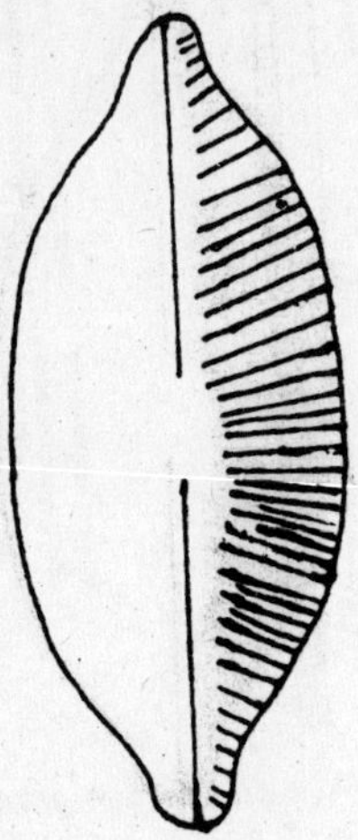

Fig. 55: *Navicula anglica*

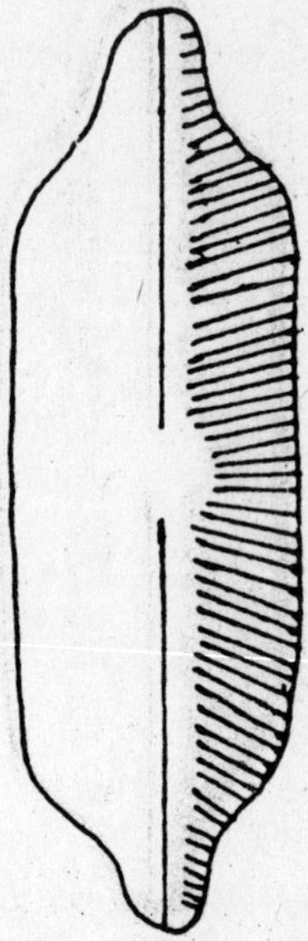

Fig. 56: *Navicula dicephala*

Valves elliptic with short rostrate and slightly capitate ends; transverse striations radial, shortened on either side of the rounded central area. (Fig. 55)

Dimensions: Cells 7.9-14.0µ broad and 19.8-40.0µ long; transverse striations 8-12 in 10µ.

Distribution: In sandy mixture on the bank of river Ganga.

Navicula dicephala **(Ehrenberg) Wm. Smith.** Tiffany and Britton, *op.cit.*, 256, Pl. 67, Fig. 768, 1952.

Valves broadly linear to linear-lanceolate with ends abruptly rostrate and somewhat capitate; transverse striations radial; central area rectangular. (Fig. 56)

Dimensions: Cells 7.6-13.0µ broad and 19.7-40.4µ long; transverse striations 14-16 in 10µ.

Distribution: In sandy material on the bank of river Ganga.

***Navicula gastrum* Ehrenberg.** Tiffany and Britton, *op.cit.*, 256, Pl. 67, Fig. 786, 1952.

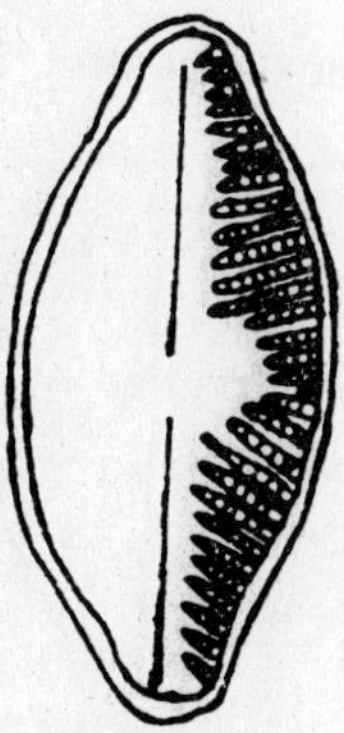

Fig. 57:
Navicula gastrum

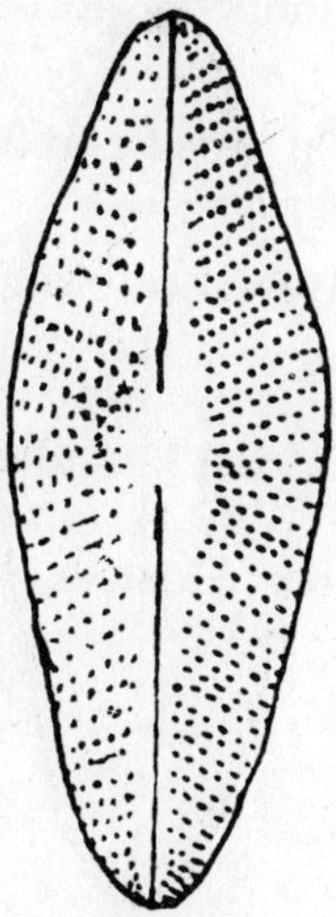

Fig. 58:
Navicula mutica

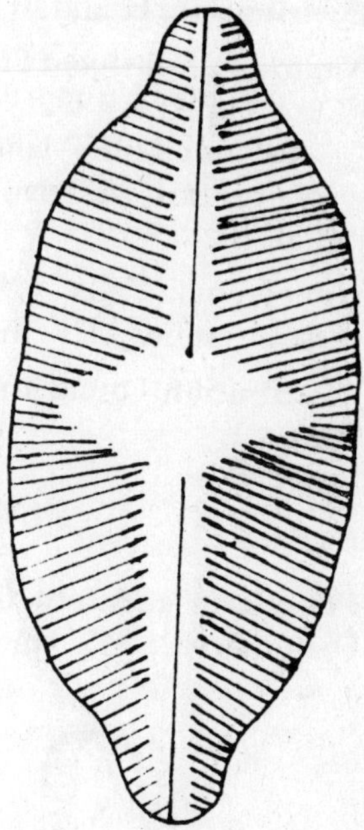

Fig. 59:
Navicula platyustoma

Valves broadly elliptic with short rostrate and broadly rounded ends; central area widened, irregular; transverse striations radial, medianly alternately long and short. (Fig. 57)

Dimensions: Cells 10.4-29.6μ broad and 25.2-63.2μ long; transverse striations 8-10 in 10μ.

Distribution: In mixture of algae on the bank of river Ganga.

***Navicula mutica* Kuetzing.** Tiffany and Britton, *op.cit.*, 254, Pl. 67, Fig. 773, 1952.

Valves lanceolate with broadly rounded ends, central area rectangular, with a single isolated puncta; transverse striations strongly punctate, radial often difficulty visible in the middle of the valve. (Fig. 58)

Dimensions: Cells 6.9-12.0μ broad and 9.8-40.0μ long; transverse striations 15-18 in 10μ.

Distribution: Found in the mixture of algae collected.

***Navicula platystoma* Ehrenberg.** Tiffany and Britton, *op.cit.*, 256, Pl. 67, Fig. 774, 1952.

Valves elliptic—lanceolate with broadly and bluntly rostrate ends; central widened, rhombo-elliptic; transverse striations radial. (Fig. 59)

Dimensions: Cells 14.9-20.0μ broad and 37.1-50.1μ long; transverse striations 14-17 in 10μ.

Distribution: Found in sandy mixture of algae on the bank of river Ganga.

PINNULARIA Ehrenberg

***Pinnularia undulata* Gregory.** Hustedt, F., A Treatise on the Diatomaceae, 315, Fig. 565, 1930.

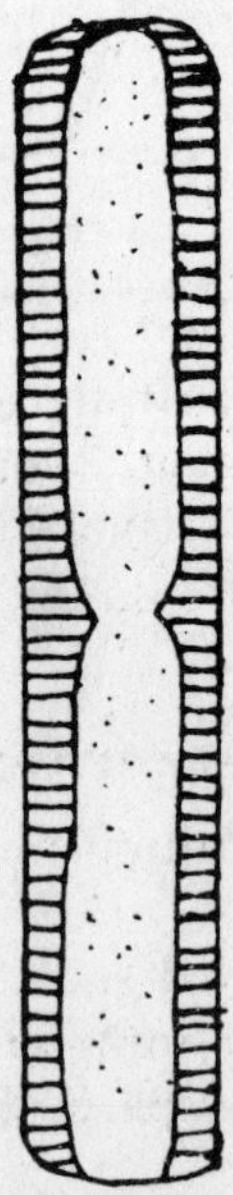

Fig. 60: Pinnularia undulata

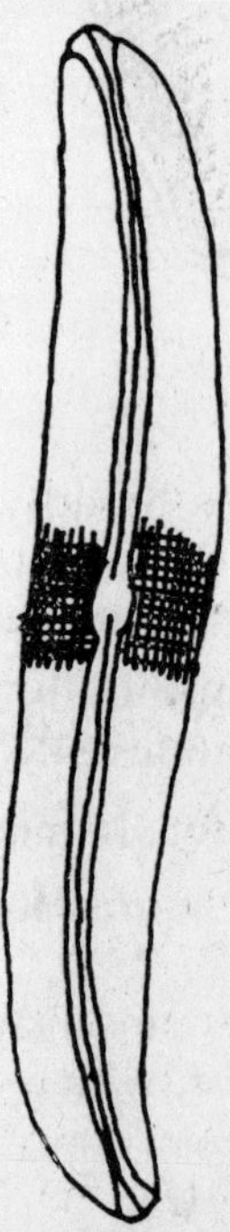

Fig. 61: Gyrosigma spencerii

Valve linear, slightly tapering to obtuse somewhat capitate ends; axial area broad; central extending to the margin; transverse striations radial.

Dimensions: Cells 33.9-38.8µ long and 6.0-7.5µ broad; transverse striations 12-15 in 10µ.

Distribution: In sandy material on the bank of river Ganga.

GYROSIGMA Hassall; emend. Cleve

***Gyrosigma spencerii* (Quekett) Cleve.** Tiffany and Britton, *op.cit.*, 269, Pl. 66, Fig. 763, 1952.

Valves sigmoid, lanceolate, tapering to narrowly rounded ends; transverse striations, parallel to the middle line. (Fig. 61)

Dimensions: Cells 11.2-25.2µ broad and 70.8-215.2µ long; transverse striations 15-18 in 10µ; longitudinal striations 16-18 in 10µ.

Distribution: In sandy mixture of algae.

GOMPHONEMATACEAE *GOMPHONEMA* C.A. Agardh

***Gomphonema constrictum* Ehrenberg.** Tiffany and Britton, *op. cit.*, 271, Pl. 72, Fig. 839, 1952.

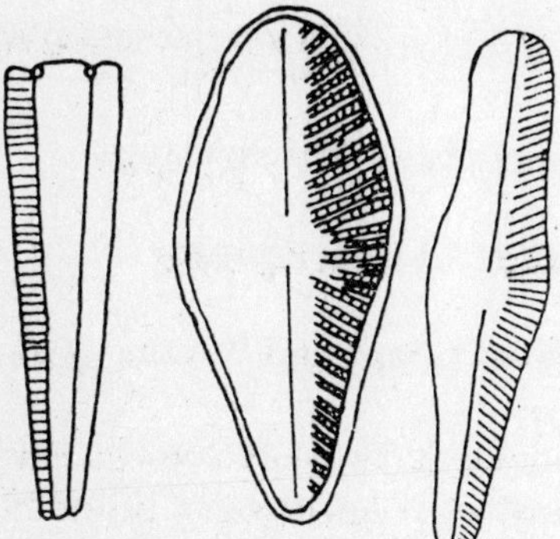

Fig. 62:
Gomphonema constrictum

Fig. 63:
Gomphonema olivaceum

Fig. 64:
Gomphonema parvulum

Valves clavate, constricted below the broad rounded apical pole, with attenuated basal pole; axial area narrow; central area broad and irregularly defined, with a dot on one side; transverse striations radial, evidently punctate, alternately long and short in the middle of the valve. (Fig. 63)

Dimensions: Cells 7.3-14.7μ broad and 25.5-63.3μ long; transverse striations 10-12 in 10μ.

Distribution: In sandy material on the bank of river Ganga.

Gomphonema olivaceum **(Lyngbye) Kuetzing.** Tiffany and Britton, *op.cit.*, 270, Pl. 72, Fig. 845, 1952.

Valves ovoid-clavate, with broadly rounded apex and acutely rounded base, axial area narrow, linear; central area widened transversely without dots; raphe straight; transverse striations radial and indistinctly punctate.

Dimensions: Cells 4.6-10.1μ broad and 14.7-40.1μ long; transverse striations 10-14 in 10μ.

Distribution: In mixture of algae.

Gomphonema parvulum **(Kuetzing) Grunow.** Tiffany and Britton, *op.cit.*, 272, Pl. 72, Fig. 838, 1952.

Valves clavate-lanceolate, with shortly constricted poles; axial area very narrow; central area small, one-sided, with isolated dot; transverse striations radial, indistinctly punctate. (Fig. 64)

Dimensions: Cells 4.0-7.3μ broad and 11.4-28.7μ long; transverse striations 13-15 in 10μ.

Distribution: In sandy material on the bank of river Ganga.

CYMBELLACEAE *CYMBELLA* C.A. Agardh

Cymbella prostrata **(Berkeley) Cleve.** Tiffany and Britton, *op.cit.*, 279, Pl. 74, Fig. 859, 1952.

Valves quite asymmetric, semielliptic, dorsally convex, ventrally straight with median expansion, poles broadly rounded; raphe straight with polar deviations; axial area narrow; central area small, round, without isolated dots; transverse striations radiate to parallel. (Fig. 65)

Dimensions: Cells 9.8-30.3μ broad and 20.4-97.8μ long; transverse striations 6-10 in 10μ.

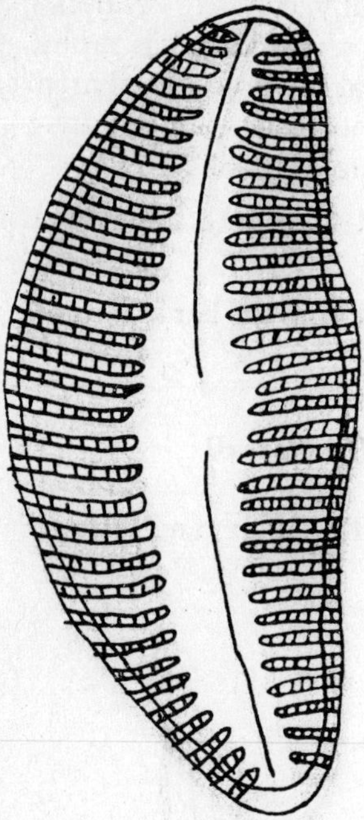

Fig. 65: *Cymbella prostrata*

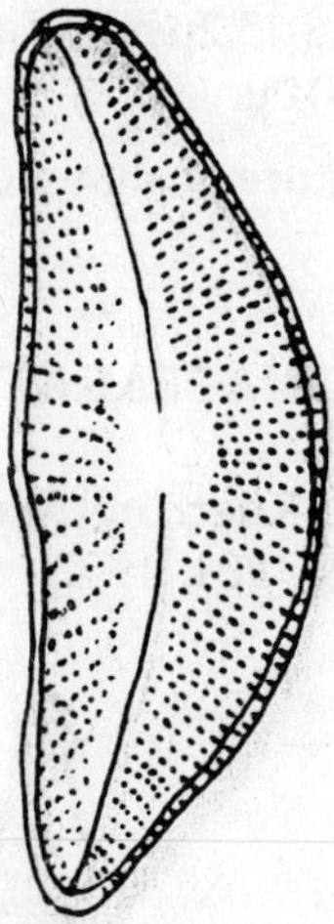

Fig. 66: *Cymbella tumida*

Fig. 67: *Cymbella ventricosa*

Distribution: In mixture of algae on the bank of river.

***Cymbella tumida* (Brebisson) Van Heurck.** Tiffany and Britton, *op.cit.*, 278, Pl. 74, Fig. 860, 1952.

Valves asymmetric and curved, broadly naviculoid, with rostrate poles, convex dorsal sides and straight or slightly convex ventral sides having a median expansion; raphe excentric; axial area narrow; central area large, round, with a ventrally placed prominent isolated dot; transverse striations punctate.

Dimensions: Cells 15.6-31.2μ broad and 39.2-105.6μ long; transverse striations 7-10 in 10μ. (Fig. 66)

Distribution: In sandy mixture of algae collected from the bank of river.

***Cymbella ventricosa* Kuetzing.** Tiffany and Britton, *op.cit.*, 279, Pl. 74, Fig. 871, 1952.

Valves somewhat semielliptic, dorsally convex, ventrally straight to slightly convex, with sharply rounded poles; raphe straight, closer to the ventral side except at the poles; axial area narrow, slightly expanded medianly, without isolated dots; transverse striations generally radiate, punctate. (Fig. 67)

Dimensions: Cells 5.6-15.2µ broad and 10.4-41.6µ long, transverse striations 10 to 16 in 10µ.

Distribution: In mixture of algae collected from the bank of river Ganga.

NITZSCHIACEAE *NITZSCHIA* Hassall

Nitzschia acicularis **(Kuetzing) Wm. Smith.** Tiffany and Britton, *op.cit.*, 286, Pl.77, Fig. 904, 1952.

Fig. 68:
Nitzschia acicularis

Fig. 69:
Nitzschia linearis

Fig. 70:
Nitzschia palea

Fig. 71:
Nitzschia dissipata

Cells with medianly spindle-shaped long attenuate poles, weakly silicified; striations very fine, scarcely visible, keel punctate. (Fig. 68)

Dimensions: Cells 2.7-4.3μ broad and 50.8-149.205μ long; striations 15-20 in 10μ.

Distribution: In sandy mixture of algae on the bank of river Ganga.

***Nitzchia linearis* (Agardth) Wm. Smith.** Tiffany and Britton, *op.cit.*, 288, Pl. 76, Fig. 891, 1952.

Cells rectangular—linear in girdle view, with median constriction; valves linear and generally with parallel sides and smaller, capitate poles; keel punctate. (Fig. 69)

Dimensions: Cells 4.8-6.8μ broad and 68.4-178.4μ long; striations 20-28 in 10μ.

Distribution: In sandy material on the bank of river.

***Nitzschia palea* (Kutz.)** W. Sm. Heurck, *op.cit.*, 401, Pl. 17, Fig. 554, 1896.

Valves more linear—lanceolate, with apices shortly rostrate; the median not distant; girdle face linear-lanceolate, with apices rounded or truncate. (Fig. 70)

Dimensions: Cells 4.2-5.8μ broad and 28.2-64.0μ long; about 10-12 carinal dots in 10μ; about 32-34 striae in 10 m.

Distribution: Planktonic in water and in wet sand of river Kalinadi. In sand and water culture.

***Nitzschia dissipata* (Kutz.) Grun.** Heurck. *op.cit.*, 394, Pl. 16, Fig. 525, 1896; Tiffany and Britton, *op.cit.*, 288, Pl. 76, Fig. 898, 1952.

Valve lanceolate, with apices very slightly rostrate; keel slightly eccentric; striae excessively faint. (Fig. 71)

Dimensions: Cells 4.2-6.8μ broad and 21.3-35.0μ long; keel punctate about 5-8 in 10μ.

Distribution: Planktonic and in wet sand on the bank of river Kalinadi and in sand culture.

CYANOPHYTA CYANOPHYCEAE
CHROOCOCCALES CHROOCOCCACEAE
GENUS *MERISMOPEDIA* Meyen

***Merismopedia convoluta* Breb.** Desikachary, T.V., Cyanophyta, 152, Pl. 29, Fig. 13, 1959; Tiffany and Britton, *op.cit.*, 334, Pl. 91, Fig. 1051, 1952.

Cells spherical to oblong, rarely longer, flat, convolute colonies; blue-green, or yellowish. (Fig. 72)

Dimensions: Cells 4.1-5.0μ broad and 8.1-9.5μ long.

Distribution: Among other algae and in sand and water culture.

***Merismopedia tenuissima* Lemm.** Desikachary, *op.cit.*, 154, Pl. 29, Fig. 7, 1959.

Cells pale blue-green, closely packed in colonies of 16-100 cells, subspherical. (Fig. 73)

Dimensions: Cells 2.6-3.1μ in diameter.

Distribution: Planktonic and invariably in association with desmids and other algae. Found in sand and water culture.

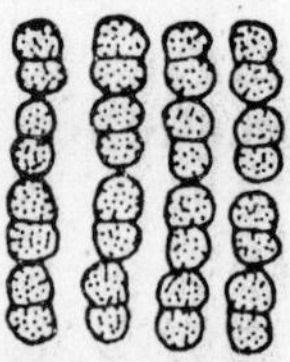

Fig. 72: *Merismopedia convoluta*

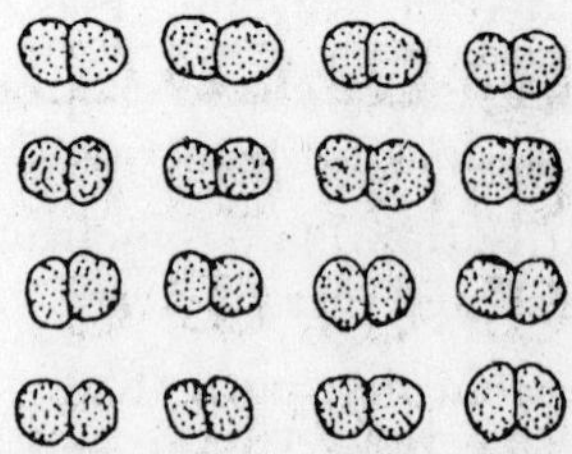

Fig. 73: *Merismopedia tenuissima*

***Merismopedia punctata* Meyen.** Desikachary, *op. cit.*, 155, Pl. 23, Fig. 5 and Pl. 29, Fig. 6, 1959.

Colonies small, 4-64 cells not closely packed, spherical or ovoid, pale blue-green. (Fig. 74)

Dimensions: Cells 2.6-3.4μ in diameter.

Distribution: Planktonic and in stagnant water of river Kalinadi or among other algae. Found in sand and water culture.

***Merismopedia glauca* (Ehrenb.) Nag.** Desikachary, *op.cit.*, 155, Pl. 29, Fig. 5, 1959, Tiffany and Britton, *op.cit.*, 334, Pl. 91, Fig. 1051, 1952.

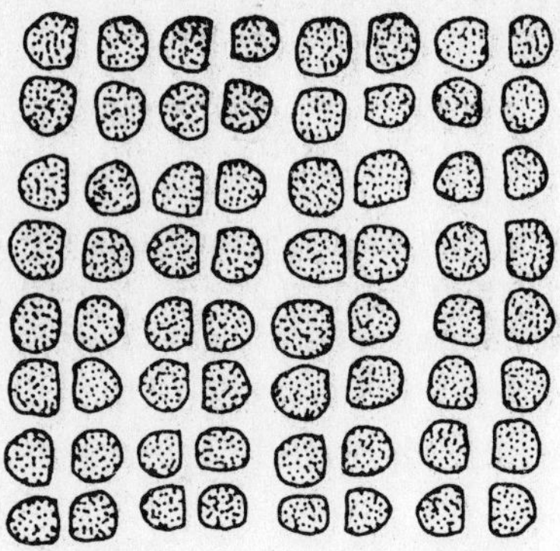

Fig. 74: *Merismopedia punctata*

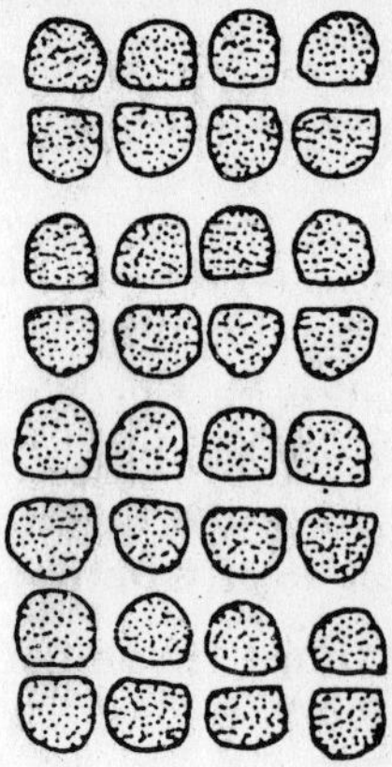

Fig. 75: *Merismopedia glauca*

Colonies mostly small with 16-64 cells, rarely more, cells oval or spherical, closely arranged pale blue-green. (Fig. 75)

Dimensions: Cells 2.9-5.8μ in diameter.

Distribution: As planktonic in stagnant waters and among other algae of river. Found in sand and water culture.

CHROOCOCCUS Nag.

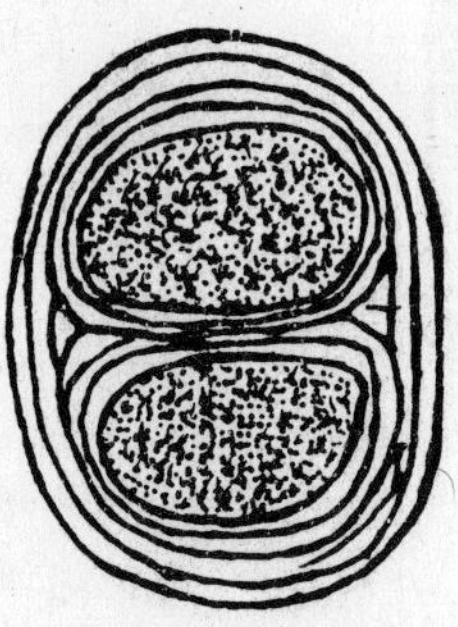

Fig. 76: *Chroococcus*

***Chroococcus tenax* (Kirchn.)** Desikachary, T.V. Cyanophyta, 103, Pl. 26, Figs. 7, 16, 1959.

Cells mostly in groups of 2-4, sometimes 8-16, blue-green or olive coloured; sheath colourless or yellow to brown, very thick, very distinctly lamellated.

Dimensions: Cells without sheath 14.9-22.7μ and with sheath 18.1-26.7μ in diameter.

Distribution: Found on the bank of river Ganga.

NOSTOCALES OSCILLATORIACEAE
OSCILLATORIA Vaucher

Oscillatoria formosa **Bory ex Gomont.** Desikachary, *op.cit.*, 232, Pl. 40, Fig. 15, 1959.

Thallus blue-green; trichome straight, slightly constricted at the cross-walls, bright blue-green, attenuated at the ends and bent; cells nearly quadrate, septa sometimes slightly granulated; end cells nearly obtuse, calyptra absent, not capitate. (Fig. 77)

Dimensions: Trichome 3.6-6.4μ broad; cells up to 1/2 as long as broad, 2.4-5.6μ long.

Distribution: Found on the bank of river Ganga.

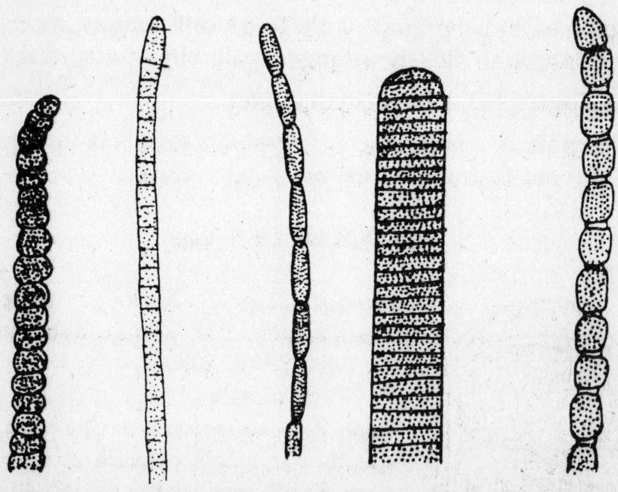

Fig. 77: *Oscillatoria formosa* Fig. 78: *Oscillatoria limnetica* Fig. 79: *Oscillatoria limosa* Fig. 80: *Oscillatoria pseudogeminata*

Oscillatoria limnetica **Lemm.** Desikachary, *op.cit.*, 226, Pl. 37, Fig. 3, 1959.

Trichome straight or slightly bent, distinctly constricted at the cross-walls, pale blue-green, not attenuated, not capitate; end cells rounded, calyptra absent. (Fig. 78)

Dimensions: Trichome 1.4-1.5μ broad, cells 3.9-12.0μ long.

Distribution: Found in floating condition on the bank of river Ganga.

***Oscillatoria limosa* Ag. ex Gomont.** Desikachary, *op.cit.*, 206, Pl. 42, Fig. 11, 1959.

Thallus dark blue-green; trichome more less straight, dull blue-green, brown or olive-green, not constricted at the cross-walls, or only slightly constricted, cross-walls, frequently granulated; end-cell flatly rounded with slightly thickened membrane. Fig. 79)

Dimensions: Trichome 12.4-17.6μ broad; cells 1/3-1/6 as long as broad, 1.6-5.2μ long.

Distribution: Found in floating condition on the bank of river Ganga.

***Oscillatoria pseudogeminata* G. Schmid.** Desikachary, *op.cit.*, 228, Pl. 41, Fig. 10, 1959.

Thallus pale or dirty blue-green, trichomes coiled, pale-green, ends not attenuated; cells as long as broad or somewhat longer or shorter than broad, not constricted at the cross walls, cross-walls thick, not granulated, end cell rounded, calyptra absent. (Fig. 80)

Dimensions: Trichomes 4.7-6.0μ broad; cell 1.0-2.3μ long.

Distribution: On the moist bank of river Ganga.

***Oscillatoria princeps* Vaucher ex Gomont.** Desikachary, *op.cit.*, 210, Pl. 37, Figs. 1, 10, 11, 1959.

Trichomes blue-green, more or less brownish, mostly forming a thallus, mostly straight, not constricted at the cross-walls, slightly or briefly attenuated at the apices and bent; end cells flatly rounded, slightly capitate. (Fig. 81)

Dimensions: Trichomes 20-56μ broad, cells 3.6-6.8μ long.

Distribution: In moist sand of river Kalinadi and mixed with other algae and in sand culture.

Oscillatoria tenuis* var. *tergestina Rabenh. Desikachary, *op.cit.*, 222, Pl. 42, Figs. 13, 14, 1959.

Differs from the type form in its diameter. (Fig. 82)

Dimensions: Trichomes 4.0-5.8μ broad; cells 2.9-4.6μ long.

Distribution: In stagnant water and in moist sand on the bank of river Kalinadi. Found in sand and water culture.

***Oscillatoria subbrevis* Schmidle.** Desikachary, *op.cit.*, 207, Pl. 37, Fig. 2, 1959.

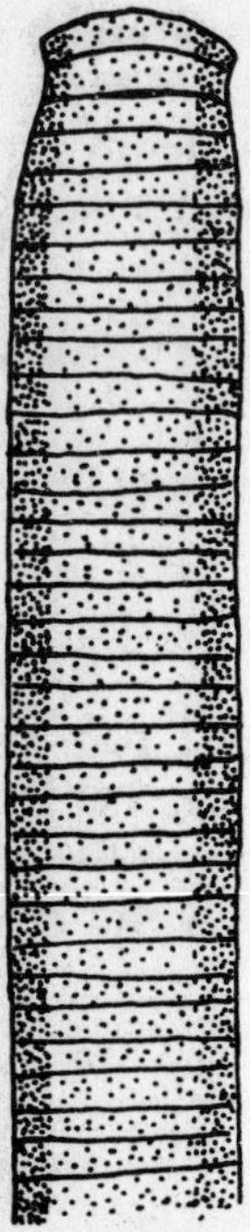

Fig. 81:
Oscillatoria princeps

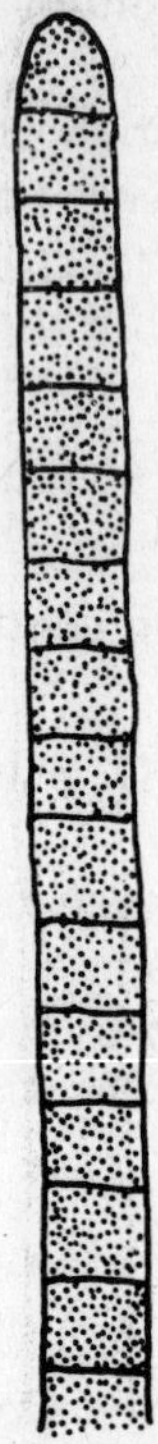

Fig. 82: *Oscillatoria*
tenius var tergestina

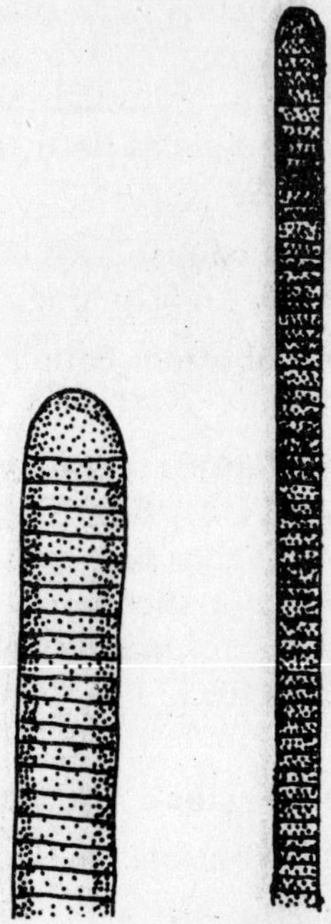

Fig. 8.3:
Ocillatoria subbrevis

Trichomes single, nearly straight, not attenuated at the apices; cells not granulated at the cross-walls; end-cell rounded, calyptra absent. (Fig. 83)

Dimensions: Trichomes 4.7-6.0μ broad; cells 1.0-2.3μ long.

Distribution: On the moist bank of river Ganga.

PHORMIDIUM Kutz.

***Phormidium ambiguum* Gomont.** Desikachary, *op.cit.*, 266, Pl. 44, Fig. 16 and Pl. 45, Fig. 6, 1959.

Thallus more or less expanded, bright blue-green, dark or yellowish green; filaments flexuous, variously entangled; trichomes slightly constricted at the cross-walls, at the ends not attenuated, not capitate, blue-green; sheath thin, firm or diffluent sometimes thick and more or less lamellated; cells shorter than broad, rarely granulated at the cross-walls, sometimes with gas-vacuoles; end cell rounded, calyptra absent. (Fig. 84)

Dimensions: Trichome 3.7-6.0μ broad, cells 1.3-2.7μ long.

Distribution: Found among other algae in Ganga water.

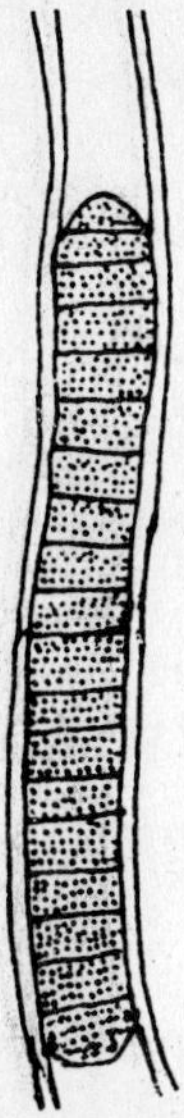

Fig. 84: *Phormidium ambiguum*

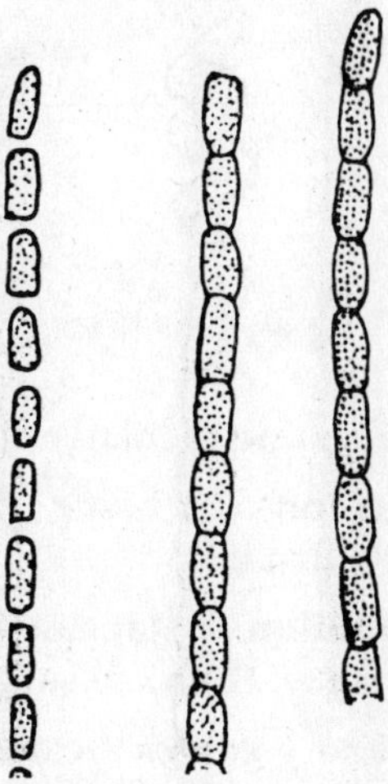

Fig. 85: *Phormidium tenue*

***Phormidium tenue* (Menegh.) Gomont.** Desikachary, *op.cit.*, 259, Pl. 43, Fig. 13 and 15, 1959.

Thallus pale blue-green, thin, membranous, expanded; trichome straight or slightly bent, densely entangled, slightly constricted at the cross-walls, attenuated at the ends, pale blue-green; sheath thin, diffluent; cell up to 3 times longer than broad, septa not granulated, cross-walls not commonly visible; end-cell acute-conical, calyptra absent. (Fig. 85)

Dimensions: Trichomes .9-2.7μ broad; cells 2.3-5.0μ long.

Distribution: Found among other algae in Ganga water.

***Spirulina major* Kutz. ex Gomont.** Desikachary, *op.cit.*, 196, Pl. 36, Fig. 13, 1959; Tiffany and Britton, *op.cit.*, 354, Pl. 97, Fig. 1124, 1952.

Fig. 86: *Spirulina major*

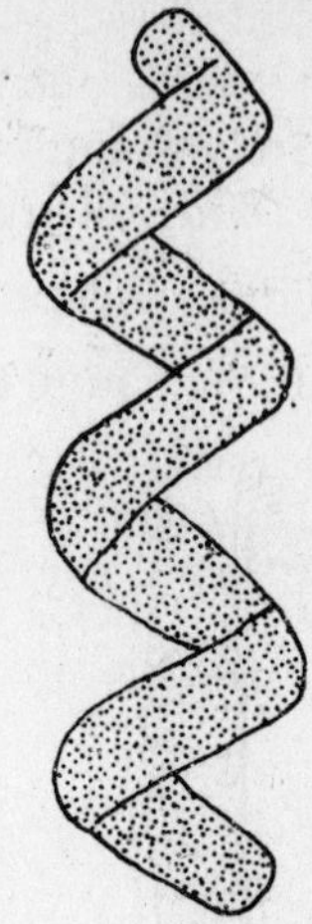

Fig. 87: *Spirulina gigantea*

Trichome regularly spirally coiled, blue-green. (Fig. 86)

Dimensions: Trichome 1.4-2.6μ broad; spirals 2.9-3.9μ broad and 2.9-4.9μ distant.

Distribution: In stagnant water of river Kalinadi and mixed with other algae. Found in sand and water culture.

***Spirulina gigantea* Schmidle.** Desikachary, *op.cit.*, 197, Pl. 36, Figs. 12, 14, 17, 1959.

Trichome deep blue-green, regularly spirally coiled, at the end conical attenuated. (Fig. 87)

Dimensions: Trichome 3.1-4.0μ broad; spirals 11.2-15.8μ broad.

Distribution: Free-floating in Kalinadi water or mixed with other algae and in sand and water culture.

NOSTOCACEAE *ANABAENA* Bory

***Anabaena oryzae* Fritsch.** Desikachary, *op.cit.*, 396, Pl. 72, Fig. 3, 1959.

Thallus soft, green, gelatinous, membranous; trichomes short, straight densely aggregated, generally parallel; cells more or less barrel-shaped; heterocysts terminal and inter calary, broader than the vegetative cells, generally barrel-shaped, sometimes spherical, single; spores rarely single next to the terminal heterocyst, commonly away from the intercalary heterocysts or in series, subspherical or short ellipsoidal. (Fig. 88)

Dimensions: Trichomes 2.6-3.0μ broad; heterocysts 3.0-3.4μ broad; spores about 5.5μ broad and 6.5μ long.

Distribution: Mixed with other cyanophycean forms in flowing water and in sand culture.

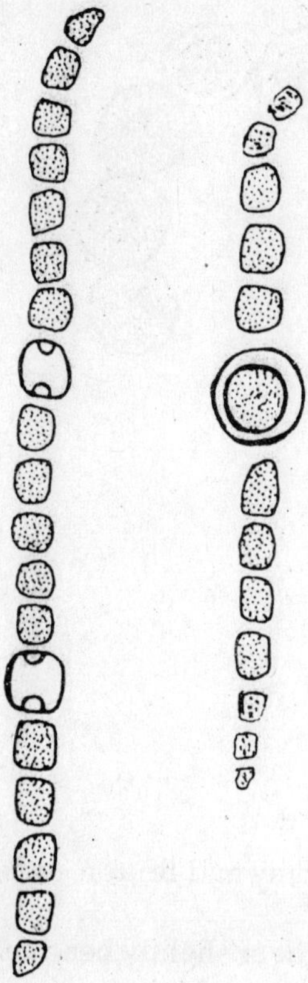

Fig. 88: *Anabaena oryzae*

***Anabaena fertillissima* Rao, C. B.** Desikachary, *op.cit.*, 398, Pl. 74, Fig. 1, 1959.

Trichomes single, straight or bent, with almost rounded end cells; cells barrel-shaped; heterocysts almost spherical; spores in long chains, often making the whole trichome sporogenous adjoining the heterocysts but formed centrifugally, almost spherical, with a smooth hyaline outer wall. (Fig. 89)

Dimensions: Trichomes 5.0-5.5μ broad; cells 4.9-7.9μ long; heterocysts 6.5-8.4μ broad, spores 5.0-7.8μ broad and 3.4-8.4μ long.

Distribution: Mixed with other algae and in sand culture.

***Anabaena iyengarii* Bharadwaja**. Desikachary, *op.cit.*, 406, Pl. 78, Fig. 2, 1959.

Trichome single or irregularly curved; end cell conical with rounded apex cells barrel-shaped; heterocysts barrel-shaped, rarely spherical; spores ellipsoidal often in long or short chains, rarely single on both sides of the heterocysts. (Fig. 90)

Dimensions: Trichome 3.5-4.3μ broad; cells 3.5-6.0μ long; heterocysts 4.8-6.5μ broad and 6.4-8.8μ long; spore 7.5-9.1μ broad and 9.1-18.9μ long.

Distribution: Mixed with other algae in stagnant water of the river Kalinadi.

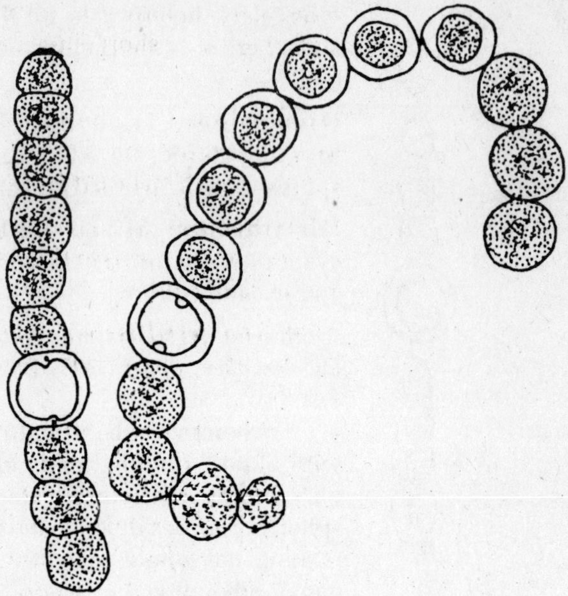

Fig. 89: *Anabaena fertillissima*

Anabaena plancktonica **Brunnthaler,** Tiffany and Britton, *op.cit.*, 357, Pl. 98, Fig. 1128, 1959.

Filaments free-floating solitary, straight or slightly bent; cells rounded or broadly ellipsoid; heterocysts spherical akinetes solitary adjacent to heterocysts. (Fig. 91)

Dimensions: Cells 9.8-14.2μ broad and 9.8-12.2μ long; heterocysts 12.4-22.4μ broad; spores 13.8-20.2μ broad and 12.6-29.8μ long.

Distribution: Mixed with other algae of river Kalinadi.

Anabaena variabilis **Kutzing ex Born. et Flah.** Desikachary, *op.cit.*, 410, Pl. 71, Fig. 5, 1959.

Thallus gelatinous, dark green; trichome without any sheath, flexuous, end cells conical, obtuse; cells barrel-shaped; heterocysts spherical or oval; spores not contiguous with the heterocyst, barrel-shaped, in series. (Fig. 92)

Dimensions: Cells 4-6.4μ broad, 2.4-6μ long; heterocysts 6-7.6μ broad, 7.6-9μ long; spores 6.8-10.8μ broad, 7.2-14μ long.

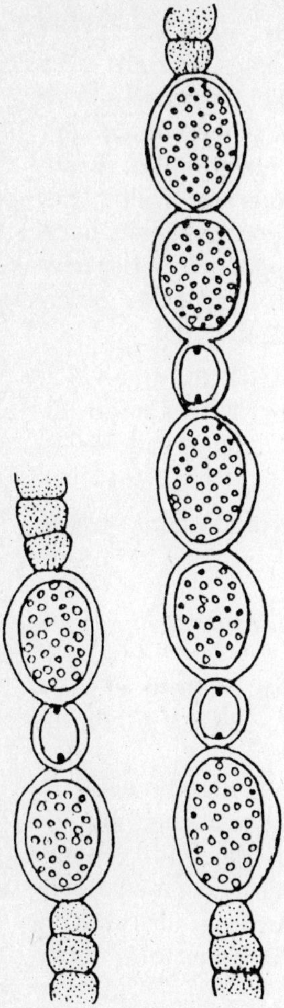

Fig. 90: *Anabaena iyengarii*

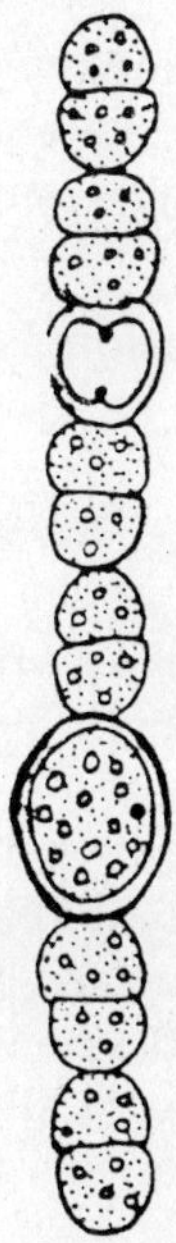

Fig. 91: *Anabaena planctonica*

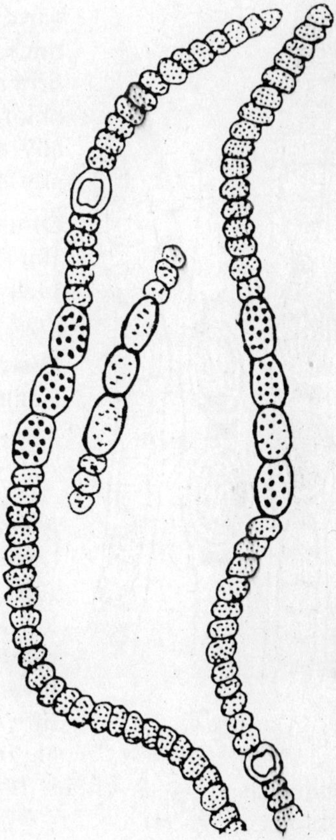

Fig. 92: *Anabaena variabilis*

Distribution: Found in floating condition at the bank of river Ganga.

AULOSIRE Kirchner

***Aulosira fertilissima* Ghose.** Desikachary, *op.cit.*, 431, Pl. 80, Fig. 6, 1959.

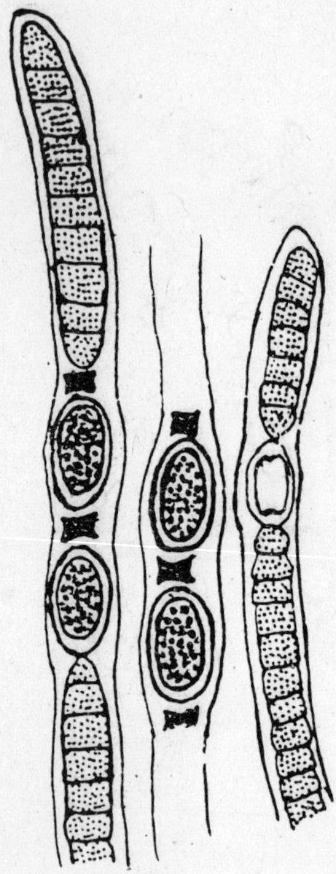

Fig. 93: *Aulosira fertilissima*

Stratum expanded, dark blue-green membranous; trichomes straight or a little flexuous, parallel or densely intricate; cells cylindrical when young, later barrel-shaped, contents granular; sheath thick, at first gelatinous and hyalin, later firm and brown; heterocysts intercalary, oblong or elliptical; spores in series usually alternating with dead cells, generally oblong-elliptical. (Fig. 93)

Dimensions: Cells 4.8-10.8μ broad, 5.2-10μ long; heterocysts 7.2-8.8μ broad, 10-14.4μ long; spores 10-12.8μ broad, 18-24μ long.

Distribution: Found on the bank of river Ganga.

NOSTOC Vaucher

***Nostoc muscorum* Ag. ex Born. et Flah.** Desikachary, *op. cit.*, 385, Pl. 70, Fig. 2, 1959.

Thallus gelatinous-membranous, irregularly expanded, attached by the lower surface, tuberculate, dull olive or brown; filaments densely entangled; cells short barrel shaped to cylindrical, upto twice as long as broad; heterocysts nearly spherical; spores oblong. (Fig. 94)

Dimensions: Trichome 2.8-4.8μ broad; cells 5.6-9.6μ; heterocysts 6.4-7.2μ broad, spores 4-8μ broad, 8-12μ long.

Distribution: Found on the bank of river Ganga.

SCYTONEMATACEAE *TOLYPOTHRIX* Kutzing

***Tolypothrix tenuis* (Kutz.) Johs. Schmidt em.**, Desikachary, 494, Pl. 102, Fig. 4, 1959.

Thallus caespitose or cushion-like, blue green or brown; filaments repeatedly branched; sheath thin, close to the trichome,

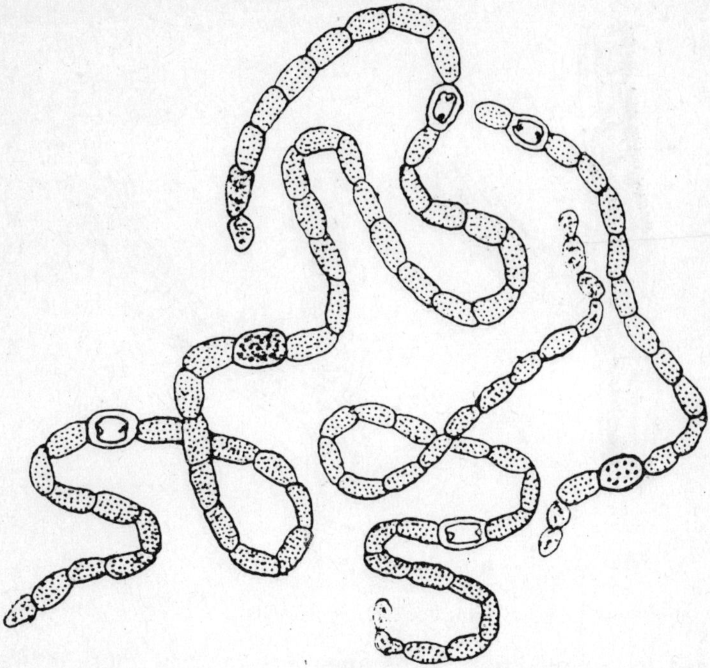

Fig. 94: *Nostoc muscorum*

at first colourless, later yellowish brown, often lamellated; cells quadrate or longer than broad blue-green, slightly or not constricted at the cross walls; heterocysts cylindrical rounded or discoid. (Fig. 95)

Dimensions: Filaments 6-17.2μ broad; cells 4-12.8μ broad; heterocysts 6.14μ broad and 2.4-6.4μ long.

Distribution : Found on the bank of river Ganga.

RIVULARIACEAE *CALOTHRIX* Ag.

***Calothrix membranacea* Schmidle.** Desikachary, *op.cit.*, 542, Pl. 106, Fig. 10, 1959.

Thallus papery, blue green; filaments long, horizontal, mostly curved, irregularly floccose, rarely branched, gradually attenuated; sheath thin, hyalin not lamelated; trichomes torulose, only seldom trichome produced into a hair; cells half as long as

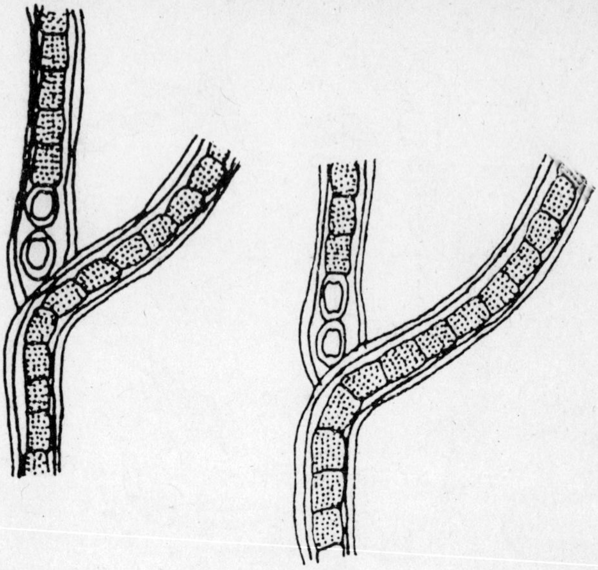

Fig. 95: *Tolypothrix tenuis*

broad to subquadrate, in the apex and the hair more or less hyalin and elongate, heterocyst present, basal. (Fig. 96)

Dimensions: Filaments 4-7.6μ broad and the heterocysts 4-6μ broad.

Distribution: Found on the bank of river Ganga.

GLOEOTRICHIA Ag.

Gloeotrichia indica **Schmidle.** Desikachary, *op.cit.*, 560, Pl. 112, Fig. 6, 1959.

Thallus spherical, soft, hollow (or solid), trichomes made of 2-3 cells, and a long hair narrow hair with cylindrical cells; spores nearly cylindrical, sheath at first hyaline later yellowish brown and close to the trichome. (Fig. 97)

Dimension: Cells 8.4-10.4μ broad; spores 59.6-71.2μ long, without sheath 17.2-20.8μ broad, with sheath 20.4-26.8μ broad.

Distribution: Found among other algae on the bank of river Ganga.

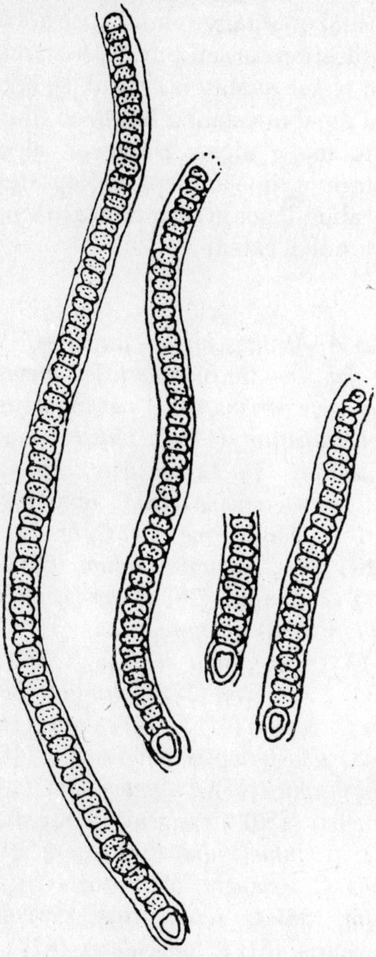

Fig. 96: *Calothrix membranancea*

Fig. 97: *Gloeotrichia indica*

Qualitative and Quantitative Assessment of Algae

A total number of 577 algal species spread over 151 chlorophyceae, 198 cyanophyceae, 223 chrysophyceae and 5 euglenophyceae were recorded from Ganga water. A very large number of planktonic algae including diatoms formed broad chunk of algae. Plankton counts show existence of 199 to 44800 cells per litre. While importance of each species in biology of Ganga water can not be denied, algae showing luxuriant growth and qualitative abundance are of major significance and their implications in self-purification, disease spread and maintenance of water quality is crucial. In here broad synoptical background of algal infestation has been illustrated with special reference to major algae. However, algal infestation shows marked variance in time and space. Reported algae may be found in numerical abundance in accordance to kind of water, its nutritional and pollutional extent.

Chlorophyceae

(1) *Actinastrum hantzschii*, (2) *A. Ahantzschii var. fluviatile*, (3) *Ankistrodesmus convolutus*, (4) *A. falcatus*, (5) *Bumilleriopsis brevis*, (6) *Botryococcus braunii*, (7) *Bulbochaete varians*, (8) *B. rectangularis*, (9) *B. reticulata*, (10) *Cosmarium granatum*, (11) *C. punctulatum*, (12) *C. nitidulum*, (13) *C. formosulum*, (14) *C. circulare*, (15) *C. depressum*, (16) *C. birectum*, (17) *C. meneghiniana*, (18) *C. reniforme*, (19) *C. sexangulare f. minima*, (20) *C. moniliforme*, (21) *C. broomei*, (22) *C. constrictum*, (23) *C. botrytis*, (24) *C. subimpressulum*, (25) *C. speciosum*, (26) *C. turgidum*, (27) *C. curtum*, (28) *C. formosulum var. nathorstii*, (29) *C. rectangulare*, (30) *C. quinarium*, (31) *C. subcrenatum*, (32) *C. cucumis*, (33) *Closterium acutum*, (34) *C. acerosum*, (35) *C. lanceolatum*, (36) *C. braunii*, (37) *C. lunula*, (38) *C. gracile*, (39) *C. turgidum*, (40) *C. venus*, (41) *C. parvulum*, (42) *C. monailiferum*, (43) *C. sp.* (44) *Closteriopsis longissima*, (45) *Characium naeqelii*, (46) *C. prinqsheimii*, (47) *C. ambiquum*. (48) *Chlorella vulgaris*, (49) *C. conductrix*, (50) *Crucigenia tetrapedia*, (51) *Chlorococcum humicola*, (52) *C. infusionum*, (53) *Chara sps.* (54) *Cladophora glomerata*, (55) *C. crispata*, (56) *Coelastrum cambricum*, (57) *C. proboscidium*, (58) *C. reticulatum*, (59) *C. sphaericum*, (60) *Coleochaete irregularis*, (61) *C. orbicularis*, (62) *C. sps.* (63) *Hydrodictyon reticulatum*, (64) *Chaetophora elegans*, (65) *Compsopogon sps.* (66) *Euastrum inermus*, (67) *Lagerheimia subsalsa*, (68) *Mougeotia calcarea*, (69) *M. robusta*, (70) *M. sphaerocarpa*, (71) *M. transeaui*, (72) *M. genuflexa*, (73) *M. elegantula*, (74) *M. boodlei*, (75) *Dedogonium gracillimum*, (76) *O. gracilius*, (77) *Hormidium subtilis*, (78) *Scenedesmus quadricauda var. bicaudatus*, (79) *Oocystis crassa*, (80) *O. spediatum*, (81) *O. elliptica*, (82) *O. solitaria*, (83) *O. borgei*, (84) *Protococcus viridis*, (85) *Pediastrum simplex*, (86) *P. simplex var. duodenarium*, (87) *P. boryanum*, (88) *P. boryanum var.*

longicorne, (89) *P. tetras*, (90) *P. tetras var. tetraodon*, (91) *P. tetras var. excisum*, (92) *P. duplex*, (93) *P. duplex var. clathratum*, (94) *Rhizoclonium hieroglyphicum*, (95) *Scenedesmus armatus*, (96) *S. acuminatus*, (97) *S. arcuatus*, (98) *S. abundans*, (99) *S. bijuga*, (100) *S. dimorphus*, (101) *S. obliguus*, (102) *S. prismaticus*, (103) *S. guadricauda*, (104) *S. guadricauda var. guadrispina*, (105) *S. guadricauda var. westii*, (106) *S. carinatus*, (107) *S. bijuga var. alternans*, (108) *S. longus var. minutus*, (109) *S. acutiformis*, (110) *Selenastrum bibraianum var. gracile*, (111) *S. gracile*, (112) *S. bibraianum*, (113) *Staurastrum oxyacanthum*, (114) *S. gracile*, (115) *Stigeoclonium tenue*, (116) *S. longipilum*, (117) *Sphaerocystis schroeteri*, (118) *Spirogyra ellipsospora*, (119) *S. punctiformis*, (120) *S. protecta*, (121) *Sirogonium illinoisense*, (122) *S. sticticum*, (123) *Spirogyre cleveana*, (124) *S. pratensis*, (125) *S. singularis*, (126) *S. sp.* (127) *Uronema confervicolum*, (128) *Ulothrix subconstricta*, (129) *U. zonata*, (130) *U. variabilis*, (131) *U. subtilissima*, (132) *Tetraedron muticum*, (133) *T. tumidulum*, (134) *T. minimum f. tetralobulatum*, (135) *Trachelomonas tambowika*, (136) *T. charkowiensis*, (137) *Tetradesmus smithii*, (138) *Westella botryoides*, (139) *Zygnemopsis decussata*, (140) *Zygnema collinsianum*, (141) *Z. cruciatum*, (142) *Z. tenue*, (143) *Cosmarium sexangulare*, (144) *Crucigenia rectangularis*, (146) *Golenkinia radiata*, (147) *Microspora quadrata*, (148) *S. guadricauda var. longispina* (149) *S. guadricauda var. maximus*, (150) *Stigeoclonium glomeratum*, (151) *Tetradesmus wisconsinense*.

Cyanophyceae

(152) *Anabaena variabilis*, (153) *A. gelatinicola*, (154) *A. planktonica*, (155) *A. fertilissima*, (156) *A. sphaerica*, (157) *A. oryzae*, (158) *A. circinalis*, (159) *A. constricta*, (160) *A. anomala*, (161) *A. orientalis*, (162) *A. oscillarioides*, (163) *A. iyengarii*, (164) *A. iyengarii var. tenuis*, (165) *A. spiroides*, (166) *A. affinis*, (167) *A. fuellebornii*, (168) *A. iyengarii var. attenuata*, (169) *Aphanocapsa montana*, (170) *A. littoralis*, (171) *A. biformis*, (172) *A. muscicola*, (173) *A. crassa*, (174) *A. elachista*, (175) *A. elachista var. conferta*, (176) *Aphanothece bullose*, (177) *A. saxicola*, (178) *A. pallida*, (179) *A. naegelii*, (180) *A. microscopica*, (181) *A. castagnei*, (182) *Aphanocapsa pulchra*, (183) *A. pulchella*, (184) *Arthrospira platensis*, (185) *A. spirulinoides*, (186) *Calothrix calvata*, (187) *C. scytonemicola*, (188) *C. marchica*, (189) *C. marchica var. intermeida*, (190) *C. bharadwajae*, (191) *C. gloeocala*, (192) *C. thermalis*, (193) *C. viguieri*, (194) *C. elenkinii*, (195) *C.*

geitonos, (196) *C. fusca*, (197) *C. montanus*, (198) *C. pallidus*, (199) *C. minor*, (200) *C. tenex*, (201) *C. macrococcus*, (202) *C. varius*, (203) *C. minutus*, (204) *C. refescens*, (205) *C. indicus*, (206) *Calothrix marchica* var. *crassa*, (207) *Chamaesiphon sideriphilus* var. *glabra*, (208) *Coelosphaerium kuetzingianum*, (209) *Cylindrospermum minutissimum*, (210) *C. stagnale*, (211) *C. muscicola*, (212) *C. licheniforme*, (213) *C. majus*, (214) *C. michailovskaense*, (215) *Dichothrix gypsophila*, (216) *Dactylococcus infusionum*, (217) *Chroococcus cohaerens*, (218) *Gloeocapsa quaternata*, (219) *G. stegophila*, (220) *G. compacta*, (221) *G. pleurocapsoides*, (222) *G. marginata*, (223) *G. magma*, (224) *G. nigrescens*, (225) *G. atrata*, (226) *G. decorticans*, (227) *G. kuetzingiana*, (228) *Gloeotrichia natans*, (229) *G. indica*, (230) *G. intermedia*, (231) *G. echinulata*, (232) *G. echinulata* var. *berhampurense*, (233) *G. raciborskii*, (234) *G. raciborskii* var. *salsettense* (235) *G. raciborskii* var. *longispora*, (236) *G. raciborskii* var. *conica*, (237) *G. longicauda*, (238) *G. ghosei*, (239) *G. raciborskii* var. (240) *Gloeothece montanus*, (241) *G. rhodochlamys*, (242) *Hydrococcus rivularis*, (243) *Lyngby infixa*, (244) *L. majuscula*, (245) *L. dendrobia f. lurida*, (246) *L. polysiphonae*, (247) *L. putealis*, (248) *L. limnetica*, (249) *L. porphyrosiphonis*, (250) *L. contorta*, (251) *L. confervoides*, (252) *L. hieronymusii*, (253) *L. epiphytica*, (254) *L. aerugineo-coerule*, (255) *Microcystis aeruginosa*, (256) *M. aeruginosa* var. *major*, (257) *M. robusta*, (258) *M. viridis*, (259) *M. flos-aquae*, (260) *M. incerta*, (261) *M. protocystis* (262) *M. pseudofilamentosa*, (263) *M. marginata*, (264) *M. lamelliformis*, (265) *Merismopedia punctata*, (266) *M. tenuissima*, (267) *M. glauca*, (268) *M. convoluta*, (269) *M. elegans*, (270) *Myxosarcina spectabilis*, (271) *Nostoc linckia*, (272) *N. linckia* var. *arvense*, (273) *N. muscorum*, (274) *M. carneum*, (275) *N. ellipsosporum*, (276) *N. verrucosum*, (277) *Nostochopsis lobatus*, (278) *Nodularia herveyana*, (279) *N. spumigena*, (280) *N. spumigena* var. *major*, (281) *Oscillatoria limosa*, (282) *O. formosa*, (283) *O. princeps*, (284) *O. pseudogeminata*, (285) *O. willei*, (286) *O. homogenea*, (287) *O. acutissima*, (288) *O. subbrevis*, (289) *O. boryana*, (290) *O. tenuis*, (291) *O. curviceps*, (292) *O. curviceps* var. *angusta*, (293) *O. earlei*, (294) *O. tanganyikae*, (295) *O. acuta*, (296) *O. subuliformis*, (297) *O. minnesotensis*, (298) *O. anguina*, (299) *O. ornata*, (300) *O. schultzii*, (301) *O. perornata*, (302) *O. corallinae*, (303) *O. nigroviridis*, (304) *O. chalybea*, (305) *P. anbiguum* var. *major*, (306) *O. subbrevis* var. *crassa*, (307) *O. okeni*, (308) *O. laetevirens*, (309) *O. laetevirens* var. *minimus*, (310) *O. limnetica*,

(311) *O. angustissima*, (312) *O. acuminata*, (313) *O. minima*, (314) *Phormidium ambiguum*, (315) *P. foveolarum*, (316) *P. tenue*, (317) *P. anomala*, (318) *P. purpurascens*, (319) *P. pachydermaticum*, (320) *P. fragile*, (321) *P. molle*, (322) *Raphidiopsis curvata*, (323) *Rivularia aguatica*, (324) *R. hansgirgi*, (325) *R. brccarina*, (326) *R. manginii*, (327) *Schizothrix lamyi*, (328) *S. muelleri*, (329) *Spirulina majar*, (330) *S. gigantea*, (331) *S. subsalsa*, (332) *S. laxissima*, (333) *S. labyrinthiformis*, (334) *S. subtilissima*, (335) *Oscillatoria calcuttensis*, (336) *O. foreaui*, (337) *Synechococcus aeruginesus*, (338) *Symploca muscorum*, (339) *Anabaena vaginicola*, (340) *Aphanocapsa grevillei*, (341) *Aphanocapsa endophytica*, (342) *Chamaesiphon sideriphilus*, (343) *Lyngbya aestuariii var. arbustiva*, (344) *Nostoc calcicola*, (345) *Oscillatoria sancta*, (346) *Phormidium lucidum*.

Chrysophyceae

(347) *Asterionella formosa*, (348) *A. gracillima*, (349) *Amphora ovalis*, (350) *A. ovalis var. pediculus*, (351) *A. ovalis var pediculus f. minor*, (352) *A. salina*, (353) *A. angusta*, (354) *Achnantes lanceolata*, (355) *A. exilis*, (356) *A. minutissima*, (357) *A linearis*, (358) *A. lanceolata var. dubia*, (359) *A. affinis*, (360) *A. parvula*, (361) *Cymbella cuspidata*, (362) *C. cistula*, (363) *C. perva*, (364) *C. affinis*, (365) *C. tumida*, (366) *C. pusilla*, (367) *C. laevis*, (368) *C. leptoceras*, (369) *C. leptoceras var. Elongata*, (370) *C. microcephala*, (371) *C. amphicephala*, (372) *C. obtusa*; (373) *C. cymbiformis*, (374) *C. turgida*, (375) *C. ventricosa*, (376) *C. prostrata*, (377) *C. delicatula*, (378) *Cyclotella meneghiniana*, (379) *C. kuetzingiana*, (380) *C. glomerata*, (381) *C. operculata*, (382) *C. comta*, (383) *C. melosiroides*, (384) *Cocconeis pediculus*, (385) *C. placentula*, (386) *C. costata*, (387) *Cyclotella ocellata*, (388) *Cymatopleura elliptica*, (389) *Diatoma elongatum*, (390) *D. hiemale*, (391) *D. vulgare*, (392) *Denticula elegans*, (393) *Cymbella subagualis*, (394) *Diploneis elliptica*, (395) *D. puella*, (396) *Epithemia gibba*, (397) *E. gibba var. ventricosa*, (398) *E. gibberula*, (399) *E. zebra*, (400) *E. sorex*, (401) *E. musculus*, (402) *E. turgida*, (403) *Encyonema caespitosum*, (404) *E. caespitosum var. Auerswaldii*, (405) *E. ventricosum*, (406) *E. caespitosum var. lata*, (407) *E. prostratum*, (408) *Eunotia pectinalis var. ventricosa*, (409) *Fragilaria capucina*, (410) *F. capucina var. acuta*, (411) *F. crotonensis*, (412) *F. construens*, (413) *F. virescens*, (414) *F. pinnata*, (415) *F. construens var, venter*, (416) *Frustulia rhomboides*, (417) *Fragilaria capucina var. acuminata*, (418) *Gyrosigma spencerii*, (419) *G. scalproides*, (420) *G. kuetzingii*,

(421) *G. acuminatum*, (422) *Gomphonema sphaerophorum*, (423) *G. montanum*, (424) *G. montanum var commutatum*, (425) *G. constrictum*, (426) *G. gracile*, (427) *G. augur*, (428) *G. olivaceum*, (429) *G. olivaceum var. vulgaris*, (430) *G. parvulum*, (431) *G. parvulum var. suncapitata*, (432) *Gomphonema parvulum var. lanceolata*, (433) *G. intricatum*, (434) *Nitzschia punctata*, (435) *Hantzschia amphioxys*, (436) *H. amphioxys var. intermedia*, (437) *Mastogloia smithii*, (438) *M. lanceolata*, (439) *M. Dansei*, (440) *M. Grevillei*, (441) *Melosira distans*, (442) *M. crenulata*, (443) *M. italica*, (444) *M. varians*, (445) *M. granulata*, (446) *M. arenaria*, (447) *Navicula ambiqua*, (448) *N. exiqua*, (449) *N. anglica*, (450) *N. sphaerophora*, (451) *N. perpusilla*, (452) *N. laevissima*, (453) *N. dicephala*, (454) *N. capitata*, (455) *N. cryptocephala*, (456) *N. protracta*, (457) *N. pupula*, (458) *N. tuscula*, (459) *N. seminulum*, (460) *N. permogala*, (461) *N. radiosa*, (462) *N. lanceolata*, (463) *N. inaequiletera*, (464) *N. cocconeiformis*, (465) *N. humerosa*, (466) *N. ventricosa*, (467) *N. mutica*, (468) *N. platystoma*, (469) *Cryptocephala var. exilis*, (470) N. brebissonii (471) *N. salinarum*, (472) *N. humilis*, (473) *N. sps.*, (474) *N. gallica*, (475) *N. kutzingiana*, (476) *N. borealis*, (477) *N. oculata*, (478) *N. Hilseana*, (479) *N. radiosa var. tenella*, (480) *N. Atomus*, (481) *N. flotowii*, (482) *N. gracilis*, (483) *N. retusa*, (484) *N. falaisensis*, (485) *N. exilis*, (486) *N. cruciculs var protracta*, (487) *N. gregaria*, (488) *N. amphisbaena*, (489) *N. mutica var. ventricosa*, (490) *N. peregrina*, (491) *N. peregrina var. menisculus, f-upsaliensis*, (492) *N. braunii*, (493) *N. pygmaea*, (494) *N. exilissima*, (495) *N. lanceolata f. minor*, (496) *N. lepidula*, (497) *N. simplex*, (498) *N. cencellata var. ammophila*, (499) *N. cincta*, (500) *N. cincta var. heulferi*, (501) *N. sculpta*, (502) *Nitzschia linearis*, (503) *N. linearis var. tenuis*, (504) *N. palea*, (505) *N. palea var. fonticola*, (506) *N. palea var. tenurostris*, (507) *N. anqustata*, (508) *N. acicularis*, (509) *N. dissipata*, (510) *N. thermalis*, (511) *N. obtusa*, (512) *N. microcephala*, (513) *N. longissima*, (514) *N. Denticula var. Delognei*, (515) *N. frustulum var. perpusilla*, (516) *N. ovalis*, (517) *N. palea var. debilis*, (518) *N. lanceolata f. minima*, (519) *N. obtusa var. nana*, (520) *N. sinuata*, (521) *N. apiculata*, (522) *N. debilis*, (523) *N. subtilis*, (524) *N. amphibia*, (525) *N. vermicularis*, (526) *N. angularis var. affinis*, (527) *N. commutata*, (528) *N. Tryblionella*, (529) *N. tryblionella var. calida*, (530) *N. Hungarica*, (531) *N. Frustulum var. minutula*, (532) *N. dissipata var. acuta*, (533) *N. Tryblionella var. levidensis*, (534) *N. dissipata var. media*, (535) *N. subtilis var. paleaceae*, (536) *Pleurosigma*

spencerii, (537) *P. spencerii var. smithii*, (538) *P. parkeri*, (539) *P. scalproides*, (540) *Pinnularia undulata*, (541) *Rhopalodia gibba*, (542) *R. gibberula*, (543) *Synedra ulna*, (544) *S. ulna var. agualis*, (545) *S. ulna var. denica*, (546) *S. vaucheriae*, (547) *S. radians*, (548) *S. amphicephala*, (549) *S. pulchella*, (550) *S. acus*, (551) *S. affinis var. parva*, (552) *Surirella robusta*, (553) *S. robusta var. parva*, (554) *S. elagans*, (555) *S. ovalis*, (556) *S. ovalis var. minuta*, (557) *Synedra vaucheriae var. parvula*, (558) *S. vaucheriae var. perminuta*, (559) *Synedra ulna var. longissima*, (560) *S. ulna var. obtusa.*, (561) *Surirella linearis*, (562) *Achnanthes biasolettiana*, (563) *Stauroneis anceps*, (564) *S. anceps var. linearis*, (565) *Stauroncis acuta*, (566) *Stauronus smithii*, (567) *Fragilaria mutabilis*, (568) *Tetracyclus repestrus*.

Euglenophyceae

(569) *Euglena spirogyara*, (570) *E. graculis*, (571) *E. viridis*, (572) *Phacus amminatus*, (573) *P. pleuronectus*, (574) *Phacus langicauda*, (575) *Euglena minuta*, (576) *Trachelomonas tambowik*, (577) *T. charkowiensis*.

Qualitative Distribution

Qualitatively algal species show sequential gradual increase from winter to summer months upto April. Various classes of algae represented during different months show increase in number of algal species from November to April. But during May-August gradual decline in three classes (chlorophyceae, bacillariophyceae and cyanophyceae) and then gradual increase from September-October could be registered.

Benthic Algae

Among chlorophycean forms 94 species (*Actinastrum hantzschii, A. hantzschii var. fluviatile, Ankistrodesmus convolutus, A. falcatus, Bulbochaete rectangularis, B. reticulata, B. varians, Cladophora glamerata, Chaetophora elegans, C. incrassata, C. pisiformis, Crucigenta tetrapedia, Chlorella vulgaris, Chlorococcum humicola, C. infusionum, Closterium acerosum, C. moniliferum, C. parvulum, C. turgidum, C. venus, Coleochaete irregularis, Coelastrum cambricum var. intermedium, C. proboscidium, Closteriopsis longissima, Cosmarium broomei, C. botrytis, C. constrictum, C. curtum, C. circulare, C. depressum, C. formosulum, C. formosulum var. nathorstii,*

TOTAL NUMBER OF ALGAE IN EACH CLASS FROM OCT TO SEPT

CHLOROPHYCEAE	CYANOPHYCEAE	BACILLARIOPHYCEAE	EUGLENOPHYCEAE	TOTAL
151	198	223	5	577

TOTAL NUMBER OF ALGAE IN EACH MONTH

Class	Oct	Nov	Dec	Jan	Feb	Mar	April	May	June	July	Aug	Sep
CHLOROPHYCEAE CYANOPHYCEAE BACILLARIOPHYCEAE EUGLENOPHYCEAE	92	126	148	168	180	221	290	296	260	61	8	1

TOTAL NUMBER OF ALGAE IN EACH CLASS AND MONTH

Class	Oct	Nov	Dec	Jan	Feb	Mar	Apr	May	June	July	Aug	Sep
CHLOROPHYCEAE	1	36	38	41	42	57	63	52	48	20	5	1
CYANOPHYCEAE	38	48	57	62	65	73	109	112	98	20	1	0
BACILLARIOPHYCEAE	53	42	51	62	73	88	119	126	120	19	2	0
EUGLENOPHYCEAE	0	-	2	3	2	3	3	5	5	2	0	0

C. granatum, C. meneghinii, C. nitidulum, C. punctulatum, C. reniforme, C. speciosum, C. subimpressulum, Cosmarium sp., Characium naegelii, Euastrum inermius, Golenkinia radiata, Hydrodictyon reticulatum, Mougeotia calcarea, M. genuflexa, M. sphaerocarpa, Oedogonium gracillimum, Oocystis eremosphaeria, O. solitaria, Pediastrum boryanum, P. duplex, P. duplex var. *clathratum, P. simplex, P. simplex* var. *duodenarium, P. tetras, P. tetras* var. *excisum, P. tetras* var. *tetraodon, Protococcus viridis, Rhizoclonium hieroglyphicum, Scenedesmus armatus, S. acuminatus, S. abundans, S. arcuatus, S. acutiformis, S. bijuga, S. bijuga var. alternans, S. carinatus, S. dimorphus, S. longus* var. *minutus, S. obliquus, S. prismaticus, S. quadricauda, S. quadricauda* var. *bicaudatus, S. quadricauda* var. *longispina, S. quadricauda* var. *quadrispina, S. quadricauda* var. *westii, Sphaerocystis schroeteri, staurastrum gracile, S. hexacerum, S. oxyacanthum, Selenastrum bibraianum, S. gracile, Spirogyra punctiformis, S. rectangularis, S. singularis, Spirogyra sp., Spirogyra sp., Spirogyra sp., Tetraedron trigonum* var. *minua, Ulothrix zonata, Zygnema collinsianum, Z. czurdae* and *Z. tenue*) were observed. The only charophycean form *Chara excelsa* was recorded.

Bacillariophycean forms including 130 species (*Amphora ovalis, A. ovalis* var. *pediculus, A. ovalis* var. *pediculus forma minor, Achnanthes exilis, Achnanthes linearis, A. lanceolata, A. lanceolata* var. *dubia, A. minutissima, Cymbella affinis, C. amphicephala, C. cistula, C. cymbiformis, C. cuspidata, C. leptoceros, C. leptoceros* var. *elongata, C. laevis, C. microcephala, C. naviculiformis, C. obtusa, C. parva, C. pusilla, C. prostrata, C. tumida, C. ventricosa, Cocconeis placentula, C. pediculus, Cyclotella glomerata, C. kuetzingiana, C. meneghiniana, C. operculata, C. ocellata, Denticula elegans, Epithemia gibba, E. gibba* var. *ventricosa, E. gibberula* var. *producta, E. sorex, E. zebra, Encyonema caespitosum, E. prostratum, E ventricosum, Fragilaria crotonensis, F. construens, Gomphonema augur, G. constrictum* var. *capitatum, G. gracile, G. gracile* var. *dichotomum, G. montanum, G. montanum* var. *commutatum, G. olivaceum, G. parvulum, G. parvulum* var. *lanceolata G. parvulum* var. *subcapitata, G. sphaerophorum, Gyrosigma kuetzingii, G. spencerii, G. scalproides, Hantzschia amphioxys, H. amphioxys* var. *intermedia, Melosira distans, M. granulata, M. granulata* var. *curvata, Mastogloia smithii, Navicula anglica, N. braunii, N. borealis, Navicula capitata, N. cryptocephala, N. cincta, N. cincta* var. *Heufleri, N. cancellata* var. *ammophila, N. dicephala, N. exigua, N. exilis, N. gastrum, N. humilis, N. hilseana,*

N. inaequilatera, N. lanceolata, N. mutica, N. oculata, N. platystoma, N. protracta, N. pupula, N. perpusilla, N. radiosa, N. sphaerophora, N. simplex, N. salinarum, N. seminulum, N. sculpta, N. tuscula, N. ventricosa, Nitzschia jacicularis, N. angustata, N. apiculata, N. dissipata, N. dissipata var. *media, N. linearis, N. linearis* var. *tenuis, N. microcephala, N. obtusa, N. obtusa* var. *nana, N. palea, N. palea* var. *tenuirostris, N. palea* var. *fonticola, N. subtilis, N. sinuata, N. thermalis, N. tryblionella* var. *calida, N. vermicularis, Pleurosigma spencerii* var. *nodifera, P. scalproides, Pinnularia undulata, Rhopalodia, gibba, R. ventricosa, Synedra acus, S. pulchella, S. radians, S. ulna, S. ulna* var. *aequalis, S. ulna* var. *danica, S. ulna* var. *longissima, S. ulna* var. *lanceolata, S. vaucheriae* var. *parvula, S. vaucheriae* var. *perminuta, Stauroneis anceps, Surirella elegans, S. robusta, S. ovalis* var. *minuta, Tetracyclus rupestris*) could be recorded. Among euglenophycean forms *Euglena gracilis, E. spirogyra* and rhodophycean *Composopogon sp.* were observed.

Cyanophycean forms spread over 120 species (*Anabaena affinis, A. circinalis* var. *crassa, A. fertilissima, A. gelatinicola, A. iyengarii, A. iyengarii* var. *tenuis, A. iyengarii* var. *attenuata, A. oscillarioides, A. oryzae, A. orientalis, A. plactonica, A. sphaerica* var. *attenuata, Anabaena sp., Anabaena sp., Aulosira bombayensis, A. prolifica, Aphanocapsa biformis, A. elachista, A. littoralis, A. montana, A. muscicola, Aphanothece bullosa, A. saxicola, Arthrospira spirulinoides, Chroococcus indicus, C. macrococcus, C. minutus, C. minor, C. montanus, C. pallidus, C. rufescens, C. tenax, C. turgidus, C. varius, Cylindrospermum muscicola, C. michailovskoense, C. stagnale, Coelosphaerium kuetzingianum, Calothrix clavata, C. geitonos, C. gloeocola, C. marchica, C. marchica* var. *intermedia, C. scytonemicola, Gloeocapsa quaternata, G. nigrescens, Gloeocapsa stegophila, Gloeotrichia echinulata* var. *berhampurense, G. ghosei, G. intermedia* var. *kanwaensis, G. indica, G. longicauda, G. natans, G. raciborskii, G. raciborskii var. bombayense, G. raciborskii* var. *longispora, G. raciborskii* var. *kashiense, G. raciborskii* var. *salsettense, Lyngbya aestuarii* var. *arbustiva, L. contorta, L. confervoides, L. porphyrosiphonis, L. polysiphonae, Merismopedia convoluta, M. glauca, M. punctata, M. tenuissima, Microcystis aeruginosa, M. elabens, M. incerta, M. protocystis, M. robusta, M. viridis, Myxosarcina spectabilis, Nostoc carneum, N. linckia, N. linckia* var. *arvense, N. muscorum, N. verrucosum, Nodularia harveyana* var. *sphaerocarpa, N. spumigena, Oscillatoria anguina, O. acuta, O. angustissma, O. acutissima, O.*

corallinae, O. chalybea, O. curviceps, O. earlei, O. formosa, O. foreaui, O. limosa, O. lacustris, O. limnetica, O. laete-virens var. *minimus, O. minnesotensis, O. minima, O. nigroviridis, O. okeni, O. ornata, O. princeps, O. pseudogeminata* var. *unigranulata, O. raoi, O. subbraevis, O. subliformis, O. tenuis, O. tenuis* var. *tergestina, O. willei, Phormidium ambiguum, P. anomala, P. purpurascens, P. tenue, Raphidiopsis sp., Rivularia hansgirgi, Spirulina gigantea, S. laxissima, S. labyrinthiformis, S. major, Synechocystis aquatilis* and *S. pevalekii*) were recorded.

Planktonic Algae

Twenty chlorophycean species comprising (*Actinastrum hantzschii, Ankistrodesmus convolutus, A. falcatus, Chlorococcum humicola, Chlorella vulgaris, Closteriopsis longissima, Crucigenia tetrapedia, C. quadrata, Coelastrum cambricum* var. *intermedium, Closterium acerosum, Golenkinia radiata, Pediastrum boryanum, Scenedesmus armatus, S. bijuga, S. dimorphus, S. quadricauda, S. quadricauda* var. *quadrispina, Selenastrum gracile, S. westii* and *Sphaerocystis schroeteri*) were recorded as planktonic forms.

Among bacillariophycean planktons 28 species (*Achnanthes exilis, Amphora ovalis* var. *pediculus, Cymbella microcephala, Cyclotella comta, C. glomerata, C. kuetzingiana, C. meneghiniana, C. operculata, Gyrosigma scalproides, Gomphonema parvulum, Mastogloia smithii, Melosira crenulata, M. distans, M. granulata, M. granulata* var. *curvata, M. varians, Nitzschia acicularis, N. linearis* var. *tenuis, N. microcephala, N. palea, Navicula dicephala, N. gregaria, N. inaequilatera, Synedra acus, S. ulna, S. ulna* var. dancia, *S. vaucheriae* var. *parvula* and *Stauroneis anceps*) were witnessed.

A numerical depletion in cyanophycean species was witnessed and only six species (*Merismopedia glauca, M. punctata, M. tenuissima, Microcystis aeruginosa, Oscillatoria limnetica* and *Phormidium tenue*) could be observed as planktons.

Distribution of Algae

Algae were found distributed as well defined communities showing major infestation and are grouped into (a) planktons (b) bottom sand algae (c) algae growing on walls, pillars of bridge or other manmade structures (d) epiphytic, benthic, floating or detached algae.

Periodicity and Succession

Diverse algal forms growing in the river exhibit a clearcut periodicity and succession. They may be broadly grouped into summer, winter and rain forms of either annuals, perennials or epimerales nature. Algae may grow at certain period and thrive rest of the time as spores or other cells capable of tiding over adverse conditions.

Certain algae (*Scenedesmus bijuga, Spirogyra punctiformis, Nitzschia palea, Synedra ulna* and *Merismopedia glauca*) grow throughout the year. Algae like *Bulbochaete rectangularis, B. reticulata, B. varians, Cosmarium, Botrytis, C. constictum, C. subimpressulum, C. speciosum, Cosmarium sp., Coleochaete irregularis, Closterium turgidum, Mougeotia genuflexa, Oedogonium gracillimum, Scenedesmus carinatus, Cymbella naviculiformis, Gomphonema sphaerophorum, Pleurosigma scalproides, Euglena spirogyra, Anabaena circinalis* var. *crassa, A. iyengarii* var. *tenuis, A. iyengarii* var. *attenuata, Coelosphaerium kuetzingianum, Gloeotrichia longicauda, Nodularia harveyana* var. *sphaerocarpa, Oscillatoria raoi, Phormidium ambiguum* and *Spirulina laxissima* were found growing distinctly in winter season; (*Ankistrodesmus convolutus, Actinastrum hantzschii* var. *fluviatile, Closterium moniliferum, C. parvulum, C. venus, Closteriopsis longissima, Cosmarium broomei, C. curtum, C. circulare, C. punctulatum, Chaetophora elegans, C. incrassata, C. pisiformis, Golenkinia radiata, Pediastrum duplex, P. tetras, P. tetras* var. *excisum, P. tetras* var. *tetraodon, P. simplex* var. *duodenarium, Protococcus viridis, Selenastrum gracile, Oocystis solitaria, Staurastrum gracile, S. hexacerum, S. oxyacanthum, Scenedesmus guadricauda* var. *bicaudatus, S. guadricauda* var. *longispina, Sphaerocystis schroeteri, Tetraedron trigonum* var. *minus, Amphora ovalis* var. *pediculus, A. ovalis* var. *pediculus forma minor, Achnanthes lanceolata* var. *dubia, A. minutissima, Cymbella affinis, C. amphicephala, C. leptoceros* var. *elongata, C. obtusa, C. parva, C. pusilla, C. ventricosa, Cyclotella ocellata, Cocconeis pediculus, Denticula elegans, Epithemia zebra, Encyonema prostratum, E. ventricosum, Fragilaria construens, Gomphonema gracile, G. gracile* var. *dichotomum, G. montanum, G. parvulum* var. *lanceolata, G. parvulum* var. *subcapitata, Hantzschia amphioxys* var. *intermedia, Melosira granulata* var. *curvata, N. borealis, N. cincta, N. cincta* var. *Heufleri, N. cancellata* var. *ammophilla, N. exilis, N. humilis, N. lanceolata, N. oculata, N. pupula, N. radiosa, N. salinarum, N. simplex, N. sculpta, Nitzschia linearis* var. *tenuis,*

N. obtusa, N. obtusa var. *nana, N. sinuata, N. vermicularis, Pinnularia undulata, Synedra Pulchella, S. radians, S. ulna* var. *longissima, S. ulna* var. *lanceolata, S. vaucheriae* var. *perminuta, Surirella ovalis* var. *minuta, S. robusta, Tetracyclus rupestris, Euglena gracilis, Anabaena gelatinicola, Anabaena sp., Aphanocapsa muscicola, A. elachista, Aulosira bombayensis, A. prolifica, Arthrospira spirulinoides, Aphanocapsa biformis, Chroococcus minor, C. montanus, Calothrix clavata, C. gloeocola, C. geitonos, C. marchica, C. marchica* var. *intermedia, Gloeocapsa nigrescens, Gloeotrichia echinulata* var. *berhampurense, G. ghosei, G. raciborskii* var. *bombayense, G. raciborskii* var. *kashiense, Lyngbya aestuarii* var. *arbustiva, L. contorta, L. confervoides, L. polysiphoniae, L. porphyrosiphonis, Microcystis elabens, M. protocystis, M. viridis, Myxosarcina spectabilis, Nostoc verrucosum, Oscillatoria acuta, O. acutissima, O. angustissima, O. curviceps, O. corallinae, O. chalybea, O. foreaui, O. lacustris, O. laete-virens* var. *minimus, O. minnesotensis, O. nigroviridis, O. okeni, O. pseudogeminata* var. *unigranulata, Phormidium purpurascens, Rivularia hansgirgi* and *Spirulina labyrinthiformis*) were observed in summer season; (*Oocystis eremosphaeria, Navicula braunii* and *N. hilseana*) were observed in rainy season. Apart from them certain forms (*Actinastrum hantzschii, Cladophora glomerata, Crucigenia tetrapedia, Chlorococcum infusionum, Cosmarium depressum, C. granatum, Characium naegelii, Hydrodictyon reticulatum, Mougeotia calcarea, M. sphaerocarpa, Pediastrum duplex* var. *clathratum, Rhizoclonium hieroglyphicum, Scenedesmus acuminatus, S. abundans, S. arcuatus, S. acutiformis, S. bijuga* var. *alternans , S. longus* var. *minutus, S. prismaticus, Selenastrum bibraianum, Spirogyra rectangularis, Spirogyra sp., Spirogyra sp., Ulothrix zonata, Zygnema czurdae, Chara excelsa, Amphora ovalis, Cymbella cistula, C. cymbiformis, C. cuspidata, C. leptoceros, C. laevis, C. prostrata, Cyclotella glomerata, Epithemia sorex, Gomphonema olivaceum, Gyrosigma kuetzingii, G. spencerii, Melosira granulata, Navicula exigua, N. mutica, N. protracta, N. platystoma, N. sphaerophora, N. seminulum, N. ventricosa, Nitzschia angustata, N. dissipata* var. *media, N. microcephala, N. palea* var. *tenurostris, N. subtilis, N. thermalis, Rhopalodia gibba, R. ventricosa, Synedra ulna* var. *danica, Surirela elegans, Compsopogon sp., Anabaena affinis, A. sphaerica* var. *attenuata, A. iyengarii, A. oscillarioides, A. planctonica, Anabaena sp., Aphanothece bullosa, Chroococcus rufescens, C. varius, Cylindrospermum muscicola, C. michailovskoense, C. stagnale,*

Gloeotrichia indica, G. natans, G. raciborskii, G. raciborskii var. *longispora, G. raciborskii* var. *salsettense, Gloeocapsa guaternata, G. stegophila, Merismopedia convoluta, Microcystis robusta, Nostoc linckia* var. *arvense, N. muscorum, Nodularia spumigena, Oscillatoria anguina, O. formosa, O. ornata, O. tenuis, O. tenuis* var. *tergestina, Phormidium anomala, Synechocystis pevalekii* and *S. aquatilis*) show growth during winter-summer seasons. Rain-winter period showed no growth of algal forms. Summer-rain seasons marked growth of *Ankistrodesmus convolutus, Closterium acerosum, Cosmarium formosulum* var. *nathorstii, C. meneghinii, C. reniforme, Euastrum inermius, Achnanthes exilis, A. linearis A. lanceolata, Cymbella microcephala, Cyclotella operculata, Fragilaria crotonensis, Gomphonema parvulum, Gyrosigma scalproides, Hantzschia amphioxys, Melosira distans, Mastogloia smithii, Navicula capitata, N. perpusilla, Nitzschia apiculata, N. palea* var. *fonticola, N. tryblinella* var. *calida, Synedra vaucheriae* var. *parvula, Aphanocapsa littoralis, Chroococcus indicus, C. macrococcus, C. minutus, C. pallidus, Gloeotrichia intermedia* var. *kanwaensis, Microcystis incerta, Oscillatoria limnetica, O. minima* and *O. subuliformis*). Certain algae (*Ankistrodesmus falcatus, Chlorella vulgaris, Chlorococcum humicola, Coelastrum cambricum* var. *intermedium, C. proboscidium, Cosmaruim formosulum, C. nitidulum, Pediastrum boryanum, P. simplex. Scenedesmus armatus, S. dimorphus, S. obliguus, S. quadricauda, S. quadricauda* var. *quadrispina, S. quadricauda* var. *westii, Spirogyra sp., S. singularis, Zygnema collinsianum, Z. tenue, Cymbella tumida, Cyclotella kuetzingiana, C. meneghiniana, Cocconeis placentula, Epithemia gibba, E. gibba* var. *ventricosa, E. gibberula* var. *producta, Encyonema caespitosum, Gomphonema augur, G. constrictum* var. *capitatum, G. montanum* var. *commutatum, Navicula anglica, N. cryptocephala, N. dicephala, N. gastrum, N. inaequilatera, N. tuscula, Nitzschia acicularis, N. dissipata, N. linearis, Pleurosigma spencerii* var. *nodifera, Synedra acus, S. ulna* var. *aequalis, Stauroneis anceps, Anbaena fertilissima, A. oryzae, Anabaena sp., Aphanocapsa montana, Aphanothece saxicola, Chroococcus tenax, C. turgidus, Calothrix scytonemicola, Merismopedia punctata, M. tenuissima, Microcystis aerutginosa, Nostoc carneum, Oscillatoria limosa, O. princeps, O. subbrevis, Phormidium tenue, Raphidiopsis sp., Spirulina gigantea* and *S. major*) were observed sporadically in various seasons but not throughout the year. While others (*Chaetophora incrassata, C. pisiformis, Closterium parvulum, Coleochaete irregularis, C. curtum,*

C. circulare, C. subimpressulum, Cosmarium sp., Golenkinia radiata, Oocystis eremosphaeria, Staurastrum gracile, S. hexacerum, S. oxyacanthum, Tetraedron trigonum var. *minus, Cymbella naviculiformis, Cyclotella ocellata, Navicula braunii, N. cancellata* var. *ammophila, N. cincta* var. *Heufleri, N. kilseana, N. simplex, N. sculpta, Nitzschia obtusa* var. *nana, Synedra ulna* var. *longissima, S. ulna* var. *lanceolata, Tetracyclus rupestris, Euglena gracilis, Anabaena gelatinicola, Aulosira bombayensis, A. prolifica, Aphanocapsa elachista, Calothrix gloeocola, Gloeotrichia ghosei, Lyngbya porphyrosiphonis, L. ploysiphoniae, Microcystis elabens, M. protocystis, Myxosarcina spectabilis, Oscillatoria lacustris* and *O. raoi.*

Benthic Algae Observed only at Certain Places

Algae observed at certain places but conspicuously absent elsewhere are given below:

Chlorophyceae (*Bulbochaete rectangularis B. reticulata, B. varians., Chaetophora elegans, C. incrassata, C. pisiformis, Closterium moniliferum, C. parvulum, C. venus, Coleochaeta irregularis, Cosmarium botrytis, C. constructum, C. circulare, C. curtum, C. formosulum* var. *nathorstii, C. subimpressulum, C. speciosum, Cosmarium sp., Golenkinia radiata, Oedogonium gracillimum, Oocystis eremosphaeria, Pediastrum duplex* var. *clathratum, P. simplex, P. simplex* var. *duodenarium, Scenedesmus acutiformis, S. bijuga* var. *alternans, S. longus* var. *minutus, S. quadricauda* var. *longispina, Staurastrum hexacerum, S. gracile, S. oxyacanthum, Spirogyra singularis, Spirogyra sp., Spirogyra sp.* and *Tetraedron trigonum* var. *minus*).

Bacillariophyceae (*Amphora ovalis* var. *pediculus forma minor, Cymbella leptoceros,* var. *clongata, C. navlcullformis, Cyclotella ocellata, Encyonema prostratum, Fragilaria construens, Hantzschia amphioxys* var. *intermedia, Melosira granulata* var. *curvata, Navicula cincta, N. cincta* var. *Ileufleri, N. cancellata* var. *ammophila, N. simplex, N. sculpta, Nitzschia obtusa, N. obtusa* var. *nana, N. sinuata, N. vermicularis, Pleurosigma scalproides, Rhopalodia ventricosa, Synedra ulna* var. *longlssima, S. ulna* var. *lanceolata, Surirella elegans,* and *Tetracyclus rupestris*).

Cyanophyceae (*Anabaena gelatinicola, A. iyengarii* var. *attenuata, A. iyengarii* var. *tenuis, Aulosira bombayensis, A. prolifica, Aphanocapsa elachista, Aphanothece saxicola, Calothrix clavata, C.*

gloeocola, Chroococccus varius, Gloeotrichia ghosei, G. raciborskii var. *bombayense, Lyngbya aestuarli* var. *arbustiva, L. porphyrosiphonis, Microcystis elabens, M. protocystis, Nostoc verrucosum, Oscillatoria corallinae, O. chalybea, O. lacustris, O. laete-virens* var. *minimus, O. nigroviridis, O. okeni, O. ornata, O. raoi, O. subbrevis, Phormidium anomala, P. purpurascens* and *Synechocystis aquatilis*).

Planktonic Algae Observed only at Certain Places

Planktonic algae observed at certain places but absent elsewhere are given below:

Chlorophyceae (*Crucigenia quadrata, Coelastrum cambricum* var. *intermedium, Closterium acerosum, Pediastrum boryanum* and *Sphaerocystis schroeteri*).

Bacillariophyceae (*Amphora ovalis* var. *pediculus, Gomphonema parvulum, Gyrosigma scalproides, Melosira granulata* var. *curvata, Navicula gregaria* and *Nitzschia microcephala*).

Cyanophyceae (*Merismopedia tenuissima* and *Phormidium tenue*).

Quantitative Distribution of Algae

For the sake of convenience quantitative distribution of algae has been grouped into (1) major algae showing profuse growth (2) minor algae exhibiting common or rare growth during different periods of the year. Quantitative assessments are based on statistical computation of Standard Deviation and Percentage Occurrence for minor and major algae. Account of the two kind of algae is given below:

Major Algae

For the sake of brevity of text the major algae has been split as under:

Winter: No alga has been recorded under this category.

Summer: No alga has been recorded under this category.

Rain: As the flow of water is very rapid and algal growth is washed out no major form could be recorded.

Winter-Summer: *Cosmarium granatum* (0.763 ± 8.522), *Nitzschia dissipata* (0.641 ± 8.13), *Navicula inaquiletera* (1.098 ± 12.677), *Gloeotrichia natans* (0.827 ± 6.323), *G. raciborskii* (0.777 ± 6.584), *G. raciborskii* var. *salsetense* (0.734 ± 7.56) were recorded under this catagory.

Summer-Rain: Under this catagory no algal forms could be recorded.

Rain-Winter: No algae could be recorded as major algae during this period.

All season algae: *Actinastrum hantzschii* (0.798 ± 6.228), *Cosmarium formossulum* (0.748 ± 10.931), *closterium acerosum* (0.677 ± 7.464), *Cladophora glomerata* (2.359 ± 17.677), *Hydrodictyon reticulatum* (1.625 ± 14.985), *Mougeotia gotlandica* (0.84 ± 8.963), *M. calcarea* (0.827 ± 7.655), *M. sphaerocarpa* (1.055 ± 10.379), *Rhizoclonium hieroglyphicum* (1.190 ± 11.317), *Scenedesmus bijuga* (0.763 ± 8.888), *S. armatus* (0.727 ± 7.657), *S. quadricauda* (0.991 ± 10. 875), *Spirogyra punctiformis* (1.019 ± 9.671), *S. species* (1.725 ± 14.270), *Chara excelsa* (1.440 ± 13.334), *Cymbella cistula* (0.755 ± 8.600), *C. tumida* (0.727 ± 8.039), *Epithemia gibba* var. *ventricosa* (0.891 ± 8.161), *Gomphonema montanum* (0.784 ± 7.872), *Navicula exigua* (0.862 ± 10.122), *N. gastrum* (0.862 ± 10.122), *N. dicephala* (0.641 ± 6.289), *N. cryptocephala* (0.656 ± 8.478), *N. cryptocephala* var. *exilis* (1.206 ± 11.878), *N. tuscula* (0.684 ± 8. 559), *N. sphaerophora* (0.988 ± 10.428), *Nitzschia palea* var. *fonticola* (0.827 ± 10.731), *N. palea* var. *tenurostris* (1.04 ± 11.175), *Pleurosigma spencerii* var. *nodifera* (0.706 ± 9.016), *Stauroneis anceps* (0.862 ± 8.073), *Synedra ulna* (1.197 ± 10.766), *S. ulna* var. *danica* (0.844 ± 8.742), *Anabaena planktonica* (0.698 ± 10.974), *A. fertilissima* (0.763 ± 8.284), *A. sphaerica* (0.696 ± 9.624), *Merismopedia punctata* (0.755 ± 7.720), *M. glauca* (0.877 ± 7.175), *M. tenuissima* (0.641 ± 6.557), *Nostoc linckia* var. *arvense* (0.926 ± 8.515), *Oscillatoria formosa* (0.969 ± 7.608), *O. princeps* (0.712 ± 7.571), *O. tenuis* (0.748 ± 8.259), *O. subbrevis* (0.741 ± 7.365), *Phormidium ambiguum* (0.691 ± 6.881), *P. tenue* (0.908 ± 7.685) and *Compsopogon species* (1.318 ± 15.150) were recorded under this category.

Algae found throughout the year: Under this category *Spirogyra singularis* (3.243 ± 22.649), *S.* species (2.202 ± 14.755), *Nitzschia palea* (1.682 ± 15.505) and *Oscillatoria tenuis* (0.748 ± 8.259) were recorded.

Minor Algae

Common Algae

Winter: During winter only *Closterium acutum* (0.135 ± 2.503) was recorded.

Summer: *Cosmarium meneghinii* (0.385 ± 5.681), *Coelastrum cambricum* var. *intermedium* (0.199 ± 2.934), *Achananthes lanceolata* var. *dubia* (0.107 ± 1.913), *Cymbella cuspidata* (0.100 ± 1.537). *C. obtusa* (0.299 ± 3.370), *Denticula elegans* (0.100 ± 1.030), *Epithemia zebra* (0.207 ± 3.825), *Gomphonema pervulum*, var. *lanceolata* (0.171 ± 3.338), *Calothrix marchica* (0.228 ± 4.043), *Chroococcus minor* (0.242 ± 4.726), *C. minutus* (0.157 ± 2.791), *Gloeocapsa stegophila* (0.157 ± 2.725), *Gloeotrichia echinulata* (0.428 ± 3.473), *G. ghosei* (0.156 ± 2.979), *G. raciborskii* var. *longispora* (0.221± 4.601), *G. intermedium* (0.149 ± 3.194), *Gloeocapsa montanus* (0.114 ± 2.146), *Microcystis viridis* (0.107 ± 1.923), *Nostoc carneun* (0.285 ± 4.830) and *Oscillatoria pseudogeminata* var. *unigranulata* (0.163 ± 2.729) were recorded under this category.

Rain: Due to rapid flow of water no alga could be recorded during rains.

Winter-Summer: During this period *Actinastrum hantzschii* var. *fluivatile* (0.299 ± 6.098), *Ankistrodesmus falcatus* (0.135 ± 2.109), *A. convolutus* (0.135 ± 2.109), *Cosmarium formosulum* var. *nathorstii* (0.178 ± 2.745), *C. nitidulum* (0.363 ± 3,980), *Closterium turgidum* (0.144 ± 2.015), *Closteriopsis longissima* (0.157 ± 1.690), *Characium naegeli* (0.107 ± 1.602), *Crucigenia tetrapedia* (0.100 ± 1.850), *Scenedesmus acuminatus* (0.128 ± 1.954), *S. quadricauda* var. *quadrispina* (0.135 ± 1.832), *S. quadricauda* var. *bicaudatus* (0.149 ± 2.598), *Selenastrum gracile* (0.199 ± 1.826), *S. westii* (0.178 ± 3.175), *Stigeoclonium tenue* (0.135 ± 1.154), *Zygnema collinsianum* (0.570 ± 5.646), *Achnanthes lanceolata* (0.292 ± 2.977), *A. exilis* (0.242 ± 2.657), *Cymbella pussila* (0.242 ± 3.157), *C. leptoceros* (0.185 ± 2.623), *Fragilaria crotonensis* (0.178 ± 2.275), *Gyrosigma spencerii* (0.413 ± 4.609), *G. scalproides* (0.199 ± 2.390), *Gomphonema augur* (0.114 ± 1.303), *G. parvulum* (0.235 ± 3.493), *G. parvulum* var. *subcapitata* (0.235 ± 3.769), *Mastogloia smithii* (0.158 ± 2.741), *Melosira granulata* var. *mizzanensis* (0.335 ± 3.919), *M. granulata* var *curvata* (0.114 ± 1.303), *Navicula platystoma* (0.157 ± 4.573), *N. lanceolata* (0.362

± 6.225), *Nitzschia microcephala* (0.157 ± 1.983), *N. angustata* (0.107 ± 2.301), *Rhopalodia ventricosa* (0.199 ± 3.085), *Synedra vaucherei* (0.235 ± 4.224), *S. vaucherei* var. *perminuta* (0.214 ± 2.969), *S. vaucherei* var. *pervula* (0.570 ± 2.791), *S. ulna* var. *obtusa* (0.314 ± 5.921), *Anabaena iyengarii* var. *tenuis* (0.378 ± 4.481), *Cylindrospermum muscicola* (0.477 ± 6.960), *C. licheniforme* (0.207 ± 3.029), *C. majus* (0.413 ± 5.638), *Gloeotrichia indica* (0.321 ± 5.242), *Microcystis flos-aquae* (0.121 ± 1.728), *Nodularia spumigena* (0.306 ± 4.718), *N. hervaeyana* (0.213 ± 4.720), *Oscillatoria willei* (0.378 ± 4.188), *O.boryana* (0.100 ± 1.697) and *O. earlie* (0.321 ± 3.388) were recorded.

Summer-Rain: Only *Chroococcus tenax* (0.335 ± 4.907) was recorded during summer rain period.

Rain-Winter: No alga was recorded.

All Season algae : *Cosmarium punctulatum* (0.485 ± 5.694), *Chlorella vulgaria* (0.599 ± 5.274), *Chlorococcum humicola* (0.456 ± 4.007), *C. infusionum* (0.214 ± 2.431), *Mougeolia viridis* (0.128 ± 1.784), *Pediastrum boryanum* (0.135 ± 1.243), *P. simplex* (0.206 ± 2.193), *P. tetras* (0.171 ± 1.477), *Scenedesmus dimorphus* (0.449 ± 5.159), *S. obliquus* (0.150 ± 2.301), *S. bijuga* var. *alternans* (0.164 ± 2.391), *S. quadricauda* var. *westii* (0.492 ± 6.890), *Spirogyra hyalina* (0.584 ± 5.076), *S. inflata* (0.513 ± 6.273), *S. species* (0.556 ± 4.092), *Zygnema tenue* (0.570 ± 5.646), *Zygnema cruciatum* (0.156 ± 2.443), *Amphora ovalis* (0.142 ± 1.557), *Asterionella formosa* (0.207 ± 1.928), *Cymbella perva* (0.100 ± 1.044), *C. laevis* (0.242 ± 3.157), *C. microcephala* (0.135 ± 1.928), *C. cymbiformis* (0.463 ± 4.461), *Cyclotella meneghiniana* (0.343 ± 3.696), *C. kuetzingiana* (0.228 ± 3.172), *C. glomerata* (0.463 ± 3. 848), *Epithemia gibba* (0.506 ± 4.907), *Encyonema caespitossum* (0.470 ± 5.681), *Gyrosigma kutzingiana* (0.178 ± 2.573), *Navicula seminulum* (0.656 ± 8.118), *N. mutica* (0.299 ± 4.100), *N. ambigua* (0.485 ± 6.513), *N. anglica* (0.363 ± 5.199), *Nitzschia acicularis* (0.584 ± 6.506), *N. subtilis* (0.449 ± 4.731), *N. linearis* (0.257 ± 3.045), *N. linearis* var. *tenuis* (0.121 ± 1.975), *Pleurosigma spencerii* (0.100 ± 1.850), *Rhopalodia gibba* (0.485 ± 4.579), *Synedra pulchella* (0.192 ± 2.301), *S. ulna* var. *aequalis* (0.328 ± 5.278), *Anabaena species* (0.363 ± 4.712), *A. iyengarii* (0.549 ± 6.719), *Aphanocapsa montanum* (0.214 ± 1.507), *Aphanothece bullosa* (0.128 ± 2.876), *Calothrix scytonemicola* (0.470 ± 5.962), *Chroococcus turgidus* (0.256 ± 3.740), *C. montanus* (0.243 ± 3.298),

C. pallidus (0.171 ± 2.863), *Calothrix fusca* var. *crassa* (0.363 ± 3.793), *Cylindrospermum stangnale* (0.435 ± 4.814), *Gloeocapsa quaternata* (0.499 ± 4.589), *Lyngbya majuscula* (0.299 ± 3.177), *Microcystis aeruginosa* (0.099 ± .589), *M. roburta* (0.420 ± 4.563), *Merismopedia convoluta* (0.527 ± 5.087), *Nostoc linckia* (0.500 ± 5.532), *Oscillatoria limosa* (0.591 ± 8.207), *O. acuta* (0.570 ± 6.631), *Spirulina gigantea* (0.285 ± 3.984) and *S. major* (0.456 ± 4.677) were observed during all three seasons sporadically but not throughout the year.

Algae found throughout the year: No alga could be recorded growing throughout the year.

Rare Algae

Winter: *Cosmarium constructum* (0.014 ± 0.577), *Oedogonium gracilius* (0.007 ± 0.289), *Sirogonium sticticum* (0.028 ± 0.651), *Fragilaria mutabilis* (0.028 ± 0.389), *Navicula salinarum* (0.021 ± 0.621), *Nitzschia subtilis* var. *paleacea* (0.036 ± 1.443), *Surirella elegans* (0.007 ± 0.289) and *Lyngbya influxa* (0.036 ± 0.996) were recorded.

Summer: Under this category *Cosmarium circulare* (0.071 ±1.642), *C. depressum* (0.099 ± 2.125), *C. birectum* (0.050 ± 1.379), *C. raineforme* (0.05 ± 1.379), *C. broomei* (0.035 ± 099), *Coelastrum sphaericum* (0.028 ± 0.651), *Coleochaete irregularis* (0.014 ± 0.389), *Closterium lanceolatum* (0.014 ± 0.058), *C. moniliforme* (0.570 ± 1.213), *C. linula* (0.007 ± 0.289)), *C. gracilae* (0.064 ± 1.055), *Characium pringsheimii* (0.043 ± 0.798), *Golenkinia radiata* (0.014 ± 0.389), *Oocystis crassa* (0.064 ± 1.215), *O. solitaria* (0.043 ± 0.798), *Scenedesmus arcuatus* (0.007 ± 0.289), *S. quadricauda* var. *longispina* (0.007 ± 0.289), *S. prismaticus* (0.021 ± 0.621), *S. acutiformis* (0.036 ± 0.793), *S. abundans* (0.036 ± 0.793), *Tetraodon muticun* (0.021 ± 0.452), *T. minimum f. tetralobulatum* (0.021 + 0.452), *Cocconeis costata* (0.014 ± 0.389), *Cymatopleura elliptica* (0.007 ± 0.289), *Diatoma vulgare* (0.014 ± 0.389), *Epithemia sorex* (0.007 ± 0.289), *Encyonema caespitosum* var. *lata* (0.036 ± 0.797), *Cymbella amphicephala* (0.028 ± 0.651), *Cyclotella operculata* (0.043 ± 0.797), *Fragilaria construens* (0.043 ± 1.168), *Gomphonema geminatum* (0.007 ± 0,289), *G. exiguum* (0.014 ± 0.389), *G. olivaceum* var. *vulgaris* (0.043 ± 0.798), *Gomphoneis herculaena* (0.014 ± 0.389), *Navicula cocconiformis* (0.028 ± 0.651), *N. peregrina* (0.036 ± 0.996), *N. ventricosa* (0.021 ± 0.797), *N. brebisonii* (0.028 ± 0.651), *N. humilis*

(0.036 ± 0.996), *N. gallica* (0.007 ± 0.289), *N. borealis* (0.036 ± 0.793), *N. ocullata* (0.014 ± 0.389), *N. amphisbaena* (0.007 ± 0.289), *N. pupula* (0.007 ± 0.289), *Nitzschia apiculata* (0.028 ± 0.797), *N. amphibia* (0.021 ± 0.621), *N. vermicularis* (0.007 ± 0.289), *N. angularis* var. *affinis* (0.021 ± 0.621), *N. Triblionella* var. *calida* (0.007 ± 0.289), *N. Hungarica* (0.014 ± 0.389), *Pleurosigma Perkerii* (0.007 ± 0.289), *P. scalproides* (0.028 ± 0.651), *Pinnularia undulata* (0.14 ± 0.577), *Stauronesis acuta* (0.028 ± 0.651), *S. smithii* (0.014 ± 0.389), *S. anceps* var. *amphicephala* (0.007 ± 0.289), *S. anceps* var. *linearis* (0.007 ± 0.289), *Surirella robusta* (0.085 ± 1.537), *S. robusta* var. *tenera* (0.007 ± 0.289), *S. ovalis* var. *minuta* (0.014 ± 0.577), *Tetracyclus rupustris* (0.007 ± 0.289), *Euglena gracilis* (0.064 ± 1.288), *E. spirogyra* (0.08810 ± 651), *Phacus pleuronoctus* (0.014 ± 0.389), *Anabaena oryzae* (0.007 ± 0.289), *A. circinalis* (0.021 ± 0.621), *A. orientalis* (0.014 ± 0.389), *Aphanocapsa biformis* (0.071 ± 1.527), *A. muscicola* (0.014 ± 0.389), *A. littoralis* (0.057 ± 1.230), *Calothrix marchica* var. *intermedia* (0.028 ± 1.155), *C. bharadwajei* (0.035 ± 0.996), *C. gloeocola* (0.007 ± 0.289), *Chroococcus macrococcus* (0.014 ± 0.577), *C. varians* (0.050 ± 1.084), *C. rufescens* (0.021 ± 0.621), *Camptylonemopsis species* (0.007 ± 0.289), *Cylindrospermum michailovskaense* (0.007 ± 0.289), *Gloeocapsa compacta* (0.050 ± 1.084), *G. pleurocapsoid* (0.007 ± 0.289), *G. atrata* (0.007 ± 0.289), *G. magma* (0.007 ± 0.289), *Gloeotrichia raciborskii* var. *kashiense* (0.071 ± 1.642), *G. longicauda* (0.007 ± 0.289), *G. echinulata* var. *berhempurense* (0.071 ± 1.992), *G. raciborskii* var. *conica* (0.007 ± 0.289), *Lyngbya putealis* (0.014 ± 0.389), *L. limnetica* (0.057 ± 1.251), *L. polysiphonae* (0.007 ± 0.289), *L. dendrobia* (0.007 ± 0.289), *L. porphyrosiphonae* (0.014 ± 0.389), *L. contorta* (0.028 ± 0.651), *Glocolhece rhodochlamys* (0.007 ± 0.289), *Dactylococcopsis raphidiodes* (0.007 ± 0.289), *Microcystis incerta* (0.007 ± 0.289), *M. protocystis* (0.21 ± 0.621), *Myxosarcina spectabilis* (0.007 ± 0.289), *Nostoc muscorum* (0.057 ± 1.614), *N. ellipsorum* (0.05 ± 1.379), *Nostochopsis lobatus* (0.007 ± 0.289), *Oscillatoria Chalybea* (0.014 ± 0.577), *O. acutissima* (0.007 ± 0.289), *O. curviceps* (0.043 ± 1), *O. tanganyika* (0.007 ± 0.289), *O. subuliformis* (0.007 ± 0.289), *O. angustissima* (0.064 ± 1.084), *O. okeni* (0.021 ± 0.621), *O. minnesotensis* (0.007 ± 0.289), *O. limnetica* (0.21 ± 0.621), *Phormidium anomala* (0.035 ± 0.992), *P. purpuresense* (0.014 ± 0.389), *P. pachydermaticum* (0.007 ± 0.289), *Rivularia aquatica* (0.085 ± 1.954), *R. hansgirg* (0.028 ± 0.651), *Sprirulina subsalsa* (0.05 ± 1.084), *S. laxissima* (0.021 ± 0.452), *Synechocystis pevalodii* (0.014 ± 0.389).

Rain: Only *Navicula peregrina* var. *Menisculus f. Upsaliensis* (0.014 ± 0.577) was recorded during rains.

Winter-Summer: *Bulbochaete rectangularis* (0.057 ± 1.231), *B. reticulata* (0.036 ± 0.793), *Characium ambiguum* (0.078 ± 0.996), *Coelastrum proboscidium* (0.05 ± 0.9), *Euastrum inermius* (0.028 ± 0.651), *Oocystis elliptica* (0.028 ± 0.651), *O. borgei* (0.036 ± 0.793), *Protoccus viridis* (0.093 ± 1.379), *Pediastrum duplex* (0.05 ± 0.793), *P. duplex* var. *clatheratum* (0.05 ± 0.793), *Selenastrum bibrainum* (0.021 ± 0.657), *Stigeoclonium glomeratum* (0.057 ± 0.779), *Spirogyra* species (0.093 ± 0.996), *Slaurastrum gracile* (0.014 ± 0.389), *Spherocystis schroeterii* (0.014 ± 0.389), *Ulothrix variahbilis* (0.078 ± 0.970), *U. subtilissima* (0.021 ± 0.452), *Zygnemopsis sphaerospora* (0.021 ± 0.452), *Achnanthes linearis* (0.078 ± 1.505), *A. minutissima* (0.093 ± 1.443), *Cymbella affinis* (0.028 ± 0.652), *Diatoma hiemale* (0.021 ± 1.505), *Encyonema ventricosum* (0.064 ± 1.055), *Gomphonema constrictum* (0.057 ± 1.073), *G. olivacium* (0.093 ± 1.240), *Melosira distans* (0.071 ± 0.937), *M. cranulata* (0.085 ± 1.414), *M. varians* (0.085 ± 1.410), *Navicula protracta* (0.064 ± 1.422), *N. capitata* (0.093 ± 1.084), *N. laevissima* (0.050 ± 0.793), *N. purpussila* (0.093 ± 1.240), *Nitzschia thermalis* (0.036 ± 0.665), *Synedra radians* (0.052 ± 1.213), *Surirella ovalis* (0.057 ±1.216), *Euglena viridis* (0.050 ± 0.792), *Phacus caudatus* var. *ovalis* (0.014 ± 0.389), *Anabaena oscillarioides* (0.036 + 0.996), *Aphanothece saxicola* (0.078 ± 0.900), *Spirulina labyranthiformis* (0.021 ± 0.452) and *Synechocystis aquatilis* (0.021 ± 0.452) were recorded.

Summer - Rainy: Under this category *Navicula radiosa* (0.085 ± 2.256) *Gloeocapsa nigrescens* (0.043 ± 0.798) were recorded.

Rain - Winter: Only *Melosira italica* var. *tenuis* (0.014 ± 0.452) was recorded.

All Seasons Algae: *Coelastrum reticulatum* (0.085 ± 1.206), *Chaetophora elegans* (0.086 ± 0.853), *Pediastrum boryanum* var. *longicorne* (0.057 ± 0.778), *P. simplex* var. *duodenarium* (0.085 ± 0.853), *P. tetras* var. *tetraedron* (0.085 ± 0.853), *P. tetras* var. *exisum* (0.078 ± 0.900), *Cocconeis placentula* (0.083 ± 0.738), *Fragilaria capusima* (0.057 ± 1.073), *Nitzschia obtusa* (0.078 ± 1.024), *N. deblis* (0.093 ± 1.379), *Synedra acus* (0.078 ± 0.848), *Arthrospira platensis* (0.093 ± 1.505), *Aphanocapsa gravillae* (0.085 ± 0.953) and *Merismopedia elegans* (0.085 ± 2.153) were recorded in different seasons.

Phycoplankton Density

Quantitative assay of phycoplanktons is highest in March and reaching its peak of 59628 cells per litre and minimum in the month of August numbering 1589 cells per litre. The data suggests that during winter months plankton count is lesser which exhibits a gradual increase as the summers approach. Out of various classes of algae bacillariophyceae comprise largest number. The average density of phycoplanktons assayed monthly is as follows:

October: The common planktonic forms were *Actinastrum* (200), *Chlorococcum* (1160), *Crucigenia* (83), *Protococcus* (666), *Merismopedia* (1750), *Oscillatoria* (1000), *Phormidium* (666), *Euglena* (83), *Amphora* (1), *Cyclotella* (1750), *Melosira* (2166) and *Synedra* (83).

November: Planktonic density in November was recorded as under - *Actinastrum* (1750), *Chlorella* (666), *Chlorococcum* (1750), *Crucigenia* (166), *Ankistrodesmus* (83), *Protococcus* (166), *Chroococcus* (2500), *Oscillatoria* (1750), *Phormidium* (1500). *Euglena* (166), *Achnanthes* (83), *Cyclotella* (5250), *Cymbella* (83), *Pleurosigma* (83), *Melosira* (2166) and *Synedra* (83).

December: Major algal planktons showed density as expressed herein *Actinastrum* (2500), *Chlorella* (800), *Chlorococcum* (300), *Crucigenia* (1750), *Ankistrodesmus* (166), *Meringospherica* (83), *Protococcus* (166), *Scenedesmus* (2166), *Selenastrum* (166), *Tetraedon* (166), *Anabaena* (800), *Chroococcus* (916), *Merismopedia* (4833), *Oscillatoria* (5250), *Phormidium* (2916), *Gloeocytis* (83), *Lepocinclis* (166), *Euglena* (666), *Achnanthes* (166), *Cyclotella* (10335), *Pleurosigma* (83), *Melosira* (10166), *Mastolgoea* (166), *Synedra* (666), *Cocconeis* (166), *Nitzschia* (3000) and *Navicula* (1750).

January: Phycoplankton density during this month was observed as follows—*Actinastrum* (1160), *Chlorella* (1167), *Crucigenia* (1333), *Closteriopsis* (555), *Ankistrodesmus* (750), *Scenedesmus* (2833), *Selenastrum* (333), *Anabaena* (1056), *Chroococcus* (500), *Merismpedia* (4500), *Oscillatoria* (3455), *Phormidium* (3279), *Euglena* (249), *Achnanthes* (666), *Amphora* (166), *Melosira* (12000), *Mastogloea* (83), *Synedra* (750) and *Navicula* (1550).

PHYCOPLANKTON DENSITY

S.No.	Genus	Oct	Nov	Dec	Jan	Feb	Mar	Apr	May	June	July	Aug	Sept
1.	*Actinastrum*	800	1750	2500	1160	1333	1554	1250	1160	750	83	—	333
2.	*Chlorella*		666	800	1580	1167	889	1750	555	800	167	—	167
3.	*Chlorococcum*	1160	1750	3000	2000	1333	999	800	666	250	83	—	—
4.	*Crucigenia*	83	166	1750	1000	417	222	333	222	83	—	—	83
5.	*Closteriopsis*				166	555	444	250	83	—	—	—	—
6.	*Ankistrodesmus*		83	166	333	750	1110	1333	1916	750	83	—	1160
7.	*Mallomonas*				83			—	—	—	—	—	—
8.	*Meringospheria*	—	—	83	—	—	—	—	—	—	—	—	—
9.	*Protococcus*	—	166	166	333	—	—	83	—	—	—	—	—
10.	*Scenedesmus*	666	1750	2166	1665	2833	3219	2500	1665	583	167	83	333
11.	*Coelastrum*	—	—	—	—	—	555	416	83	—	—	—	—
12.	*Selenastrum*	—	—	166	333	1916	2331	1916	666	417	167	83	—
13.	*Tetraedon*	—	—	—	—	83	222	166	—	—	—	—	—
14.	*Cosmarium*	—	—	166	—	—	—	83	83	—	—	—	—
15.	*Anabaena*	—	666	800	1056	1750	1332	1916	1583	750	167	83	—
16.	*Aphanothece*	—	83	—	500	667	1221	1554	750	333	167	—	—
17.	*Chroococcus*	—	166	916	—	555	999	555	333	167	83	—	—
18.	*Merismopedia*	1750	2500	4833	4500	2321	2667	1833	1750	666	167	83	333
19.	*Oscillatoria*	1000	1750	5250	3455	3552	5328	4833	2416	1083	666	83	333
20.	*Phormidium*	666	1500	2916	3150	3219	3552	2833	1750	750	167	83	167
21.	*Gloeocystis*	—	—	83	—	—	—	—	—	—	—	—	—
22.	*Phacus*	—	—	—	166	83	—	83	—	—	—	—	—
23.	*Lepocinclis*	—	—	166	500	—	—	—	—	—	—	—	—
24.	*Euglena*	83	166	666	830	249	—	167	83	—	—	—	
25.	*Achnanthes*	—	83	166	666	250	—	333	167	250	167	83	83

S.No.	Genus	Oct	Nov	Dec	Jan	Feb	Mar	Apr	May	June	July	Augs	Sept
26.	*Amphora*	—	—	—	83	—	—	83	—	—	—	—	—
27.	*Cyclotella*	1750	5250	10335	13650	12074	13889	10335	6883	5250	667	167	667
28.	*Cymbella*	—	83	—	166	83	—	1660	83	—	—		83
29.	*Pleurosigma*	—	83	83	—	83	—	—	—	—	—	—	—
30.	*Gyrosigma*	—	—	—	166	83	22	—	—	—	—	—	—
31.	*Melosira*	2116	4500	10166	12000	13916	13098	10166	5250	1750	416	250	333
32.	*Mastogloea*	—	83	166	83	—	—	167	83	83	—	—	—
33.	*Synedra*	83	750	666	750	1083	1553	1250	1166	583	416	83	83
34.	*Cocconeis*	—	—	166	—	—	—	—	83	—	—	—	—
35.	*Nitzschia*	—	666	3000	3500	2917	3667	2321	1333	1166	583	167	250
36.	*Navicula*	—	166	1750	1550	55	333	1550	1250	1083	333	83	167
37.	*Pinnularia*	—	—	—	—	83	222	167	83	83	—	—	—

February: The commonly occurring algal planktons were *Actinastrum* (1333), *Chlorella* (1167), *Chlorococcum* (1333), *Crucigenia* (417), *Closteriopsis* (555), *Ankistrodesmus* (750), *Scenedesmus* (2833), *Selenastrum* (1916), *Tetraedon* (83), *Anabaena* (1750), *Aphanothece* (667), *Chroococcus* (555), *Merismopedia* (2321), *Oscillatoria* (3552), *Phormidium* (3219), *Phacus* (83), *Euglena* (249), *Achnanthes* (250), *Cyclotellas* (12074), *Cymbella* (83), *Gyrosigma* (83), *Melosira* (13916), *Synedra* (1083), *Nitzschia* (2917), *Navicula* (555) and *Fragillaria* (83).

March: Plankton density during March is as follows: *Actinastrum* (282), *Chlorella* (222), *Closteriopsis* (444), *Chlorococcum* (999), *Crucigenia* (222), *Ankistrodesmus* (1110), *Scenedesmus* (3219), *Coelastrum* (255), *Selenastrum* (2331), *Tetraedon* (222), *Anabaena* (1332), *Aphanothece* (1221), *Chroococcus* (999), *Merismopedia* (2667), *Phormidium* (5328), Cyclotella (13889), Gyrosigma (22), Melosira (13098), Synedra (1553), *Nitzschia* (3667), *Navicula* (333) and *Fragilaria* (222).

April: Planktonic forms during this month comprise *Selenastrum* (800), *Scenedesmus* (1200), *Chlorella* (200), *Ankistrodesmus* (400), *Sphaerocystis* (200), *Actinastrum* (1200), *Merismopedia* (200), *Oscillatoria* (2866), *Nitzschia* (3466), *Synedra* (2466), *Navicula* (333) and *Epithemia* (66).

May: The density of phycoplanktons recorded during this month was as follows: *Selenastrum* (266), *Scenedesmus* (5066), *Chlorella* (66), *Sphaerocystis* (133), *Actinastrum* (666), *Oscillatoria* (400), *Cyclotella* (6466), *Nitzschia* (6200), *Synedra* (2133) and *Navicula* (333).

June: Algal phycoplanktons recorded during this month were *Selenastrum* (66), *Scenedesmus* (933), *Ankistrodesmus* (200), *Sphaeocystis* (66), *Actinastrum* (533), *Anabaena* (133), *Oscillatoria* (200), *Cyclotella* (4533), *Nitzschia* (1800), *Synedra* (4055), *Navicula* (200) and *Pleurosigma* (200).

July: Phycoplankton density recorded is as follows : *Scenedesmus* (133), *Actinastrum* (133), *Anabaena* (66), *Merismopedia* (133) and *Cyclotella* (133).

August: Algal phycoplanktons recorded were as follows : Scenedesmus (66), *Ankistrodesmus* (66), *Merismopedia* (133) and *Cyclotella* (133)

September: Following phycoplanktons were recorded during this month : *Scenedesmus* (200), *Chlorella* (66), *Scenedesmus* (133), *Anabaena* (133) and *Cyclotella* (200).

Observations suggest that numerically algal infestation is larger during summers than winters. Monthly qualitative density of various classes recorded indicates that cyanophycean forms are largest numerically followed by bacillariophyceae and chlorophyceae during winter months from November to January. Thereafter bacillariophycean forms mark numerical abundance over cyanophycean and chlorophycean algae, respectively. Euglenophyceae remains the least represented of all the other classes.

Occurrence of large number of 51 algal species comprising 14 chlorophyceae (*Hydrodictyon reticulatum, Rhizoclonium hieroglyphicum, Spirogyra singularis, Chara sp., Mougeotia calcarea, M. sphaerocarpa, Zygnema collinsianum, Z. tenue, Closterium acerosum, Cladophora glomerata, C. crispiata, Mougeotia robusta* and *Spirogyra sp.*) 13 cyanophycean (*Anabaena iyengerii, Gloeotrichia natans, G. raciborskii, Merismopedia punctata, M. glauca, Nostoc linckia, Oscillatoria princeps, O. pseydigebunata, O. formosa, O. limosa, Phormidium tenue, Spirulina major,*) and 24 bascillariophyceae (*Achnanthes linearis, Cymbella cistula, C. affinis, C. cymbiformis, C. obtusa, Epithemia gibba, E. gibba* var. *Ventricosa, E. gibberula, E. musculus, Gyrosigma kuetzingii, G. spencerii Gomphonema montanum, G. parvulum, Navicula gastrum, N. cryptocephala, N. sphaerophora, N. exiqua, Nitzschia palea. N. linearis, Rhopalodia gibba, Pleurosigma spencerii, Stauroneis anceps* and *Synedra ulna*) has been observed as major algae showing massive growth in Ganga water. There is only a marginal variance in cyanophyceae and chlorophyceaen members but bacillariophyceae outnumbered the other two classes of algae.

Water Culture Algae

Ninty three algal species were recorded from Ganga water culture. They may be grouped into two catagories:

Algae observed in nature and culture

Algae observed in nature as well as in culture are : *Actinastrum hantzschii, Ankistrodesmus falcatus, A. convolutus,*

QUANTITATIVE ASSESSMENT OF ALGAE IN GANGA WATER CULTURED

S.No.	Species	Percentage Occurrence	Standard Deviation	Coefficient of Variation
1.	*Actinastrum hantzschii*	0.247	± 0.608	156.267
2.	*Ackistrodesmus falcatus*	4.871	± 1.534	20.007
3.	*A. convolutus*	5.577	± 2.184	24.885
4.	*Chlorococcum humicola*	1.164	± 1.150	62.749
5.	*C. infusionum*	0.564	± 1.183	133.042
6.	*Closterium acerossum*	0.176	± 0.461	165.924
7.	*Chlorella valgaris*	0.953	± 1.295	86.320
8.	*Crucigenia tetrapedia*	0.176	± 0.401	165.924
9.	*Coelastrum cambricum* var. *intermedium*	0.247	± 0.608	156.266
10.	*C. microporum*	0.388	± 0.698	114.185
11.	*Franceia ovalis*	0.035	± 0.236	424.260
12.	*Oocystis elliptica*	0.105	± 0.383	230.100
13.	*Pediastrum tetras*	0.211	± 0.594	178.230
14.	*P. duplex*	0.211	± 0.594	178.230
15.	*P. simplex*	0.176	± 0.461	165.924
16.	*P. boryanum*	0.035	± 0.548	246.735
17.	*Scenedesmus quadricauda*	4.271	± 1.274	18.958
18.	*S. quadricauda* var. *longispina*	0.494	± 0.878	112.991
19.	*S. quadricauda* var. *westii*	1.235	± 1.162	59.744
20.	*S. quadricauda* var. *bicaudatus*	0.776	± 1.166	95.400
21.	*Scenedesmus quadricauda* var. *quadrispina*	0.212	± 0.485	145.530
22.	*S. quadricauda* var. *maximus*	0.035	± 0.428	192.510
23.	*S. armatus*	5.224	± 2.289	27.845
24.	*S. bijuga*	5.365	± 1.293	15.318
25.	*S. dimorphus*	2.929	± 1.754	38.029
26.	*S. bijuga* var. *alternans*	0.458	± 1.127	156.115
27.	*S. arcuatus*	0.352	± 0.784	141.084
28.	*S. acuminatus*	0.882	± 1.092	78.638

S.No.	Species	Percentage Occurrence	Standard Deviation	Coefficient of Variation
29.	*S. obliquus*	1.341	± 1.231	58.329
30.	*S. brasieliensis*	0.035	± 0.236	424.260
31.	*Selenastrum gracile*	5.788	± 1.641	18.010
32.	*S. westii*	5.788	± 2.083	22.864
33.	*Sphaerocystis schroeteri*	0.176	± 0.461	165.924
34.	*Spirogyra species*	2.259	± 4.475	125.862
35.	*Achnanthes exilis*	0.564	± 1.079	121.392
36.	*A. linearis*	0.282	± 0.616	138.532
37.	*Cyclotella meneghiniana*	2.789	± 1.195	27.228
38.	*C. glomerata*	1.906	± 0.907	30.233
39.	*C. kutzingiana*	1.306	± 0.007	97.652
40.	*C. o erculata*	0.035	± 0.548	246.735
41.	*Epithemia gibba*	0.317	± 0.707	141.420
42.	*Gomphonema montanum*	0.070	± 0.323	291.000
43.	*Gyrosigma scalproides*	0.106	± 0.383	230.100
44.	*Nitzschia palea*	3.980	± 0.895	14.253
45.	*N. palea* var. *fonticola*	1.176	± 0.900	28.935
46.	*N. palea* var. *tenuirostris*	1.094	± 1.227	71.268
47.	*N. microcephala*	0.635	± 0.091	90.750
48.	*N. angustata*	1.094	± 1.178	68.429
49.	*N. subtilis*	0.600	± 1.110	41.621
50.	*N. dissipata*	1.659	± 1.037	39.711
51.	*N. acicularis*	0.247	± 0.608	156.266
52.	*Navicula inaquiletera*	0.176	± 0.461	165.960
53.	*N. seminulum*	0.670	± 1.162	110.056
54.	*N. dicephala*	1.580	± 1.386	55.308
55.	*N. platystoma*	0.282	± 0.784	176.355
56.	*N. cryptocephala*	1.129	± 1.145	64.389
57.	*N. cryptocephala* var. *exilis*	1.023	± 1.334	82.831
58.	*N. anglica*	0.105	± 0.383	230.100
59.	*Melosira distans*	0.388	± 0.858	16.302
60.	*M. granulata* var. *angustissima*	0.458	± 0.958	132.688

S.No.	Species	Percentage Occurrence	Standard Deviation	Coefficient of Variation
61.	*Melosira varians*	0.105	± 0.383	230.100
62.	*M. cranulata*	0.423	± 0.840	126.015
63.	*Pleurosigma spencerii*	0.635	± 1.085	108.460
64.	*Mastogloia smithii*	0.211	± 0.594	178.230
65.	*Synedra ulna*	0.811	± 1.127	88.239
66.	*S. ulna* var. *aequalis*	0.424	± 0.767	115.050
67.	*S. ulna* var. *lanceolata*	0.423	± 0.970	145.515
68.	*S. vaucherei* var. *perminuta*	0.282	± 0.616	138.532
69.	*S. ulna* var. *denica*	0.388	± 0.777	127.227
70.	*Euglena acus*	0.211	± 0.594	178.230
71.	*E. spirogyra*	0.282	± 0.616	138.532
72.	*E. gracilis*	0.211	± 0.594	178.230
73.	*Phacus pleuronectus*	0.176	± 0.574	206.820
74.	*Aphanocapsa montanum*	0.070	± 0.323	291.060
75.	*A. biformis*	0.352	± 0.874	141.084
76.	*Anabaena orientalis*	0.847	± 1.940	145.522
77.	*A. fertilissima*	0.423	± 1.188	178.230
78.	*Chrooccus varius*	0.317	± 0.707	141.420
79.	*C. minutus*	0.035	± 0.428	192.510
80.	*Gloeocapsa quaternata*	0.247	± 0.608	156.266
81.	*Lyngbya contorta*	0.529	± 1.339	160.730
82.	*Merismopedia glauca*	0.458	± 1.018	140.926
83.	*M. tenuissima*	1.164	± 1.200	65.482
84.	*M. convoluta*	0.882	± 1.461	105.178
85.	*M. punctata*	1.164	± 1.201	65.482
86.	*Microcystis aeruginosa*	0.247	± 0.698	179.434
87.	*M. robusta*	0.176	± 0.461	165.924
88.	*Phacus orbicularis*	0.176	± 0.574	206.820
89.	*Oscillatoria angustissima*	1.447	± 2.740	120.284
90.	*O. limnetica*	4.624	± 3.121	42.884
91.	*Phormidium tenue*	1.976	± 2.055	66.047
92.	*Spirulina gigantea*	0.811	± 0.826	64.675
93.	*S. major*	0.811	± 1.127	88.239

Chlorococcum humicola, C. infusionum, Closterium acerosum, Chlorella vulgaris, Crucigenia tetrapedia, Coelastrum cambricum var. *intermedium, Oocystis elliptica, Pediastrum tetras, P. simplex, P. duplex, P. boryanum, Scenedesmus quadricauda, S. quadricauda* var. *longispina, S. quadricauda* var. *westii, S. quadricauda* var. *bicaudatus, S. quadricauda* var. *quadrispina. S. armatus, S. bijuga, S. bijuga* var. *alternans, S. dimorphus, S. arcuatus, S. acuminatus, S. obliquus, Selenastrum gracile, S. westii, Spherocystis schroeteri, Spirogyra species, Achnanthes exilis, A. linearis, Cyclotella meneghiniana, C. glomerata, C. kuetzingiana, C. operculata, Epithemia gibba, Gomphonema montanum, Gyrosigma scalproides, Nitzschia palea, N. palea* var. *fonticola, N. palea* var. *tenuirostris, N. microcephala, N. angustata, N. subtilis, N. dissipata, N. acicularis, Navicula inaquiletera, N. seminulum, N. dicephala, N. platystoma, N. cryptocephala, N. cryptocephala* var. *exilis, N. anglica, Melosira distans, M. varians, M. cranulata, Pleurosigma spencerii, Mastagloia smithii, Synedra ulna, S. ulna* var. *aequalis, S. ulna* var. *lanceolata, S. vaucherei* var. *perminuta, S. ulna* var. *denica, Euglena spirogyra, E. gracilis, Phacus pleuronectus, Aphanocapsa montanum, A. biformis, Anabaena orientalis, A. fertilissima, C. minutus, Gloeocapsa quaternata, Lyngbya contorta, Merismopedia glauca, M. tenussima, M. convoluta, M. punctata, Microcystis aeruginosa, M. robusta, Oscillatoria angustissima, O. limnetica, Phormidium tenue, Spirulina gigantea* and *S. major.*

Algae observed only in culture

Algae found growing in culture of the water alone but conspicuously absent in nature are: *Coelastrum microporum, Franceia ovalis, Scenedesmus quadricauda* var. *maximus, S. brasieliensis, Melosira granulata* var. *angustissima, Euglena acus, Phacus orbicularis.*

Sand Culture Algae

One hundred one algal species were recorded from sand culture of river Ganga. They may be grouped into two catagories

Algae observed in nature and culture

Algae recorded in sand culture as well as in nature are: *Ankistrodesmus falcatus, Chlorococcum humicola, C. infusionum, Chlor-*

QUANTITATIVE ASSESSMENT OF ALGAE IN GANGA SAND CULTURE

S.No.	Species	Percentage Occurrence	Standard Deviation	Coefficient of Variation
1.	*Ankistrodesmus falcatus*	3.164	± 1.278	32.870
2.	*Chlorococcum humicola*	0.949	± 0.786	67.363
3.	*Chlorococcum infusionum*	0.497	± 0.784	128.258
4.	*Chlorella vulgaris*	0.994	± 0.149	73.864
5.	*Crucigenia tetrapedia*	0.226	± 0.461	165.916
6.	*Closteriopsis longissima*	0.180	± 0.428	192.510
7.	*Closterium acerosum*	0.226	± 0.574	206.820
8.	*Golenkinia radiata*	0.135	± 0.383	230.100
9.	*Golenkinia paucispina*	0.135	± 0.383	230.100
10.	*Pediastrum simplex*	0.180	± 0.428	192.510
11.	*Protococcus virdis*	0.316	± 0.608	156.265
12.	*Scenedesmus bijuga*	2.215	± 1.822	66.934
13.	*Scenedesmus quadricauda*	1.356	± 1.283	77.004
14.	*Scenedesmus quadricauda* var. *bicaudatus*	0.632	± 0.895	115.04
15.	*Scenedesmus quadricauda* var. *quadrispina*	0.542	± 0.840	126.015
16.	*Scenedesmus armatus*	1.537	± 1.323	70.068
17.	*Scenedesmus dimorphus*	0.994	± 0.923	75.560
18.	*Scenedesmus obliquus*	0.678	± 1.043	125.184
19.	*Selenastrum bibraianum* var. *graciile*	2.757	± 2.004	59.134
20.	*Selenastrum westii*	3.209	± 1.392	35.293
21.	*Achnanthes exilis*	1.491	± 1.043	56.902
22.	*Achnanthes linearis*	0.406	± 0.786	157.180
23.	*Amphora ovalis*	0.180	± 0.428	192.510
24.	*Cyclotella meneghiniana*	0.678	± 0.786	94.308
25.	*Cyclotella glomerata*	1.039	± 1.074	84.052
26.	*Cocconeis placentula*	0.723	± 0.936	108.382
27.	*Cymbella microcephala*	0.316	± 0.608	156.266
28.	*Encyonema ventricosa*	0.497	± 0.777	127.227
29.	*Encyonema caespitosum*	0.406	± 0.707	141.420
30.	*Gyrosigma spencerii*	0.497	± 0.777	127.227
31.	*Gyrosigma scalproides*	0.678	± 0.786	94.308
32.	*Gyrosigma kutzingiana*	0.180	± 0.428	192.510
33.	*Gomphonema montanum*	0.135	± 0.383	230.100
34.	*Hantzschia amphioxys*	0.090	± 0.323	29.106
35.	*Mastogloia smithii*	0.858	± 0.998	94.585
36.	*Pleurosigma spencerii*	0.632	± 0.878	75.274
37.	*Pinnularia braunii*	0.180	± 0.428	192.510
38.	*Navicula tuscula*	0.452	± 0.784	141.084
39.	*Navicula inaquiletera*	0.994	± 1.114	91.178
40.	*Navicula purpussila*	0.271	± 0.594	178.230

ella vulgaris, Crucigenia tetrapedia, Closteriopsis longissima, Closterium acerosum, Golenkinia radiata, Pediastrum simplex, Protococcus virdis, Scenedesmus bijuga, Scenedesmus quadricauda, S. quadricauda var. *bicaudatus, S. quadricauda* var. *quadrispina, S. armatus, S. dimorphus, S. obliquus, Selenastrum gracile, S. westii, Achnanthes exilis, A. linearis, Amphora ovalis, Cyclotella meneghiniana, C. glomerata, Cocconeis placentula, Cymbella microcephala, Encyonema ventricosum, E. caespitosum, Gyrosigma spenceri, G. scalproides, G. kuetzingiana, Gomphonema montanum, Hantzschia amphioxys, Mastogloia smithii, Pleurosigma spencerii, Navicula tuscula, N. inaquiletera, N. purpussila, N. peregrina, N. cryptocephala, N. seminulum, N. exigua, N. mutica, N. dicephala, N. gastrum, N. pupula, Nitzschia palea, N. palea* var. *fonticola, N. palea* var. *tenurostris, N. tryblionella* var. *calida, N. angustata, N. acicularis, N. microcephala, N. subtilis, N. dissipata, Synedra acus, S. vaucherei* var. *perminuta, S. ulna, S. ulna* var. *denica, Stauroneis anceps, Euglena gracilis, E. spirogyra, Phacus pleuronectus, Aphanocapsa montona, Anabaena planktonica, A. fertilissima, Chroococcus tenax, C. minutus, C. varius, C. turgidus, Gloeocapsa quaternata, G. pleurocapsoid, Lyngbya contorta, L. porphyrosiphouae, Microcystis aeruginosa, Merismopedia tenuissima, M. punctata, Phormidium tenue, Nostoc linckia, Oscillatoria tenuis, O. okeni, O. limnetica, O. minima, O. acuta, O. subliformis, O. princeps, O. subbrevis, O. angustissima, O. psedogeminata* var. *unigranulata, Phormidium ambiguum, spirulina laxissima, S. gigantea, S. labyranthiformis, S. major* and *Synecocystis aquatilis.*

Algae observed only in culture

Algae observed in culture of sand alone but conspicuosly absent in nature are: *Golenkinia paucipsina, Euglena acus, Phacus acuminatus, Tr·chleomonas hispida* and *Aphanothece nidulans.*

MACROPHYTES

A total number of 36 macrophyte species were recorded during different months of the year. Their presence shows some minor variance as given below:

October: Macrophytes like *Ceratophyllum domersum, Gnapholium indicum, Majus delavayi* and *Ranunculus sceleratus* were common during this month. However, *Aponogeton natans, Gnapholium indicum, Eichhornia crassipes, Hydrilla verticillata, Potentilla supina, Polygonum glabrum, P. crispus, Jussiaea repens, Sagittaria guayanensis, Grangea maderaspatana, Cyperus platystylis, Phyla nodiflora, Burgia ammannioides, Eclipta alba, Amaranthes indicum, Launia nauchali, Salvia plebia, Salix tetrasperma, Cyathocline purpurea, Mazus pumilus, Xanthium stramonium* and *Potentilla indica* rarely grew during this month.

November: Among the macrophytes recorded *Ceratophyllum domersum, Gnapholium indicum, Polygonum glabrum, Potamogeton nodosus, Eclipta alba* and *Potentilla indica* were common during this month. Second category of species comprising *Aponogeton natans, A. crispum, Eichhornia crassipes, Hydrilla verticillata, Potentilla supina, Polygonum orientale, Potamogeton crispus, Jussiaea repens, Sagittaria sagittifolia, S. guayanensis, Grangea maderaspatana, Campanula banthamii, Cyperus platystyles, Phyla nodiflora, Burgia ammannioides, Amaranthes indicum, Launia nauchali, Salix tetrasperma, Salvia plebia, Mazus pumilus, M. delavayi, Xanthium stramonium, Cyathocline purpurea* and *Calamus tenuis* grew rarely during this month. However, *Ranunculus sceleratus* grew abundantly.

December: Macrophytes like *Gnapholium indicum* and *Ranunculus sceleratus* exhibit abundant growth. *Aponogeton crispum, Campanula banthamii, Ceratophyllum domersum, Potentilla indica, P. supina, Polygonum glabrum, Potamogeton crispus, Grangea maderspatana, Phyla nodiflora, Bergia ammannioides, Salvia plebia,* and *Cyathocline purpurea* were commonly observed. However, *Aponogeton natans, Eichhornia crassipes, Hydrilla verticillata, Polygonum orientale, Jussiaea repens, Saqittaria quayanensis, S. saqittifolia, Rumex dentatus, Cyperus platystyles, Launia nauchali Mazus pumilus, M. delavayi* and *Xanthium stramonium* grew rarely.

January: Among macrophytes *Gnapholium indicum* and *Ranunculus sceleratus* show abundant and profuse growth respectively. *Aponogeton crispum, Potentilla supina, P. indica, Polygonum glabrum, Potamogeton nodosus, P. crispus, Jussiaea repens, Campanula banthamii, Grangea maderaspatana, Bergias ammannioides, Eclipta alba, Amaranthes indicum, Launia nauchali, Salvia plebia, Mazus pumilus, Xanthium stramonium, Salix tetrasperma, Cyathocline* *purpurea* were commonly growing macrophytes of this month. However, *Aponogeton natans, Ceratophyllum domersum, Hydrilla verticillata, Polygonum orientale, Chenopodium ambrosioides, Rumex dentatus, Sagittaria sagittifolia, S. gyayanensis, Cyperus platystyles, Phyla nodiflora* and *Veronica anagallis-acquatica* were rarely observed during this month.

February: Macrophytes like *Gnapholium indicum,* and *Ranunculus sceleratus* show profuse growth while *Potentilla indica* grew abundantly during this month. *Aponoqeton crispum, Polyqonum glabrum, Potamogeton crispus, Granqea maderaspatana, Cyperus platystyles, Burgia ammannioides, Eclipta alba, Amaranthese indicum, Launia nauchali, Veronica anagallis-aquatica, Mazus pumilus, Xanthium stramonium, Cyathocline purpurea, Potentilla indica* and *P. supina* were commonly observed macrophytes of this month. *Ceratophyllum domersum, Hydrilla verticillata, Cephalanthus occidentalis, Polygonum orientale, Jussiaea repens, Sagittaria guayanensis, S. sagittifolia, Campanula banthamii, Rumex dentatus, Chenopodium imbrosioides, Salix tetrasperma, Salvia plebia* and *Phyla nodiflora* were rarely recorded during this month.

March: Among macrophytes recorded *Gnapholium indicum* and *Ranunculus sceleratus* show profuse growth during this month. Other macrophytes like *Aponoqeton crispum, Ceratophylium domersum, Potamogeton crispus, Grangea maderaspatana, Sagitteria sagittifolia, Cyperus platystyles, Phyla nodiflora, Eclipta alba, Amaranthes indicum, Chenopodium imbrosioides, Cephalanthus occidentalis, Mazus pumilus, Xanthium stramonium* and *Potentilla indica* were commonly observed. *Hydrilla verticillata, Salvia plebia, Potentilla supina, Polygonum orientale, Campanula banthamii Jussiaea repens, Saqittaria quayanensis, Veronica angallis-aquatica, Burgia ammannioides, Barringtonia acutangula, Cyathocline purpurea* and *Launia nauchali* were rarely growing mcrophytes during this month.

April: Macrophytes like *Ceratophyllum domersum, Gnapholium indicum, Polygonum glabrum, Potamogeton crispus, Ranunculus sceleratus, Grangea maderaspatana, Chenopodium imbrosioides, Cyperus platystyles, Burgia ammannioides, Launia nauchali, Cephalanthus occidentalis, Baringtonia acutangula* and *Potentilla indica* were common during this month. Other species like *Aponogeton crispum,*

Hydrilla verticillata, Potentilla supina, Potomogeton, nodosus, Polygonum orientale, Jussiaea repens, Sagittaria guayanensis, Rumex dentatus, Sagittaria sagittifolia, Phylanodiflora, Eclipta alba, Amaranthus indicum, Veronica anagolis-acquatica, Mazus pumilus, Cyathocline purpurea and *Xanthium stramonium* grew rarely during the month of April.

May: Among macrophytes *Gnapholium indicum, Ranunculus scleratus, Burgia ammannioides, Barringtoina acutangula* and *Potentilla indica* grew commonly during this month. *Aponogeton natans, A. crispus, Ceratophyllum domersum, Hydrilla verticillata, Potentilla supina, Polygonum glabrum, Potamogeton nodosus, P. crispus, Polygonum orientale, Jussiaea repens, Rumex dentatus, Sagittaria sagittifolia, S. guayanensis, Grangea madaraspatona, Cyperus platystyles, Phyla nodiflora, Eclipta alba, Amaranthus indicum, Cepualanthus occidentalis, Chenopodium imbrosioides, Launic nauchali, Mazus pumilus* and *Xanthium stramonium* were rare species of this month.

June: Macrophytes like *Aponogeton crispum, Ceratophyllum domersum, Gnapholium indicum, Hydrilla verticillata, Potentilla supina, P. indica, Polygonum glabrum, Potamogeton nodosus, P. crispus, Ranunculus sceleratus, Grangea maderaspatana, Cyperus platystyles, Burgia ammannioides, Eclipta alba, Amaranthes indicum, Launia nauchali, Xanthium stramonium, Barringtonia acutangula* and *Cephalanthus occidentalis* grew rarely during this month.

July: Only macrophytes like *Ceratophyllum domersum, Cephalanthus occidentalis, Amaranthes indicum* and *Potentilla indica* could be observed during this month with rare growth.

August: No macrophytic flora was observed during this month in river water as well as along the river bank.

September: Among macrophytes *Aponogeton crispum, Ceratophyllum domersum, Gnapholium indicum, Hydrilla verticillata, Polygonum glabrum, P. orientale, Potamogeton nodosus, P. crispus, Jussiaea repens, Sagittaria guayanensis, Rumex dentatus, Ranunculus sceleratus, Sagittaria sagittifolia, Grangea maderaspatana, Phyla nodiflora, Durgia ammannioides, Eclipta alba, Salvia plebia, Calamus tenuis, Mazus pumilus, M. delavayi, Potentilla indica* and *P. supina* were rarely observed during this month.

AERO-ALLERGENIC POLLEN SYSTEMATIC ENUMERATION OF AERO-ALLERGENIC POLLEN

Ranunculaceae

Pollen grains show considerable variation in the family (2 - 3 - colpate, 3 - zonicolpate, 6 - polyrugate, non-aperturate or polyforate, panporate. Sometimes spiraaperturate grains are found with irregular spiroid aperture, may be sub-oblate, prolate, sub-prolate. The exine pattern is mostly reticulate and few forms provided with small spinules.

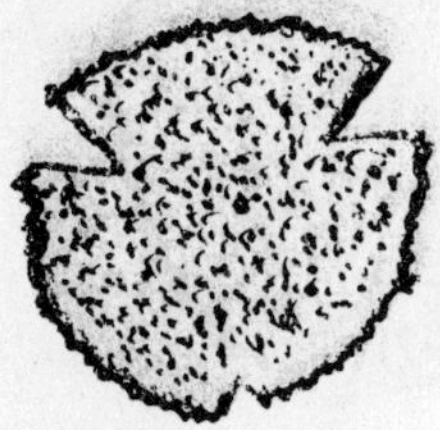

Fig. 98: *Ranunculus*

Pollen grains 3 - zonicolpate

Ranunculus Grains 3 colpate, prolate, spheroidal, exine finely reticulate, (colpi membrane granular) pollen 22-22 μm diam.

Papaveraceae

Pollen grains are 3 (or more) - zonicolpate (may be faintly colporoidate), panporate, rupate, prolate. Aperture symmetrical, rarely irregular. Exine granulate or faintly reticulate.

Fig. 99 *Argemone mexicana*

Pollen grains panporate

Argemone mexicana Grains 3 - zonocolpats, subprolate, exine 1.5 um thick, seroine thicker than nexine, few grains trisyncolpate, exine finely reticulate pollen grain 30-23 μm diam.

Cruciferae

Pollen grains are 3 (4) - zonicolpate (obate spheroidal, spheroidal subprolate). Exine distinct, sexine thicker than nexine, commonly reticulate sometime laculate or granulate.

Pollen grains 3 - zonicolpate

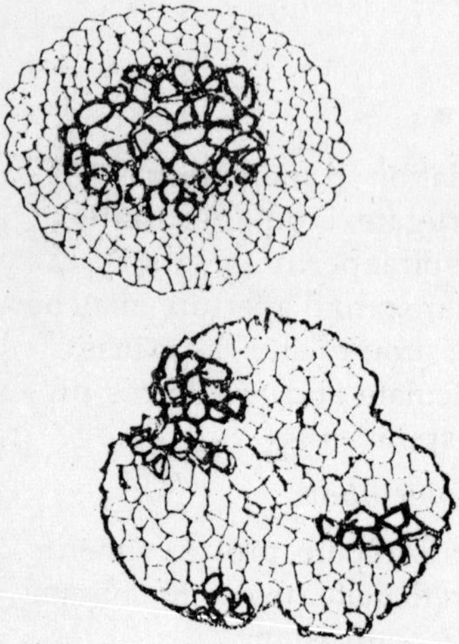

Fig. 100: *Brassica nigra*

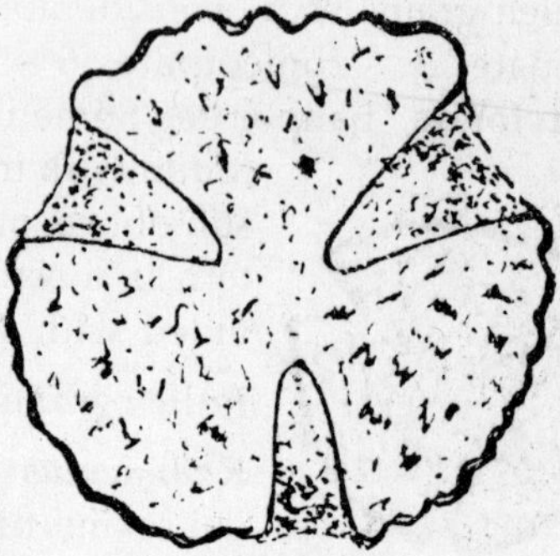

Fig. 101: *Brassica campestris*

Brassica nigra Grains 3 (4) zonicolpate, subprolate (39 x 33 μ), colpi 8-12 μ wide in the middle ends pointed, tenui-marginate (exine thinner near margins). Apocolpium distinct, diameter 4 μ. Exine 15 μ thick. Ectine thicker than end line, reticulate, columella distinct Lumina large. (Fig. 100)
Brassica campestris 3 colpate, subprolate exine ratipilate, colpi broad, ornamented, grain size 34 x 30 μm. (Fig. 101)

Capparidaceae

Pollen grains 3 - zonicolporate or 3-4 zonicolporate, subprolate or prolate. Colpi ends free or syncolpate. Ectine psilate or granulate, tegillate, sometimes showing minute spinules. Pollen grains 3 - zonicolpate.
Capparis grandis Grains 3 (4) zonocolporate, subprolate, endocolpium circular (4 μm diam.) exine 15 μm thick, sexine thicker than nexine, Psilate, pollen grain 19 x 15 μm diam. (Fig. 102)
Cleome viscosa Grains 3 - zodocolporate, suboblate, 27 x 32 μm. Endocolpium circular Exine 18 x 5 μm, Exine 1 μm thick exine surface regulate. (Fig. 103)

Fig. 102: *Capparis grandis*

Fig. 103: *Cleome viscosa*

Cariyophyllaceae

Pollen grains are generally panporate or 3 - zonicolpate. In porate forms the mature, size of pores vary considerably. Exine distinct, granulate or reticulate with irregular reticulations.

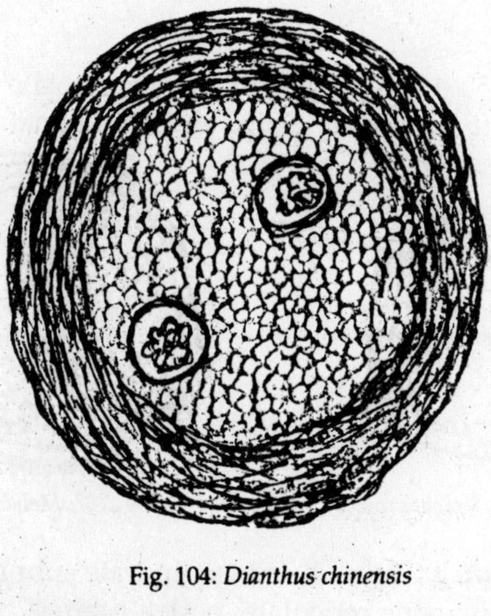

Fig. 104: *Dianthus chinensis*

Pollen grains - Pan porate

Dianthus chinensis Grains 5 - 14 - porate; Pores 3-8 μ wide, annulus 1 μ thick; exine 2.5 μ thick. Ectine thicker than endine, family set; culate, collumella in distinct.

Malvaceae

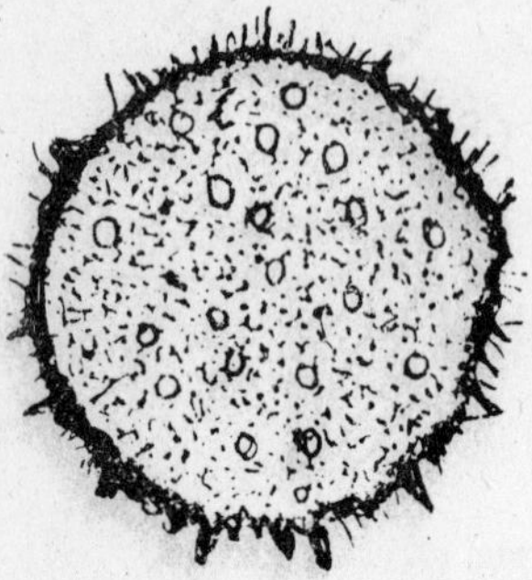

Fig. 105: *Hibiscus roso sinensis*

Pollen grains are zoniporate, zonicolporate or panporate. Exine highly ornamented with spines or baculae. On most of the forms columella at the base of spines constituting the ectine are larger and form the Ectinous cushion. The basal cushion is often denoted as Excrescence system. The lip of excrescences are pointed, rounded or flat. The base is bulbous or straight. The basal cushion may be high or low.

Pollen grains 3 - zoniporate

Hibiscus rosa sinensis Grains pantaporate, spheroidal, exine spinulose spines dimorphic length of spine 19 μm pollen 135 μm diam. Exine thick 7.5 μm thick.

Meliaceae

Pollen grains 3-5 zonicolporate, oblate or suboblate or subprolate. Sexine thicker than nexine. Colpi ends indistinct.

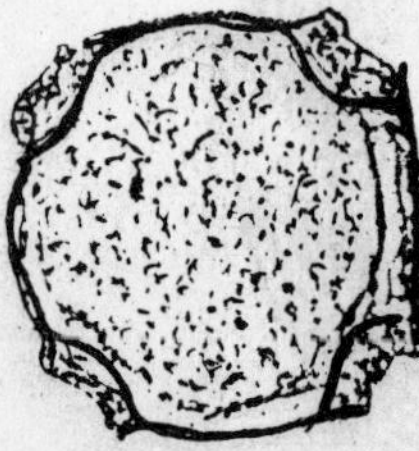

Fig. 106: *Azadirachta indica*

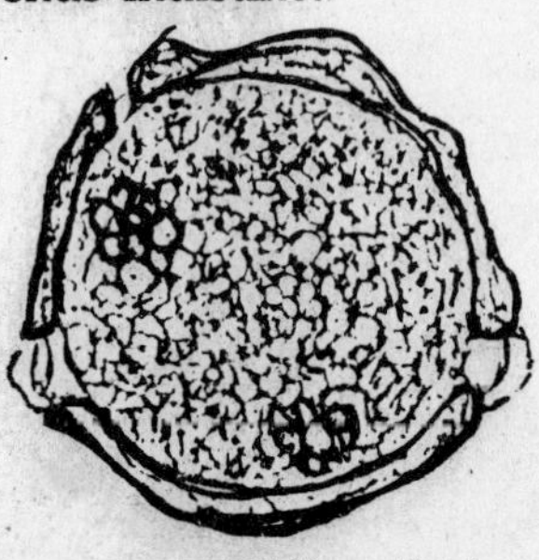

Fig. 107: *Melia azadirach*

Azadirachta indica Grains 4 - zonocolporate subprolate, circular ora 7.5 μm diam; exine reticulate, pollen grain 42 x 34 μm diam. Exine 2.5 μm thick, ectexine thicker than endexine. Exine tectate, surface psilate (LM) and punctate (SEM). (Fig. 106)

Melia azadirach Grains 4 colporate, prolate spheroidal ora circular or slightly lalongate, exine granular, pollen grain 42 x 39 μm. diam. (Fig. 107)

Rhamnaceae

Pollen grains 3 - zonicolporate. The family show homogenity in the apertural structures of pollen. The pollen may be subprolate, prolate, prolate spheroidal, suboblate and rarely triangular. Exine stratification is indistinct, but reticulations are marked. Exine near the colpi may be thick or thin.

Fig. 108: *Zyzyphus*

Pollen grains 3 - zonicolporate

Zyzyphus Grains 3 zonocolporate, oblate, brevicolporate, ora granular, exine psilate pollen less than 30 μm diam. 19-25 μm amb. Triangular exine thick 1.5 μm, vestibulate exine surface faveolate.

Anacardiaceae

Fig. 109: *Mangifera*

Pollen grains 3 - 4 zonicolporate. The aperture types sometimes vary considerably in structure and size *i.e.*, the grains may be faintly porate or poroid, prolate, suboblate spheroidal; spheroidal.

Pollen grains 3 - zonicolporate

Mangifera indica Grains tricolporate, prolate, exine striate (exine margin psilate, ora simple), pollen grain 25 x 24 μm diam.

Leguminosae

Pollen grains in this family show considerable variation in apertural characters. It includes large number of representatives with single or united or compound grains. These compound grains are common in minosaceae as tetrads and polyads. Grains may be 3 -6 zonicolporate, 3 - zonicolpate, 3 - zoniporate and prolate, subprolate, oblate, oblate spheroidal or prolate spheroidal.

Paplionaceae

3 - zonicolporate, sexine thicker than nexine, faint to pattern.

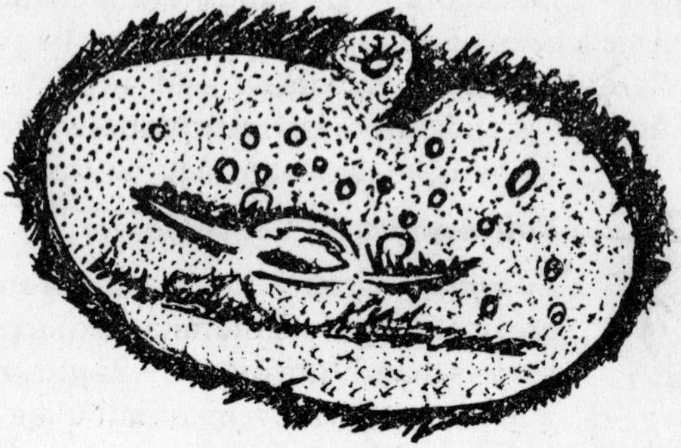

Fig. 110: *Lathyrus odoratus*

Lathyrus odoratus 3 - zonicolporate, subprolate (24 x 21 μ). Colpi ends free pointed. Few grains syncolpate. Endocolpium clear, lalongate (8 x 6 μ), lateral ends rounded. Exine pattern faintly granulate. Both ecline and endine are thick.

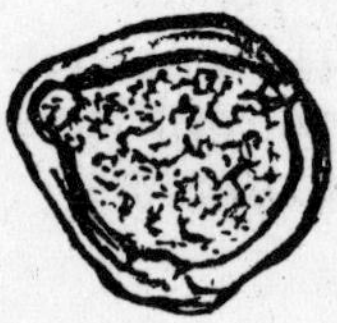

Fig. 111: *Dalbergia sisso*

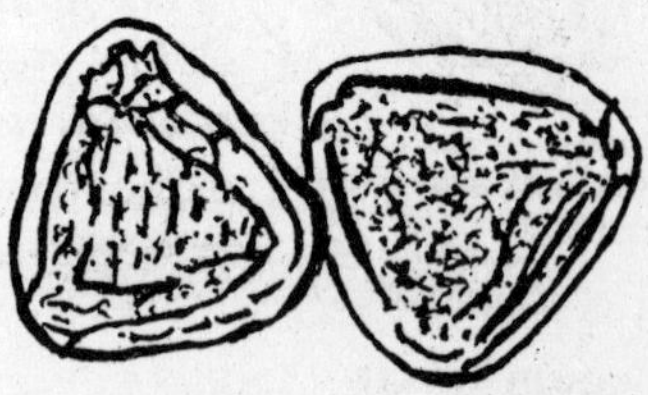

Fig. 112: *Indigofera cordifolia*

Dalbergia sisso 3 - colporate, subrolate, exine psilate ora free simple, endocolpium lalongate, lateral ends rounded, pollen 24 x 27 μm diam. (Fig. 111)

Indigofera cordifolia Grain 3 colporate, subprolate, exine granulate endo-colpium faint constricted in the middle, pollen 39 x 30 μm diam.

Caesalpinaceae

3 - zoniaperturate (colpate, colporate or porate), the exine ornamentation varies considerably ranging from striate, granulate, reticulate or aerolate in different forms.

Pollen grains 3 - zonicolporate

Cassia fistula Grains 3 zonocoporate, suborblate, oblate speroidal (Delonik) Amb. triangular exine psilate sexine as thick as nexine, 27 x 34 μm diam. (Fig. 113)

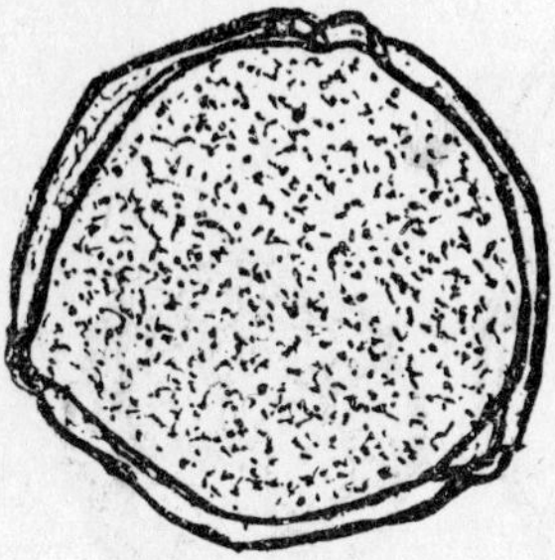

Fig. 113: *Cassia fistula*

Caesalpinia pulcherrima Grains 3 - zonoaperturate, spheroidal, pseudocolpi united at poles forming synpseudocolpic condition, exine shows gradation of pattern in order, areolate, regulate and reticulate apart from psilate pattern on pseudocolpium, pollen 85 μm diam., 77-91 μm. (Fig. 114)

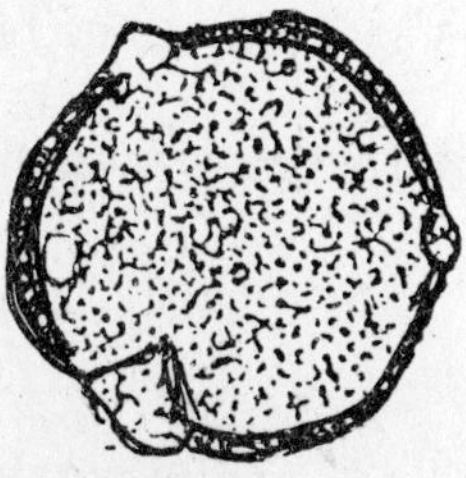

Fig. 114: *Caesalpinia pulcherrima*

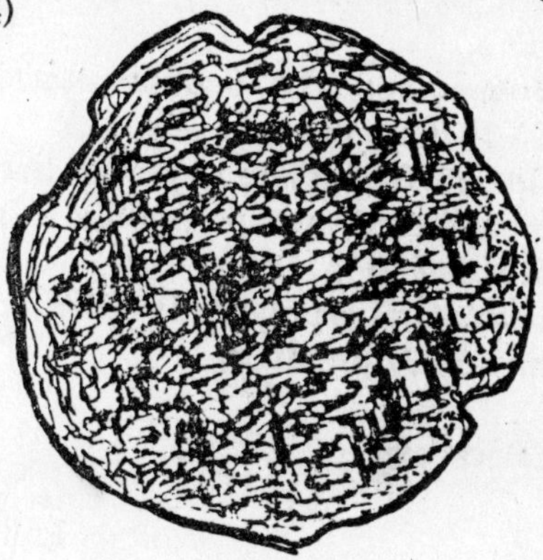

Fig. 115: *Delonix elata*

Delonix elata Grains 3 - colporate, prolate, colpi constricted in the Tip pollen grain 70 x 51 μm diam.

Mimosaceae

Pollen grains are compound, mostly in polyads, except for very few forms as Neptunia, Prosopis, Leucaena, Dichrostachys. These forms constitute common appearance in the locality under investigation. The apertures in polyads type are obscure. Therefore, perhaps the grains may be colpate or porate. Exine pattern varies from psilate, foveolate, granulate or arcolate.

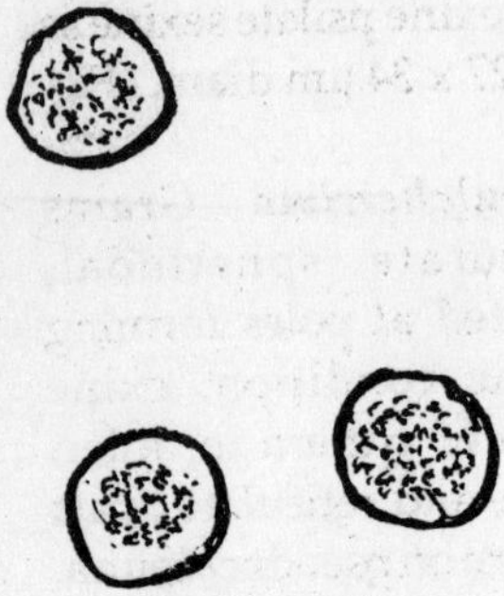

Fig. 116: *Mimosa*

Fig. 117: *Albizzia lebbek*

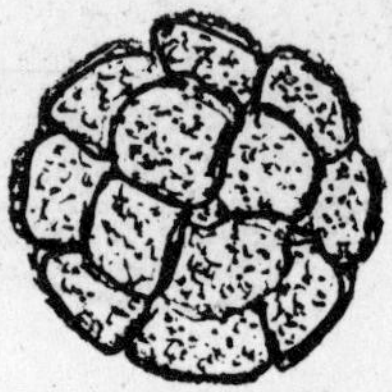

Fig. 118: *Acacia*

Mimosa polyad 4 - 8 grains, syncolpate exine thin psilate pollen 12 - 19 diam µm size. (Fig. 116)

Albizzia lebbek Grains polyad 16 called, average size of polyad 72 µm apertures not discernible, exine thick (1.4 µm) ectine faintly reticulate. (Fig. 117)

Acacia Grains polyads 16 celled, spheroidal oblate, exine granular thick 36-37 µm diam.

Caricaceae

Pollen grains 3 - zonicolporate and suboblate. Exine is faintly baculate or reticulate.

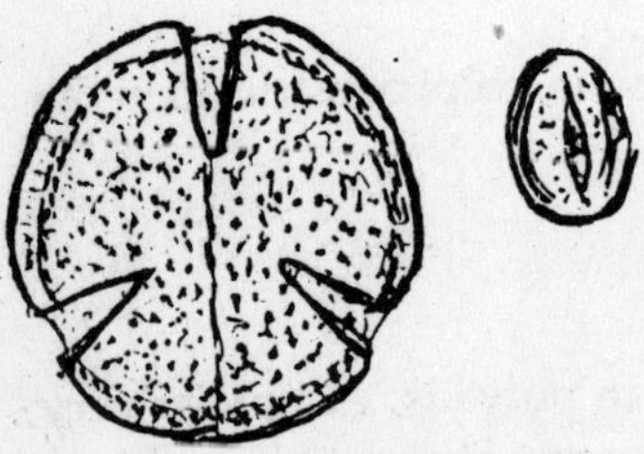

Fig. 119: *Carica papaya*

Pollen grains 3 - zonicolporate

Carica papaya Grains 3 - zonocolporate, subprolate, rarely spheroidal, endocolpium lalongate, sexine almost as thick as nexine surface faintly reticulate 33 x 29 µm diam (range 32 - 35 x 28 - 32 µm).

Myrtaceae

In general pollen are isopolar, rarely subiropolar or radio symmetric, 2-4 zonicolporate, 3-4 aperturate, para-syn colpate, Syn colpate 3 - colporoidate, peroblate, oblate or prolate spheroidal or oblate spheroidal. Exine distinct, sexine mostly as thicker as nexine.

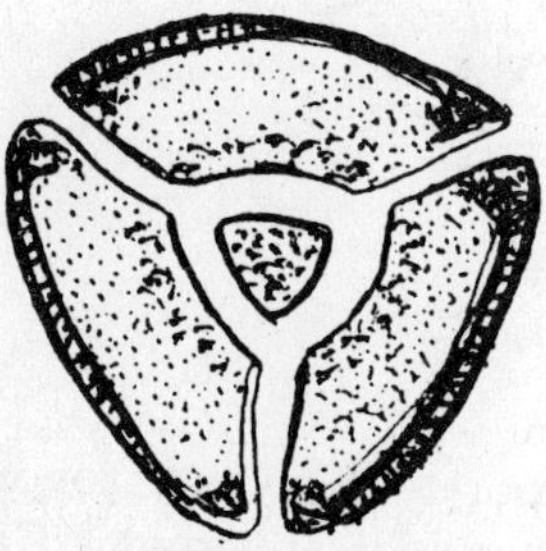

Fig. 120: *Eucalyptus*

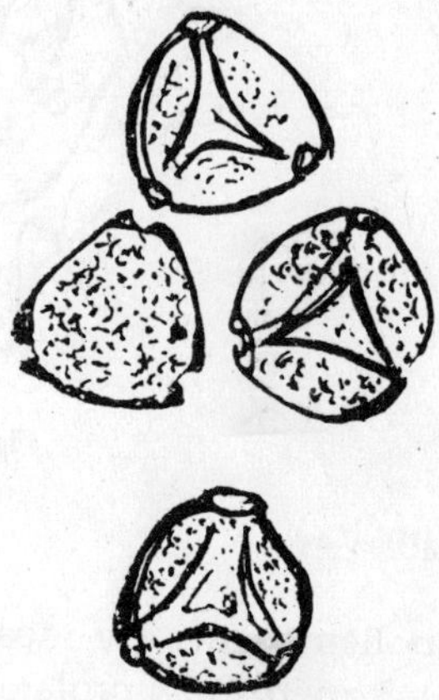

Fig. 121: *Syzigium cumini*

Eucalyptus Grains 3 – colporate or syncolporate; oblate 15-18, 22-28 µ amb semi lobate; or circular. (Fig. 120)
Syzigium cumini Grains 3 - syncolporate (rarely 4 colporate), pollen 10-17 µm diam. Amb. Triangular Endocolpium lalongate colpus 20 µm long Exine 1 µm thick. Exine surface granulate. (Fig. 121)

Umbelliferae

This family is stenopalynous, almost all the grains studied are of colporate type. The shape is subprolate, perprolate or prolate. The grains in different genera show variation in size. The exine pattern is granulate, foveolate or areolate. Exine is thin in most of the forms.

Pollen grains 3 - zonicolporate

Dacus carota Size 31 x 16 µ (range 28 - 32 x 16 µ). Endocolpium lalongate (14 x 6 µ) lateral ends rounded. (Fig. 122)

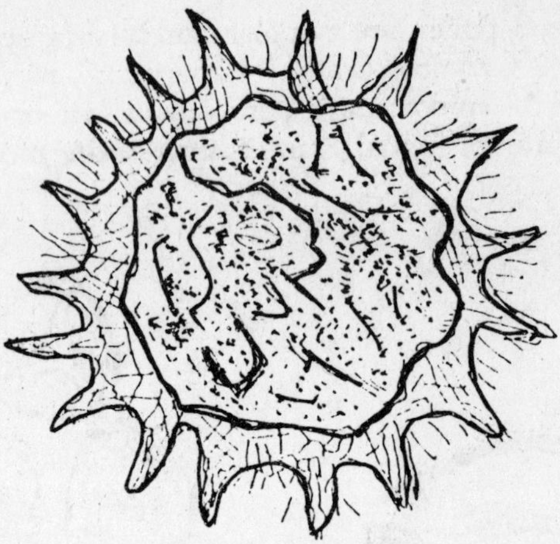

Fig. 122: *D. carota*

Compositae

The pollen grains are mostly 3 - colporate, 3-colpate zonorate (rare), 3 aperturate, prolate, oblate spheroidal, spheroidal. Colpi and pores may be distinct or indistinct. Exine tegillate or granulate. The sexine is distinct, tactate, spiniferous, baculate (ranibaculate), homogenous or demarcated into two parts, outer and inner. Endo and ecto sexine. Ecto-sexine is comparatively much thicker than endosexine. Neine may be baculate in most of the form. Sexine may be thrown into ridges forming regular symmetrical patterns, characteristic of genus vernoria. Wood house described such ridges as lophate. Echinolophate grains possess slightly spinulose characters, whereas prilolophate has smooth rides (wood house). Rarely the area adjacent to ridges are perforated (Erdt. 1952). Such grains appear to be more or less reticulate.

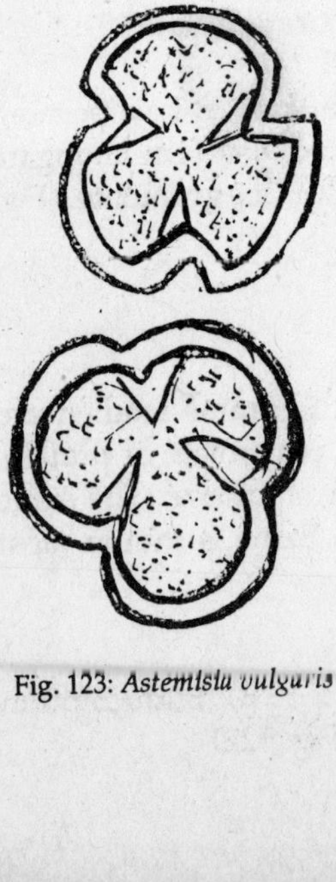

Fig. 123: *Astemisia vulgaris*

Pollen grains 3 - zoniporate

Astemisia vulgaris Grains 20.25 x 20-24 μ; colpi 18-21 x 3-4 μ; or a transversally elliptic, 1-2 x 1-2 μ; exine 1 μ thick. (Fig. 123)

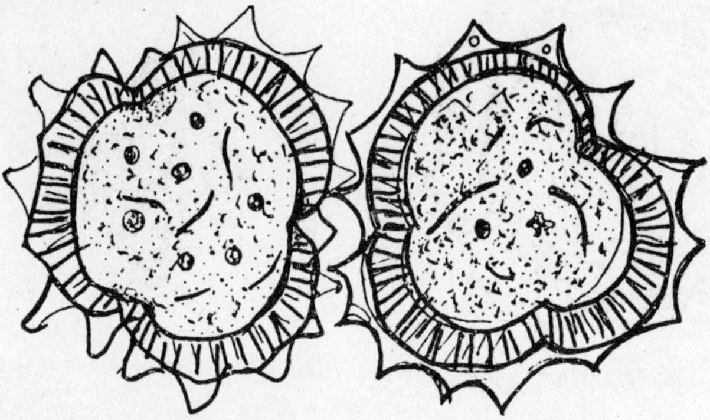

Fig. 124: *Chrysanthemum leucanthemum*

Chrysanthemum leucanthemum Grains 3 - colporate; spheroidal to oblate spheroidal; 29-31 x 30-36 μ; amb circular colpi 22-25 x 4-5 μ; or a circular, 5 μ wide; exine 2.5 μ thick ; ectine 3 x 2.5 μ. (Fig. 124)

Dahlia cultorum Grains 2-4 colporate spheroidal; 42-47 μ wide colpi 10 μ long; or a circular, 5 μ wide exine 2-3 μ thick; ectine 5-6 x 3 μ. (Fig. 125)

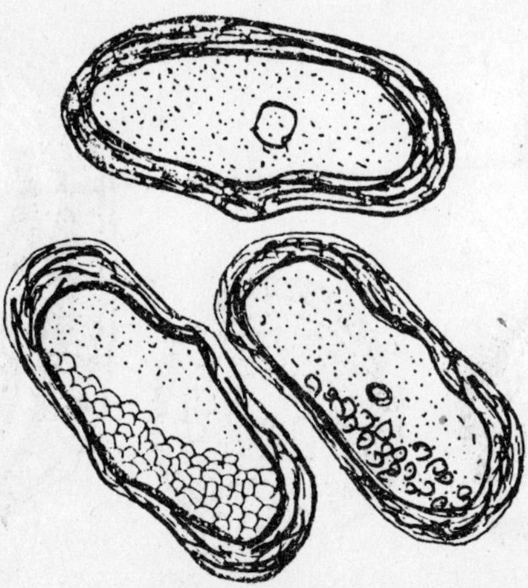

Fig. 125: *Dahlia cultorum*

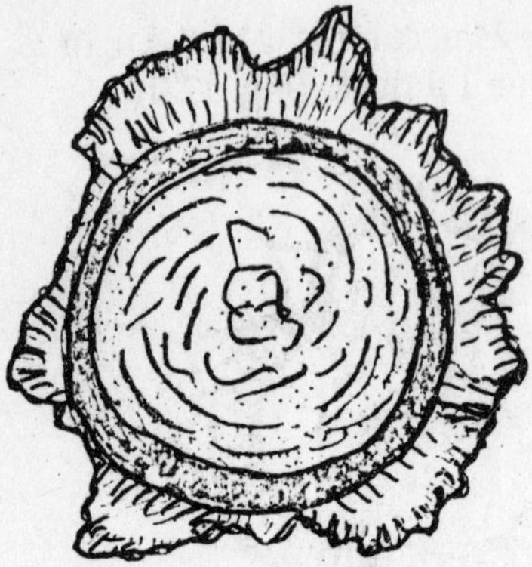

Fig. 126: *Taraxacum vulgare*

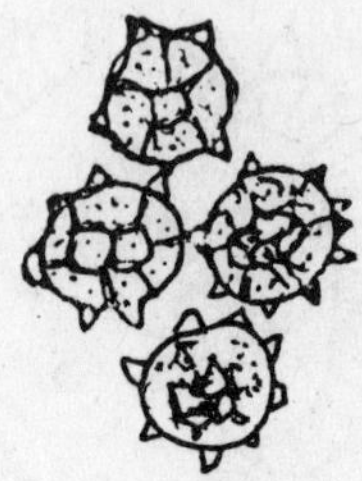

Fig. 127: *Adenostemma viscosum*

Taraxacum vulgare Surface of pollen grain showing inter poural sexine elements at low focus (black dots represents the basal part of baculla; some bright dots near the margin of the grain represent the upper part of bacula) presence of small spinules. (Fig. 126)
Adenostemma viscosum Grains 3 - colporate, amb spheroidal, planaperturate pore central protuding, pore membrane smooth excine spinulose (small 22-24 x 20-22 μm spines included, 19 x 15 μm spine excluded).

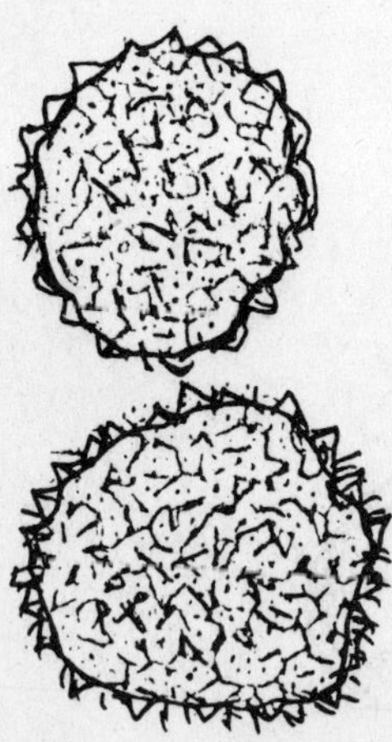

Fig. 128:
Helianthus annus

Fig. 129:
Xanthium strumarium

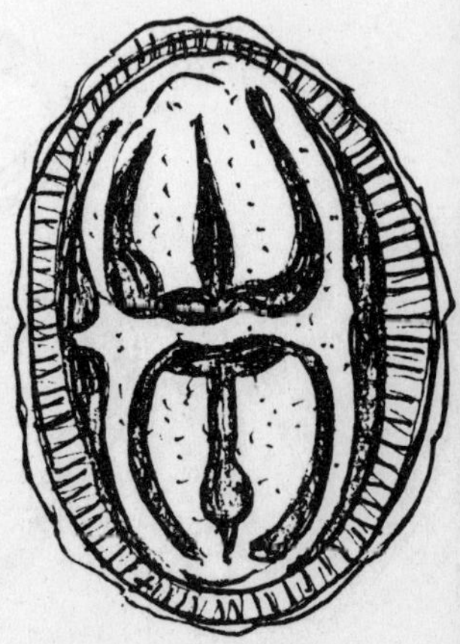

Fig. 130:
Centaurea cyans

Helianthus annus Grains 3 colporate, spheroidal, amb spheroidal plana perturate, exine echinate pollen medium, 35 x 33 diam. (Fig. 128)

Xanthium strumarium Grains zonoporate spheroidal, exine subechinate 22-25 um diam Endocolpium la-longate (3-9 μm) colpus short faint, Exine thick, spinules longer than 1 μm. (Fig. 129)

Centaurea cyans UV - micrograph 25 36 A, about x about 2900 shi x sect. (Fig. 130)

Solanaceae

Grains 3-4 zonicolporate, colporoidate, non-aperturate (few grains syncolpate), syncolporate and Parosyncolporate, 3-4 zonicolpate. Exine granulate, Psilate, prolate, subprolate, Prolate spheroidal, reticulate.

Pollen grains 3 zonicolpate

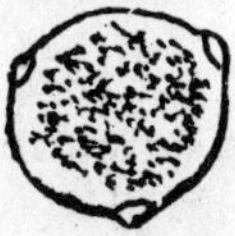

Fig. 131: Datura alba

Datura alba Grains 3 - colporate, prolate amb circular exine striato reticulate, faintly granular, pollen 41 x 45 μm diam. (Fig. 131)

Verbenaceae

Grains are 3-4 zonicolporate or 3-4 zonicolporate or 3 zonicolpate, spheroidal, prolate spheroidal, subprolate. Exine surface is spinulate, spinate, granulate, verrucate or prilate with faint LO - pattern.

Pollen grains zonicolporate

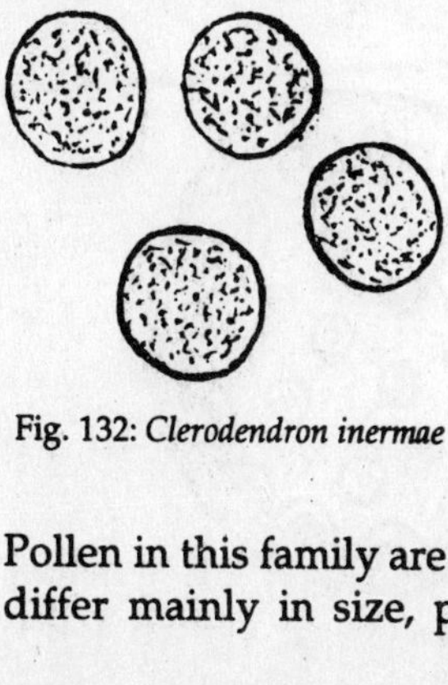

Fig. 132: *Clerodendron inermae*

Clerodendron inermae Grain 3 - zonocolpate, subprolate, and circular, exine 2.6 um thick, sexine thicker than nexine, sphinulate (spinule length 2 μm basal diam. 1 μm) pollen grain 72 x 56 μm (range 68-76 x 46-64 μm). (Fig. 132)

Amranthaceae

Pollen in this family are either panta-porate or panporate. Grains differ mainly in size, pore diameter and exine ornamentation which constitute basis for the categorisation of this family in 3 sub-families, namely Celosieae, Amranthoidae and Comphreneae. For convenience only one member from each tribe has been considered in detail.

Exine pattern is granulate, areolate or sestipilate.

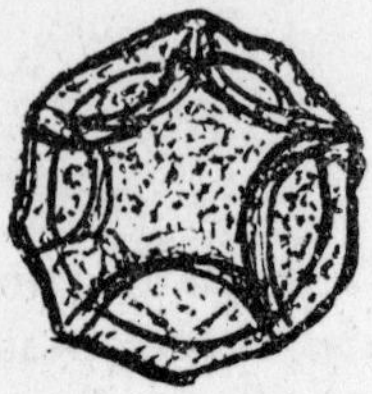

Fig. 133: *Alternenthera sessilis*

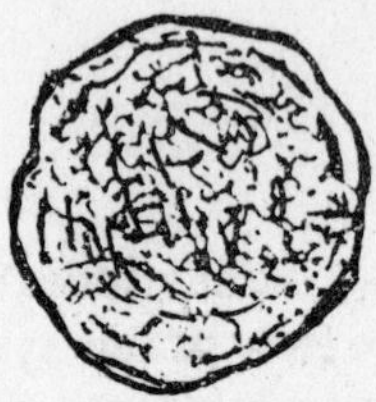

Fig. 134: *Celosia argenta*

Fig. 135: *Amaranthus spinosum*

Alternenthera sessilis Grains pantoporate, spheroidal, exine granulate, pollen 19 µm diam. (range 12-24µm). (Fig. 133)

Celosia argenta Grains periporate, prolate, spheroidal, exine thick with vestigial spinules, pollen grains 23 µm diam. (Fig. 134)

Amaranthus spinosum Grains pantaporate prolate, spheroidal exine thick with vestigial spinules, pollen 20-26 µm. (Fig. 135)

Chenopodiaceae

The grains are predominately pantoporate (Pantapororate, Stephanoporate), except for *Basella rubra* in the present collec-

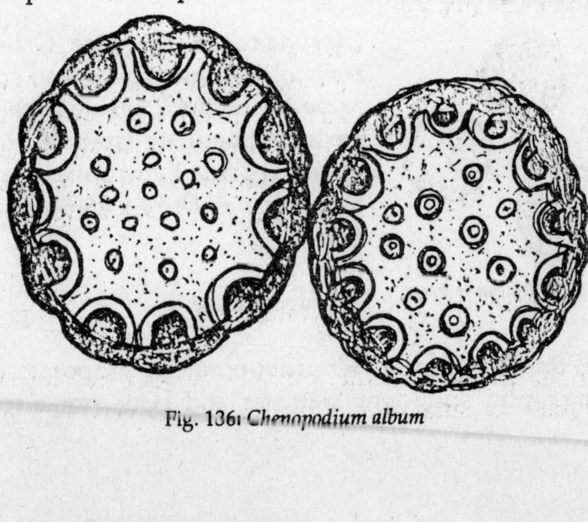

Fig. 136: *Chenopodium album*

tion, which have been observed as 6 parcolpate with square shape amb. Normally in other forms the grains are spheroidal. Exine is faintly granulate or arcolate or psilate.

Pollen grains pan porate

Chenopodium album Panporate, spheroidal, diameter 30 μ; range 21-35 μ pore distinct, spherical diameter about 3-4 μ, tenuumargenate the distance between the pore is about 4 μ. Exine is about 3 μ thick. Ectine is slightly thicker than endine, granulate. (Fig. 136)

Euphorbiaceae

Euphorbiaceae is a large family showing variations in the type of colpi and exine pattern. Grains are 3 to many zonicolporate and mostly spheroidal or oblate spheroidal. Exine is prolate, arcolate or seticulate.

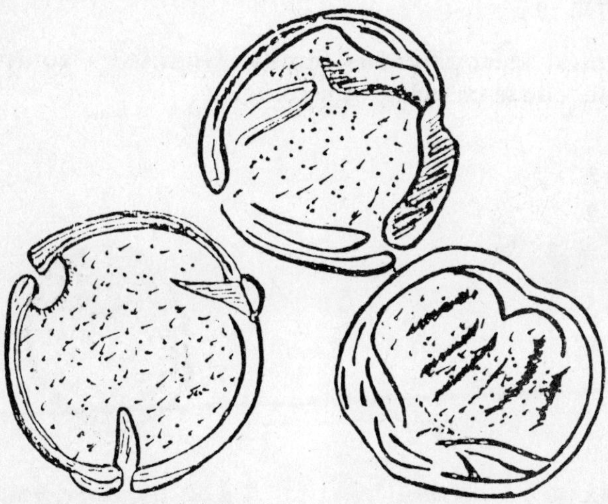

Fig. 137: *Ricinus communis*

Pollen grains Inoperturate

Ricinus communis Grains 3 - colporate; prolate - spheroidal to subprolate; 26-31 x 23-28 μ; amb. circular; colpi as long as P axis; aperture common type; or transversally elliphic; exine finely raticulate, with OL - pattern; nexine thinner than sexine. (Fig. 137)

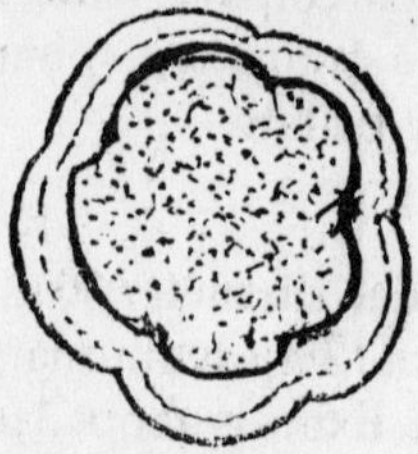

Fig. 138: *Cicca acida*

Fig. 139: *Acalypha*

Cicca acida Grains 5 - colporate, prolate spheroidal, exine psilate. (Fig. 138)

Acalypha Grains tricolpate, peroblate, 3-4-5 colpate 12 μm in diam. Sexine as thick as nexine. (Fig. 139)

Urticaceae

It is almost stenopalynous family. Grains 3 - zoniporate, suboblate, oblate or oblate spheroidal.

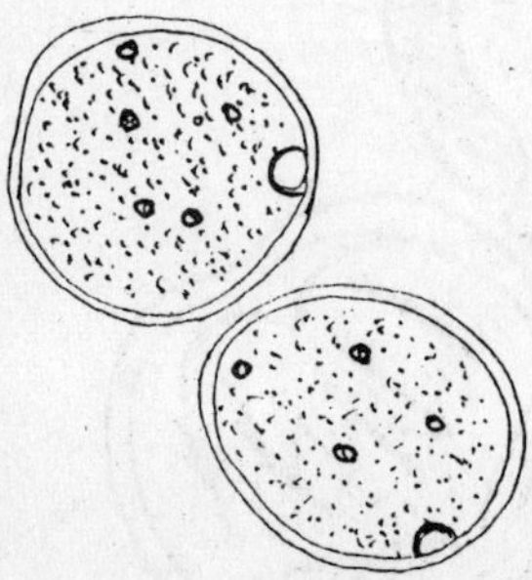

Fig. 140: *Cannabis sativa*

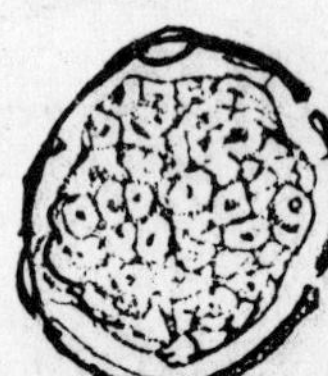

Fig. 141: *Helptelia integrifolia*

Pollen grains 3 - zoniporate

Cannabis sativa Grains 3 - zoniporate, suboblate (size 22.4 x 28.1 μ) at apertures sexine is thicker. Exine stratification indistinct, LO - pattern. (Fig. 140)

Holptelia integrifolia Grains 5 porate rarely 4 or 6, spheroidal exine slightly thickened near the pores (graniyar 20.5-25.5 μm diam) (20-23 x 23-26 μm). (Fig. 141)

Casurinaceae

Pollen grains 3 - zoniporate, oblate or suboblate. Ectine thicker than endine, tegillate.

Pollen grains zoniporate

Casurina equistifolia Grains 3 - zonoporate, oblate, exine not thickened near the pores, pores capped, exine psilate or with faint LO pattern (granular), pollen 18 x 23 μm diam. (Fig. 142)

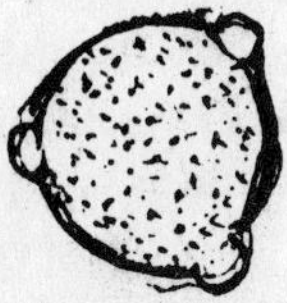

Fig. 142: Casurina equistifolia

Cannaceae

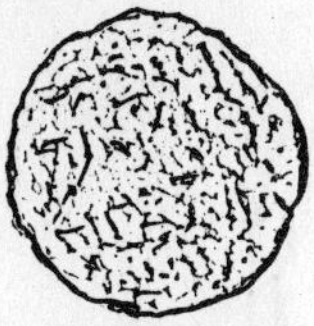

Fig. 143: *Canna indica*

Grains inaperturate, spheroidal. Ectine and endine indistinct, columella not clear. Exine stratification varies as vesrucate, psilate or baculate.

Canna indica Grains inaperturate, spheroidal spinulate, ornamented pollen 68 μm diam (range 62-71 μm). (Fig. 143)

Cyperaceae

Grains 1-4 aperturate; obovoidal (apple-shaped), rectangular, sub-spheroidal or triangular; 21-60 x 25-95 x 23-70 μ furcon in 1 aperture grains usually as long as '6' axes being 25-95 x 23-70 μ distal aperture in 4 - aperturate grains 4-10 μ wide. Sexine as thick as nexine.

Eriophorum vaginatum Pollen 1 - colpate, elongate (size 36 x 26 x 32 μ). Colpus not well defined, about 24 μ in length. Exine 1 μ thick, surface faintly granulate. (Fig. 144)

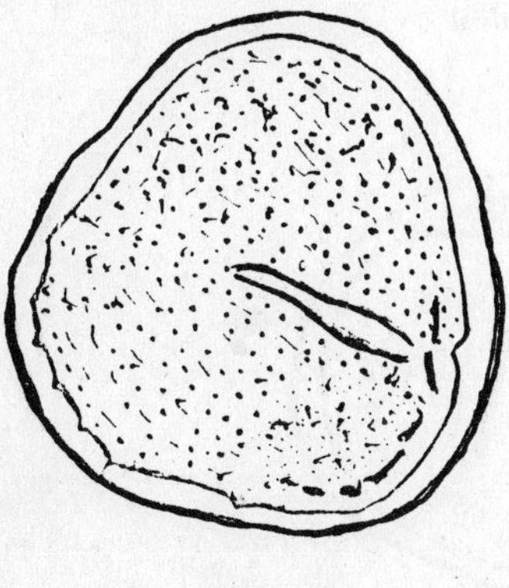

Fig. 144: Eriophorum vaginatum

Gramineae

The family is stenopalynous, having grains with single pore (1-porate) with an annular thicking, spheroidal (rarely ovoid spheroidal), average diameter 20-55 µ. Annulus faint, exine faintly granulate. Sexine as nexine, prilate or raticulate. Besides setarea glauca many species of gramineae were collected and observed but apart from variance in size of pollen grains in different species, there is little difference in other features to support a reliable means of plant identification.

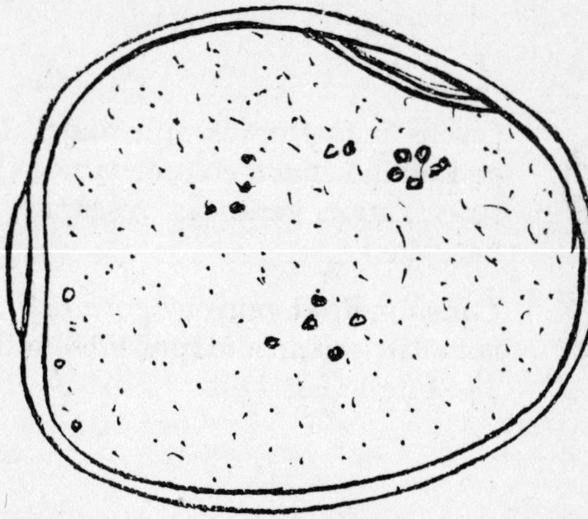

Fig. 145: *Avena sativa*

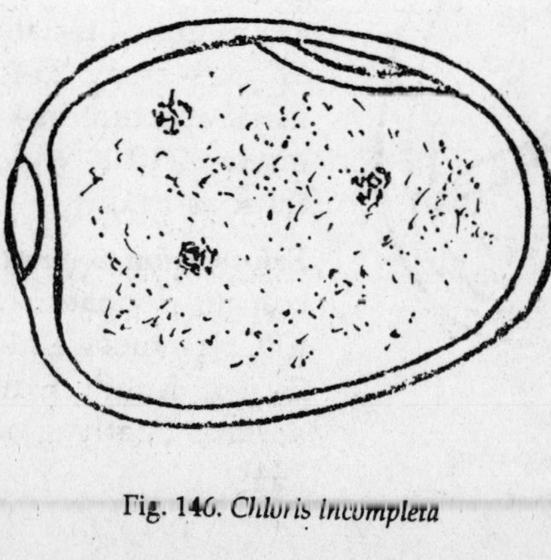

Fig. 146. *Chloris incompleta*

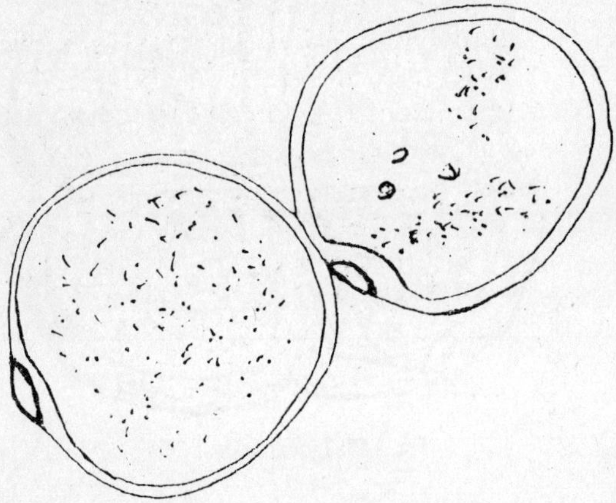

Fig. 147: *Cynodon dactylon*

Pollen grains 1 - Porate

Avena sativa Grains 1 - porate (ulcerate), spheroidal; 14-97 µ wide; aperture drop tube, pores annulate 1.5 - 7 µ wide 11-28 in wide including thickness of annulus, the annulus elevated, 2-6 (-12) µ thick exine 1-1.5 µ thick, nexine thinner than sexine. (Fig. 145)

Chloris incompleta Grains monoporate, spheroidal, pore spheroidal (2.8 µm diam) conspicuous protruding, operculate, operculum spheroidal, exine coarsely granular, pollen medium 34 x 30 µm diam. (Fig. 146)

Cynodon dactylon Single porate grains and exine is thin walled, grains are round in shape. (Fig. 147)

Typhaceae

Grains 1 - porate; 24-29 µ wide ; aperture common type, pores 3-4 x 3-4 µ; exine 1-1.5 thick; tectum with scabrate processes; sexine reticulate, with OL pattern.

Typha angustifolia Grains 1 - porate, exine 1-1.5 µ thick, sexine reticulate with OL pattern. (Fig. 148)

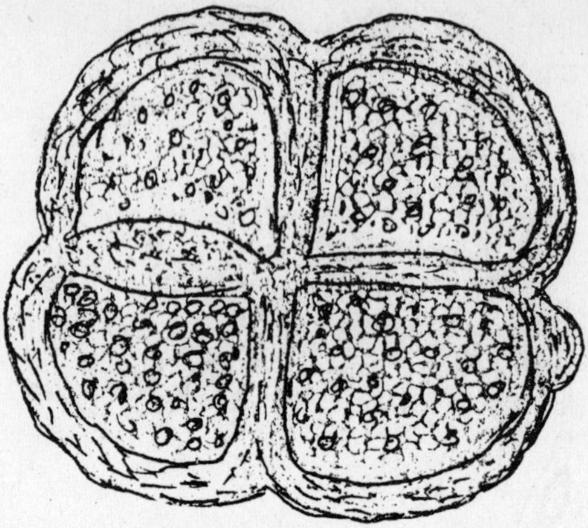

Fig. 148: *Typha angustifolia*

FISH

Fish Diversity and Density

There is a good deal of diversity in quantitative and qualitative distribution and occurrence of major and minor carps. The fishes migrate and transmigrate in time and space, therefore, results are indicative of basic trends alone. A total number of 25 fish species were recorded during different months of the year and their presence shows some variance as given below:

October: During this month *Mystis tingra* (9.52), *Bagarius bagarius* (9.95), *Eutropichtyes vacha* (24.67), *Mastocanabatus species* (1.3), *Chela laubuca* (34.20), *Labeo baata* (9.52), *L. rohita* (0.86), *L. calabasu* (0.86), *Paleomon species* (2.16), *Cirrhan mrigala* (0.43), *Clarius batoachus* (2.59), *Wallago attu* (2.16) and *ophicembelus strituss* (0.86) were recorded.

November: During this month Mystis tingra (8.07), *Bagarius bagarius* (13.8), *Eutropichtys vacha* (36.2), *Mastocanabatus species* (0.52), *Chela laubuca* (22.13), *Labeo baata* (8.07), *L. calabasu* (0.26), *Palemon species* (8.07), *Labeo rohita* (0.78), *Clarius batoachus* (1.04), *Wallago attu* (0.78), *Katla katla* (0.78), *Ilisha motius* (0.26), *Notopterus notopterus* (0.52), *Rita rita* (0.26), *Ailia coila* (0.26) and *Channa gachua* (4.42) were recorded.

December: During this month *Mystis tingra* (22.24), *Bagarius bagarius* (21.04), *Eutropichtyes vacha* (18.25), *Mastocanabatus species* (0.18), *Nangra punctata* (0.37), *Chela laubuca* (24.58), *Chanda randa* (0.74), *Labeo baata* (4.09), *L. rohita* (1.12), *Palemon species* (1.67), *Cirrhana mrigala* (0.37), *Mystis seenghala* (0.93), *Mastocembelus armatus* (1.49), *Clarius batoachus* (0.55), *Wallago attu* (0.74), *Ophicembelus stritus* (0.56) and *Channa gachua* (3.16) were recorded.

January: During this month *Mystis tingra* (17.5), *Bagarius bagarius* (23.1), *Eutropichtyes vacha* (12.0), *Mastocanabatus species* (0.5), *Chela laubuca* (22.8), *Chanda ranga* (1.2), *Labeo baata* (15.2), *L. rohita* (0.8), *Paleomon species* (1.0), *Cirrhana mrigala* (0.1), *Punctius tincto* (0.3), *Clarius batoachus* (0.3), *Wallago attu* (0.3), *Rhinonugil corsula* (0.1), *Ophicembelus stritus* (0.3), *Rita rita* (0.5) and *Channa gachua* (3.7) were recorded.

February: During this month *Mystis tingra* (19.6), *Bagarius bagarius* (24.7), *Mystis seenghala* (0.4), *Eutropichtyes vacha* (17.7), *Mastocanabatus species* (0.3), *Nangra punctata* (0.4), *Chela laubuca* (25.2), *Labeo baata* (2.7), *Paleomon species* (2.7), *Cirrhana mrigala* (0.4), *Punctius species* (0.1), *Clarius batoachus* (1.1), *Mastocembelus armatus* (1.4), *Wallago attu* (1.1) and *Channa gachua* (3.7) were recorded.

March: During this month *Mystis tingra* (20.2), *M. seenghala* (0.8), *Bagarius bagarius* (17.8), *Eutropichtyes rache* (17.0), *Nangra punctata* (0.5), *Mastocanabatus species* (0.4), *Chela laubuca* (26.5), *Labeo baata* (5.5), *L. rohita* (1.0), *Chanda ranga* (0.4), *Paleomon species* (1.6), *Wallago attu* (0.2), *Mastocembelus armatus* (2.3), *Rhinonugil corsula* (0.2), *Katla katla* (0.2), *Rita rita* (0.2), *Ilisha motius* (0.1), *Ailia coila* (0.1) and *Channa gachua* (4.2) were recorded.

April: *Mystis tingra* (20.8), *M. seenghala* (0.3), *Bagarius bagarius* (19.8), *Eutropichtyes vacha* (7.8), *Mastocanabatus species* (0.3), *Chela laubuca* (33.0), *Ailia coila* (0.4), *Chanda ranga* (0.3), *Labeo baata* (0.7), *L. rohita* (5.4), *L. calabasu* (0.1), *Paleomon species* (1.3), *Cirrhana mrigala* (1.0), *Punctius tincto* (0.1), *Mastocembelus armatus* (0.3), *Wallago attu* (0.3), *Rhinonugil corsula* (0.1), *Ophicembelus stritus* (0.1), *Ilisha motius* (0.3), *Notopterus notopterus* (0.1), *Clarius batoachus* (0.3), *Rita rita* (0.3), *Ailia coila* (0.1), *Channa gachua* (5.5.) and *Katla katla* (0.3) were recorded during this month.

May: *Mystis tingra* (25.2), *M. seenghala* (0.6), *Bagarius bagarius* (21.1), *Eutropichtyes vacha* (6.1), *Mastocanabatus* species (0.1), *Chanda ranga*

FISH DENSITY AND DIVERSITY OF RIVER GANGA

Species	Percentage Occurrence	Standard Deviation	Coefficient of Variation
Mystis tingra	20.326	+52.039	58.307
Bugurius bugarius	19.643	+49.809	57.749
Eutropichtyes vacha	14.158	+45.109	36.281
Mystis seenghala	00.398	+2.221	126.914
Mastocanabatus sp.	00.380	+1.131	73.860
Nangra punctata	00.361	+1.621	102.379
Chela laubuca	26.589	+62.056	53.153
Chanda ranga	00.380	+2.229	133.740
Labeo ba ta	04.327	+24.506	128.979
Labeo rohita	02.752	+12.479	103.274
Paleomon sp.	01.480	+4.583	70.508
Cirrhana mrigala	00.645	+3.326	117.388
Punctius tincto	00.114	+0.674	134.800
Clarius batoachus	00.493	+1.946	89.815
Mastocembelus armatus	00.759	+5.033	150.990
Wallago attu	00.569	+2.067	107.843
Rhinonugil corsula	00.247	+1.379	127.292
Ophicembelus stritus	00.190	+1.030	123.600
Catla catla	00.030	+1.193	102.514
Ilisha motius	00.114	+1.435	287.000
Notopterus notopterus	00.171	+1.055	140.667
Labeo calabasu	00.133	+0.793	135.943
Rita rita	00.285	+1.138	91.040
Ailia coila	00.133	+0.669	114.686
Channa gachua	04.897	+12.717	59.149

(0.5), *Ailia coila* (0.3), *Labeo baata* (5.4), *L. rohita* (5.4), *Chela laubuca* (31.9), *Paleomon species* (1.5), *Clarius batoachus* (0.5), *Cirrhana mrigala* (1.5), *Mastocembelus armatus* (0.1), *Wallago attu* (0.1), *Katla katla* (0.5), *Notopterus notopterus* (0.5), *Ilisha motius* (0.6), *Rita rita* (0.5) and *Channa gachua* (6.1) were recorded during this month.

June: During this month *Mystis tingra* (26.1), *Bagarius bagarius* (18.9), *Eutropichtyes vacha* (8.9), *Mastocanabatus species* (0.5), *Chela laubuca* (26.3), *Chanda ranga* (0.1), *Nangra punctata* (0.7), *Mystis seenghala* (0.1), *Labeo baata* (0.7), *L. rohita* (0.8), *Paleomon species* (0.3), *Cirrhana mrigala* (1.5), *Punctius species* (0.1), *Clarius batoachus* (0.7), *Mastocembelus armatus* (0.3), *Wallago attu* (0.5), *Rhinonugil corsula* (0.8), *Ophicemdelus stritus* (0.1), *Katla katla* (0.1), *Ilisha motius* (0.7), *Notopterus notopterus* (0.1), *Rita rita* (0.1), *Ailia coila* (0.3) and *Channa gachua* (7.5) were recorded.

July: *Mystis tingra* (26.9). *Bagarius bagarius* (16.9), *Eutropichtyes vacha* (7.3), *Chelalaubuca* (4.3), *Chanda ranga* (0.4), *Labeo baata* (0.4), *L. rohita* (6.0), *Ilisha motius* (0.4), *Ailia coila* (0.4), *Paleomon species* (0.4), *Nangra punctata* (0.4) and *Channa gachua* (6.0) were recorded.

August: *Mystis tingra* (21.2), *Bagarius bagarius* (19.1), *Eutropichtyes vacha* (7.0), *Chela laubuca* (27.2), *Labeo rohita* (7.0), *L. calabasu* (1.0), *Cirrhana mrigala* (2.6), *Punctius tincto* (1.0), *Wallago attu* (1.0), *Ilisha motius* (1.0), *Mastocembelus armatus* (1.0), *Ophicembelus stritus* (1.0), *Katla katla* (1.0), *Ailia coila* (1.0) and *Channa gachua* (9.0) were recorded during this month.

September: During this month *Mystis tingra* (26.4), *Bagarius bagarius* (24.3), *Chela laubuca* (18.2), *Labeo rohita* (5.2), *L. baata* (5.2), *Paleomon species* (4.3), *Clarius batoachus* (1.7), *mastocembelus armatus* (0.8), *Katla katla* (1.7), *Notopterus notopterus* (1.7), *Rita rita* (0.8) and *Channa gachua* (11.3) were recorded.

Gut Content Algae of Certain Fish

Gut contents examination of fish *Eutropichtyes vacha* and *Chela laubuca* netted from Ganga water periodically showed presence of algae present in identifiable form as listed below:

Algae in gut content of Eutropichtyes vacha

Results of gut contents examination showed presence of twelve species of algae in gut contents of *E. rache*. Out of them

ALGAE IN GUT OF FISHES

S. No.	Species	Percentage Occurrence	
		Eutropichtyes vacha	*Chela laubuca*
1.	*Ankistrodesmus falcatus*	8.82	11.22
2.	*Actinastrum hantzschii*	-	3.06
3.	*Chlorella vulgaris*	11.76	9.18
4.	*Scenedesmus dimorphus*	7.84	16.33
5.	*S. obliquus*	17.64	7.14
6.	*Selenastrum gracile*	13.75	18.38
7.	*Microcystis aeruginosa*	2.94	-
8.	*Gloeocapsa quaternata*	3.92	-
9.	*Cycloglomerata*	8.80	-
10.	*Cymbella laevis*	-	3.06
11.	*Spirogyra species*	6.86	-
12.	*Navicula lanceolata*	4.90	-
13.	*N. gastrum*	-	1.02
14.	*N. cryptocephala*	-	6.12
15.	*Nitzschia palea*	7.84	10.20
16.	*N. palea* var. *fonticola*	-	7.14
17.	*Synedra ulna*	4.90	6.12
18.	*S. ulna* var. *aequalis*	-	4.08

five species (*Ankistrodesmus falcatus, Chlorella vulgaris, Scenedesmus dimorphus, S. obliguus, Selenastrum gracile*) of chlorophyceae, five species (*Cyclotella glomerata, Navicula lanceolata, Nitzschia palea, N. palea* var. *fonticola, Synedra ulna*) of bacilariophyceae and two species (*Gloeocapsa quaternata, Microcystis aeruginosa*) of cyanophyceae could be recorded.

Algae in gut content of *Chela laubuca*

Presence of 13 algal species in gut contents of *C. laubuca* was observed. Out of them 7 species (*Ankistrodesmus falcatus, Actinastrum hantzschii, Chlorella vulgaris, Scenedesmus dimorphus, S. obliquus, Selenastrum gracile, Spirogyra species*) belonging to chlorophyceae and six species (*Cymbella laevis, Nitzschia palea, N. palea* var. *fonticola, Navicula cryptocephala, Synedra ulna, S. ulna* var. *aequalis*) of bacillariophyceae could be recorded.

EFFECT OF *Scenedesmus dimorphus* ON GROWTH OF *Eutropichtyes vacha*

(Average of three replicates)

S. No.	Concentrations	No. of fish	Weight		
			Initial (in mg)	Final (in mg)	% Increase
1.	Control	10	148.382	150.272	1.274
2.	5 gram	10	151.123	158.653	4.983
3.	10 gram	10	154.982	174.010	12.277

EFFECT OF *Selenastrum gracile* ON GROWTH OF *Eutropichtyes vacha*

(Average of three replicates)

S. No.	Concentrations	No. of fish	Weight		
			Initial (in mg)	Final (in mg)	% Increase
1.	Control	10	151.274	154.060	1.842
2.	5 gram	10	153.710	159.280	3.625
3.	10 gram	10	149.820	166.950	5.427

EFFECT OF *Scenedesmus dimorphus* ON GROWTH OF *Chela laubuca.*

(Average of three replicates)

S. No.	Concentrations	No. of fish	Weight		
			Initial (in mg)	Final (in mg)	% Increase
1.	Control	10	154.642	159.227	2.965
2.	5 gram	10	151.123	165.573	9.562
3.	10 gram	10	148.382	173.967	17.243

Pisciculture

In vitro cultural experiments conducted with *Scenedesmus dimorphus* and *Selenastrum gracile* on two common carps *Eutropichtyes vacha* and *Chela laubuca* as feed showed:

Eutropichtyes vacha

Scenedesmus dimorphus as *E. vacha* feed

Results indicated a marked increase in growth of fish expressed in terms of weight. Fish fed with 10 grams of algae per week showed maximum percentage increase in growth of fish.

***Selenastrum gracile* as *E. vacha* feed**

Observations indicated a marked increase in growth of fish. Fish fed with 10 grams of algae shows maximum percentage increase in weight of fish.

Chela laubuca

***Scenedesmus dimorphus* as *C. laubuca* feed**

A perusal of data showed maximum increase in weight of fish with 10 grams of algae. An increase in concentration of alga Scenedesumus dimorphus promoted the rate of increase in weight of *C. laubuca*.

***Selenastrum gracile* as *C. laubuca* feed**

Results emphasized that an increase in weight of fish was observed with increase of concentration of algae. Maximum increase in weight of fish was recorded with 10 grams of algae.

EFFECT OF *Selenastrum gracile* ON GROWTH OF *Chela laubuca*.

(Average of three replicates)

S. No.	Concentrations	No. of fish	Weight		
			Initial (in mg)	Final (in mg)	% Increase
1.	Control	10	152.456	155.285	1.856
2.	5 gram	10	148.762	159.494	7.214
3.	10 gram	10	149.253	169.065	13.274

Rene Dubos (1975) rightly stated on Quote "Now and then in my dreams, I see a place where one might initiate and publicize programme to give environmental movement a positive and constructive philosophy that would compliment the present defensive attitude of environmental conservation and protection." Nevertheless, this dream of Dubos is not singular, the global environmental community is aspiring to develop a technology and a system that could control and co-ordinate use and reuse of flowing river water for drinking, domestic, industrial and agricultural purposes. But there are severe handicaps in transforming this dream into reality. The problems causing impediment in development of an unpolluted, safe and hygenic water are many and

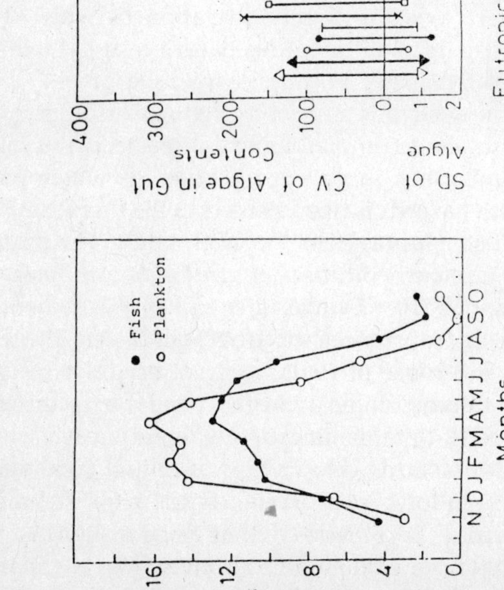

Correlative fish and plankton density in Ganga water

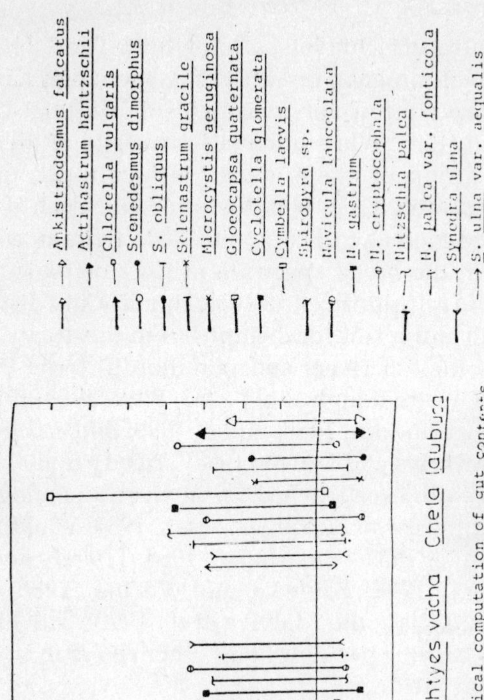

Statistical computation of gut contents algae of *Eutropichtyes vacha* and *Chela laubuca*.

beyond control of technological knowhow. The defensive exploration and environmental research movement emphasizes creative aspects of riverine water management. Hillman (1961) rightly pointed out that "While adequate coverage of all work with a particular group of plants must perforce touch most fields of botanical research, it is impossible to consider each of the problems in its general context, to do so would be to write an encyclopedia." However, in the broad spectrum of riverine biology and its involvement in self-purification, water quality and disease spread is the strength and forms focal emphasis in this treatise.

The ecology of rivers and pollution in India have been reviewed elsewhere (Singh *et al.*, 1982; Patil, 1983; Bilgrami, 1986; Ajmal and Raziuddin, 1988; Sinha, 1988; Sinha, 1989; Sharma *et al.*, 1990; Shrotriya and Dubey, 1990; Trivedy *et al.*, 1990; Trivedy, 1990). There are excellent works on riverine biology (Saxena *et al.*, 1966; Parmasivam and Sreenivasan, 1981; Venkateswarlu and Sampat Kumar, 1982; Somashekar, 1984; Trivedy and Goel, 1984; Dubey *et al.*, 1986; Kudesia and Verma, 1986; Reddy and Vekateswarlu, 1987 and Malviya *et al.*, 1990). The literature suggests that Ganga is peculiar of all other rivers in the country for its biological attributes.

Perhaps there exists a direct co-relation between chemical constituents of organic and inorganic nature coupled with physical attributes of the water body and microbe profile. The biology of a riverine system is subject to changes in physico-chemical characteristics, and complementary impacts of microbes. Involvement of microbes in self-purification of water pollution and public health hazards is documented in the literature (Shukla, 1989; Nigam, 1986; Gupta, 1990; Tripathi, 1990). The problem of monitoring, abatement, pollution, control of health hazards and resultant biology of river Ganga *inter alia* physico-chemical attributes in unison constitutes environmental priorities for exploitation, use and reuse of river water. It becomes incumbent, therefore, that microbiological profile need be recorded with special reference to organisms growing luxuriantly.

Public health hazards due to high levels of total and fecal coliform bacteria in lotic water of the river Ganga are of pivotal importance. It may be remarked that fecal coliforms form a natural intestinal flora of man and animals. The coliforms pass out with fecal matter into the flowing waters. During present investigation both total and fecal coliforms do not exceed the limit of less than 5000 MPN/100ml for drinking raw water source prescribed for conventional treatments and less than 500 MPN/100ml for swimming purpose. Observations reveal a great deal of variance in MPN/100ml of total and fecal coliforms throughout the calendar year. This is due to continuous influx of sewage water and unending human activity which directly or indirectly makes the river polluted coupled with certain kind of pollutant input load affecting coliform levels at various places. Total coliforms include fecal, enteric and other intestinal gram negative bacteria. Their existence in water beyond limit of pollutional hazard in the water causes dysentry, gastroentritis, septimecia and enteric fever in both human beings and animals. Coliforms are ideally rated as indicators of fecal pollution in lotic waters. Assessment of fecal coliforms in the river revealed that the water is unfit for direct consumption by human beings and animals.

Despite pioneering works of Cooke (1977); Cooke and Matsuura (1969) on fungal composition of fresh water ecosystem, little is known about fungi of fresh water bodies. Literature supports elsewhere isolation of fungi from Ganga waters (Bilgrami and Dutta Munsi, 1979; Shanker and Prasad, 1983). Herein, observed species were also earlier recorded from waste and polluted waters (Cooke, 1970, 1977; Rai, 1981; Vittal and Aravazhi, 1981). Ganga is known for its variance in biological fabric within short distances, and totality of fungal infestation in entire river is unknown. Results of present investigation emphasize importance of fungi of Ganga waters in self-purification and public health hazards. Ten genera spread over 21 species were isolated from the river water. The occurrence of fungal spores and hyphae in Ganga waters indicates that they are capable of utilizing the nutrients from the polluted waters. Cooke and Matsuura (1969) reported an increase of 5-200 fold in the fungal and yeast populations within a year in a stabilization pond, spaking thereby of the mileu of pond as a favourable habitat. In water ecosystem fungi occupy the same functional status as bacteria, bringing about effective degradation. Bacteria induce B.O.D. reduction alone while fungi also set in degradation of phosphates and ammonia-N spilled through domestic waste (Patil, 1979; Heremith, 1984). Fungi possess capability of reducing B.O.D. as well as phosphates and ammonia-N (Thanh and Simard, 1973; Patil 1979; Heremith 1984). Therefore, fungi are better degraders as compared to bacteria. Presence of fungi and yeasts in the waters receiving organic enrichment is of pivotal importance in self-purification of water and can not be simply ignored as contaminants (Cooke, 1973).

The bottom sand of river Ganga supports 12 genera spread over 26 species. Literature supports isolation of soil fungi (Raillo, 1928; Takahashi, 1919; Mishra, 1986; Akhtar *et al.*, (1987) from different parts of the world. Species of genera *Alternaria, Aspergillus, Chaetomium, Cladosporium, Cunninghamella, Curvularia, Fusarium, Mucor, Penicillium, Phoma* and *Trichoderma* comprise a group of allergenic fungi isolated during present investigation. *Aspergillus* sp. and *Penicillium* sp. cause athlete's foot. *Aspergillus* sp. *Penicillium* sp., *Mucor* sp. and *Candida* sp. cause otomycosis. Diseases like aspergillosis, mucormycosis and penicilloisis are caused by *Aspergillus* sp., *Mucor* sp. and *Penicillium* sp. respectively. *Aspergillus fumigatus, A. flavus, A. nifwe* and *A. terreus* cause aspergillosis. *Cunninghamella elegans* and *Mucor* sp. induce *zygomycosis*. Certain allergenic diseases like allergenic rhinitis, bronchial asthama, hypersensitivity and pneumonitis are caused by fungal organisms like *Cunningghamella* sp., *Curvularia* sp., *Aspergillus* sp. *Trichoderma* sp. and *Fusarium* sp., Presence of such fungi in Ganga is of pivotal significance from view point of infections in both animals and human beings, thereby posing public health problems.

The fungi are also involved in self-purification of water. During investigations four fungal species including *Bipolaris* sp., *Drechslera bicolor, Botrytis Terrestris* and *Cephalosporium acremonium* were recorded and were involved in self-purification of water.

The Ganga bottom sand was also studied with special reference to keratinophilic fungi and dermatophytes. Isolation of some dermatophytes from soil has aroused interest in the keratinophilic fungi of soils during last two decades (Ajello, 1953; Ajello, 1960; De-Vries, 1962; Stockdale 1963; Ajello, 1963; Padhye and Carmichael, 1971; Ajello and Alpert, 1972; Kumert, 1972; Cabrita and Figueiredo, 1973; Ulfig and Krzysztof, 1983; Ajello and Padhye, 1974; Verma *et al.*, 1982; Jain, 1986). Recently work on distribution of such fungi has been carried out in Lyon's region (Michel-Brun and Monier, 1981); Egypt (Abdel Fattach *et al.*, 1982); Poland (Ulfig and Koroz, 1983); Rome (Mercentini *et al.*, 1983) and Spain (Calvo *et al.*, 1984). In India, keratinophilic fungal studies were initiated by Dey and Kakoti (1953). They isolated *Microsporum gypseum* from a soil sample of animal house in Dibrugarh, Assam. Present Study unravel the presence of three genera including 6 species namely *Chrysosporium evolceanui*, *C. indicum*, *C. tropicum Microsporum fulvum*, *M. gypseum* and *Irichophyton simii* from bottom sand of river Ganga at Kanpur. Dermatophytes are essentially related to diseases of cutinized and keratinized parts of animals and human beings, however, keratinophilic organisms possess proteolytic enzymes capable of degrading keratinized substrates, and keratinophilic organisms explicit competitive ability of degradation of keratin substrate in natural conditions.

The qualitative abundance of cyanophyceae has been recorded elsewhere (Fritsch, 1905; Gupta, 1957; Singh, 1961; Nair, 1965 and Shukla, 1971; 1983). However, bacillariophycean forms outweighed even cyanophyceae. Facts suggest larger number of bacillariophyceae as sequel to cater increased planktonic forms. However, number of species are not constant within fresh water environments (Morrison and Fair, 1966).

The algal flora of the river Ganga shows a distinct periodicity and succession of algae. Among algal forms belonging to various classes, some of them grew throughout the year, others during larger span of time or short-lived and only recorded sporadically. Algae exhibit preferences for periodicity and succession in different seasons. Despite seasonal preferences they may continue to thrive in the transitory periods of seasons. Qualitatively bacillariophyceae outweigh cyanophyceae, chlorophyceae, euglenophyceae, charophyceae and rhodophyceae.

Major algal infestation in lotic water exhibits qualitative diversity in time and space. On the basis of periodicity and distribution, algae may be grouped into winter, summer and rain, winter-summer, summer-rain, rain-winter, all three seasons and throughout the year as annuals, perennials or ephimerals. But during present observations major algae were observed during winter-summer, all three seasons and throughout the year. Majority of them come up at certain period, thrive and endure unfavourable conditions in the form of spores or other means of perennation. Apart from them, there are certain algae not only confined to distinct season but exist as planktons and mud or bottom forms. Epiphytic algae except planktons remain attached temporary or permanently. Predictably, attachment is through mucilage of specialized cells. There is also sequential periodicity and variance in different layers of water, making representative study more complex.

It would be of interest to mention role of minor algae in pickets and pockets of Ganga water. Certain algae observed rarely are suggestive of specialized quality of water in a particular area and such algae could be used as indicator organisms. There are certain other rare forms (*Ankistrodesmus falcatus, Coelastrum microporum, Pediastrum boryanum, P. duplex, P. simplex, Scenedesmus acuminatus, S. arcuatus, S. abundans, S. prismaticus, Stigeoclonium glomeratum, S. tenue, Achnanthes minutissima, Cymbells affinis, Cymatopleura solea, Cocconeis placentula, Hantzschia amphioxys, Melosira granulata, M. varians, Nitzschia obtusa, Anabaena iyengarii* var. *tenuis, Lyngbya putealis, Oscillatoria chalybea, Microcystis flos-aquae* and *Spirulina laxissima*) found growing in a specified place. Such algae are unknown for indication of water quality reported elsewhere in the literature (Palmer, 1969). The growth of such forms needs to be studied for their utilization as indicator of the kind and quality of the water they inhabit.

The annual changes in phycoplankton abundance and community structure occurs with some degree of predictability in temperate waters. Probably this is the best studied aspect of algal ecology. However, such findings are basic for lotic river waters in tropics. During winter season, with low temperature, low light intensities and short days phycoplankton biomass and productivity are generally low in spite of elevated nutrient concentrations. As the water warms and stratification begins to set in, the mixed layer becomes shallower and the algae are able to take full advantage of the nutrient stockpile. The algal bloom continues until nutrient depletion and increased predation signal the summer decline in phycoplankton production. However, present observations suggest that due to shallower waters during summer and winter phycoplankton growth increases for larger availability of light and nutrient concentrate. In fresh water systems, diatoms are usually the dominant species of the spring blooms and are succeeded by chlorophyceae and cyanophyceae in more eutrophic situations.

There is an evidence of seasonal variation in phycoplankton from other rivers (Kopizynska, 1980; and Nandan and Patel, 1983). The seasonal succession of algae involves the factors such as light, nutrients, temperature and grazing pressure of the water. It is likely that the decline in diatom abundance is related to the slow regeneration of silicate in the mixed layer at the metabolic level, the decline in the phycoplankton abundance during some months in calendar year may be due to decrease in photosynthesis/respiration ratio. The preceding overview, while studied in a broad sense, does not seem to explain extensive changes in abundance of species within the annual cycle. Success of any given species is likely to be unique to that species and will depend on its physiological capabilities and tolerances on the influence of other organisms preceding it and coexisting with it in the successional sequence.

Literature abounds in local floras and exhaustive texts like those of Eardtman, 1952; Woodhouse, 1935; Nair, 1962; Rao and Shukla, 1975 and others. Basic palynology concerned with morphological importance of pollen features in taxonomy has been described earlier by Nair, 1970; Rao and Shukla, 1975; Lakhanpal, 1958; Vishnumitre, 1962. However, there is good deal of advance in understanding of palynological significance of air-brone pollen. The topic on palynological importance of these small structures in spread of allergenic diseases in both human beings and cattle has been reviewed elsewhere (Nair, 1964; Tilak, 1982, 1989 and others). However, present investigation has emphasized allergenic significance of Ganga basin pollen laying stress on germinational physiology and morphological peculiarities of pollen in spread of allergenic diseases. Some of the earlier reports have described aeropollen grains of allergenic nature from Delhi, Lucknow, Jaipur, Mysore, Trivandrum, Waltair, Hyderabad, Calcutta, Bhopal, Shillong, Aurangabad, Bombay, Pune and Nagpur. There are also reports on contribution of allergenic shrubs grasses of Unnao Local flora (Shukla and Mishra, 1978). Aero-pollen allergenic calendar of Ganga basin has been prepared for the first time as described earlier. It is needless to emphasize that aero-allergins have direct implications in spread of respiratory disorders. Other allergenic pollen may cause skin or eye infections through direct contact with pollen while they are carried by water and used in human beings and cattle. In the wake of such significance aero-allergins rank very high. Recent international Aero-biological Conference (1990) has emphasized focal theme of global aero-allergenic pollen spora as of considerable significance in spread of public health hazards and control measures and number of species were observed in air spora of Ganga basin.

Aero-palynological studies have indicated that a large number of plants are important in studies of allergy. Out of a total of 900 species of plants only approximate 50-60 plants are of allergenic significance (Nair, 1964). These plants include the grasses (wild or cultivated), harbaceous weeds and trees. The grasses are the most widely distributed plants and are wind pollinated. Other species are seen flowering for a major part of the year. Although about 100 anemophilous species occur in Delhi, only about half a dozen are found to grow in abundance, produce large amounts of pollen and may be considered the important causes of pollinosis. It is believed that allergenic existence in various species of grasses is common, but not a rule. *Cynodon dactylon, Sorghum vulgare, Panicum austroasiaticum*, and others are allergenic grasses. There are several weeds, significant among which belong to the families Amaranthaceae, Chenopodiaceae and Compositae. Although, the true rag weed *Ambrosia* which is the main cause of hay fever is not found in Ganga basin. Two other members of the rag weed family (Compositae), namely *Xanthium strumarium* and *Artemisia scoparia* occur in this area. While the latter species is uncommon, the former is quite abundant and frequently causes pollinosis. In the family Amaranthaceae, *Amranthus spinosus* is perhaps the most abundant and wild spread weed. The family Chenopodiaceae is represented here by nine species hailing from six genera, but *Chenopodium album* is perhaps the most common. In the family Cyperaceae there are many species found occurring the basin of which *Cyperus rotundus* is most common. Another important member of the family is *Scirpus mautimus*, though less common, has considerable pollen output per plant.

Other important weeds which cause pollinosis are *Cannabis sativa* (hemp) and *Argemone mexicana* having abundant pollen production. The latter is quite common, produces abundant pollen and has been found responsible for pollinosis in some patients.

The rag weeds are the most important plants from the stand point of allergy and are said to account for 75 per cent of hay fever cases in the country. At the same time, it is known that in only a very small proportion pollen hypersensitiveness is exclusively restricted to rag weed.

Compared to the pollinosis caused by grasses and weeds, the one caused by trees is milder and of shorter duration. In Ganga basin cases of allergy due to tree pollen are observed in the spring and early summer (middle of Feb - middle of May) and it does not last long. The patients sensitive to tree pollen may be sensitive to two or more genera. Although certain trees causing pollinosis may not be abundantly distributed, the amount of pollen produced from a single tree is so profuse, that it may be equivalent to that produced by weeds covering a fairly large area. The height of the tree is another factor aiding the dissemination of pollen. Therefore, it is important to take into consideration those trees in the immediate environment of the patient.

Thus, the most important criterion entitles a pollen or a spore to be listed as allergenic in a certain locality in its ability to give rise to immunological response in the form of antigen-antibody reaction. The reaction is the basis of a positive skin reaction. In other words, although the abundance of a plant in a locality and frequency of a pollen in the atmosphere provide very useful clues as to the types of pollen which should be collected for preparing their extracts, it is chiefly the result of direct or indirect skin test with their extracts which indicate the causative agents in respiratory allergy in a given locality.

A total number of 51 allergenic pollen recorded from Ganga basin provide a basis of therapeutic treatment of allergy patients and also useful in recognition of allergenic pollen in other localities.

Riverine eco-system is a highly integrated ecologically complex system where the role of macrophytes is an important factor helping in maintenance of ecological balance. The macrophytes growing in river and along the banks were recorded (Shukla, 1989). Aquatic macrophytes play a pivotal role in primary productivity of river eco-system. Bilgrami and Data-Munshi (1979) have observed that biomass was directly related to productivity. Biomass was maximum at the centre where the primary productivity was greater and it was minimum where the productivity lower. Thirty six macrophytic species were recorded from river Ganga eco-system. However, the diversity of the species and their presence or absence varies from place to place. The assessment of macrophytes during entire calendar year marks the abundance of certain species in some months and shows rare appearance in other times. Comparatively, July-August show rare appearance of macrophytes with thinner growth and in February-March growth was observed maximum. Results indicate that the presence of macrophytes is witnessed throughout the stretch of river Ganga (Sinha, 1988 and Shukla, 1989). Macrophytic organisms proliferate where organic matter is present in the eco-system. Several workers (Gupta and Shukla, 1990) have recorded the worst condition of polluted river water in the stretch down stream Kannauj. The river once synonym of purity and reverence is now polluted due to industrial effluents and domestic wastes increasing with passage of time.

The macrophytic growth causes problem of organic pollution and also interacts with the riverine system modulating its biological attributes. Significance of macrophytes as a source of subsidiary protein food was reported earlier (Das, Shukla and Gupta, 1967; Shukla, 1983). Besides this macrophytic growth increase in the river is likely to cause hindrance to bathers and nagivators. Aquatic environments supporting growth of hydrophytes are best with typical conditions due to interaction of macrophytes and environment. Low oxygen content linked with high organic constituent simulate ideal conditions for growth of algae (Rao, 1953; Hicks, 1937; Shukla and Pandey, 1979).

There is considerable variance in diversity and density of fishes in river Ganga in time and space. However, twenty-five fish common species were recorded. Common edible fish of commercial value comprised *Eutropichtyes vacha, Chela laubuca, Labeo baata, L. rohita, L. calabasu, Punctius tincto, Wallago attu, Ophicembellus stritus* and *Mystis tingra*. Maximum density of fishes were also recorded during April and minimum in August. Results emphasize that algal growth is closely related to fish growth as density of fishes increases with an increase in density of algae and decrease with decline in number of algae.

Role of algae in fisheries has been emphasized by Randhawa (1960) and Patil (1960). Utilization of algae as source of fish food in fresh water fish culture has also been emphasized (Ross,

1955; Moitra, 1956; Alikunhi, 1958; Philipose, 1959; Fish, 1951; Singh, 1961; Gupta and Ahmad, 1964, 1966; Tilden, 1968; Venkataraman and Becker, 1985).

Conclusively, the totality of biological fabric of Ganga ecosystem is very complex and rich. It could be exploited in multiple ways for the dwellers of Indo-Gangetic Plains. River has a tremendous capacity to assimilate and digest enormous amounts of pollutants. While it may be anticipated that there is an increase in pollution of river as it decends from mountains to plains (Shukla, 1989), typical increase in number of species with increase in pollution marvels a fact negating usual norms of eutrophication.

The planktonic growth is inversely proportional to benthic growth with former exceeding the latter. Changes in benthic and planktonic growth along the bank or mid-stream are enormous. Surprisingly, the total number of 577 algal species observed is very large by any standards, and explicits Ganga as a vast resource of algal genetic stock.

Chapter 6

Utilization of Algae of Ganga Water

The multiple uses of algae have been described herein to focus the significance. The aim is to pin-point prospective utilization of algal flora in socio-economic, public health, domestic and industrial development of the Ganga basin.

A SOURCE OF GENETIC STOCK OF ALGAE

Ganga water shows, the presence of a very large number of algae. The genetic stock of Ganga can be used in multiple ways. The algae may serve as inoculum for raising the mass cultures used for various commercial, industrial, medicinal and agricultural purposes.

PHYCOHORMONES IN RELATION TO AGRICULTURE

Existence of growth promoting substances in various plants have been known. In Ganga water the alga like *Phormidium tenue* has been observed. It can be isolated for raising mass culture of algae for extraction of growth hormones.

A SOURCE OF ANTIBIOTIC SUBSTANCES

Out of a vast array of algae recorded some of them are known to yield antibiotic substances. Observations are suggestive of the fact that with known pockets of algal growth and time of growth during a calendar year, algae with antibiotic substances may be collected and utilized as a source for isolation of material for *in vitro* culture and subsequent extraction of antibiotic substances. Chlorellin is extracted from *Chlorella* which inhibits growth of certain bacteria. Alga like *Nitzschia palea* is known to reduce the growth of bacteria *Escherichia coli*. It has also been reported that extracts of *Cladophora* and *Lyngbya* possess antiviral properties and kill the strains of certain bacteria (*Pseudomonas* and *Mycobacterium*). *Mycrocystis* is popularly known for its inhibitory action to *Staphylococcus* and *Closteridium*. Algae like *Hydrodictyon reticulatum*, *Chlorella vulgaris* and *Oscillatoria princeps* can be also used for extraction of antibiotics.

A SOURCE OF SUBSIDIARY PROTEIN FOOD

Some of the green and blue-green algal species of digestive nature can be exploited for extraction of proteins. *Chlorella vulgaris*, *Scenedesmus arcuatus*, *S. abundans*, *S. acuminatus*, *S. bijuga*, *S. dimorphus*, *S. obliquus*, *S. prismaticus*, *S. quadricauda* var. *bicaudatus*, *S. quadricauda* var. *westii* and *Spirulina major* are the forms which could be used as a source of subsidiary protein food for human consumption because of their large protein contents.

A SOURCE OF VITAMINS

The algae are considered rich in vitamin A, B, C, D, E, K and B_{12}. The vitamins A and D are commercially obtained from the livers of shark and similar fish. They were originally synthesized from the planktonic algae particularly forming the food of the fish. *Nitzschia* rich in vitamin A, *Chlorella* in vitamin B and *Euglena gracilis* in vitamin B_{12} are found in Ganga waters and can be exploited for extraction of corresponding vitamins.

ALGAE FIXING NITROGEN

One of the roles ascribed to algae is the fixation of atmospheric nitrogen. Certain blue-green algae like *Anabaena ambigua, A. fertilissima, A. iyengarii* var. *attenuata, A. oryzae, A. spiroides, Cylindrospermum licheniforme, C. muscicola, Calothrix elenkinii, Mastigocladus laminosus, Nostoc calcicola, N. ellipsosporum, N. linckia, N. linckia* var. *arvense, N. muscorum, N. piscinale* and *Nodularia harveyana* var. *sphaerocarpa* have the capacity to fix atmospheric nitrogen. They can be used as inoculum to raise mass cultures of such algae for application in fields of rice, wheat and other crops to boost productivity.

In general, the majority of heterocystous forms are capable of nitrogen fixation. Of these, three genera of Nostocaceae (*Nostoc, Anabaena* and *Cylindrospermum*) and a few forms of Rivulariaceae are of great importance. However, only in recent years some non-heterocystous forms such as *Oscillatoria princeps* and *Gloeocapsa sp.* have also been reported to fix atmospheric nitrogen. Such algae have been recorded from Ganga water.

A SOURCE OF FISH FOOD

Role of algae as a food for fish has been well emphasized in the literature. It is known that phycoplanktons, certain filamentous and multicellular algae serve as a source for fish food. Fish exhibit preferential algal feeding. The fish may feed on variety of algae but it is not likely to digest all algae. It has been recorded that perhaps diatoms are most easily disgested by fish and serve as ideal food. Following algae growing in the river Ganga may be used in fish farming.

Ten chlorophycean (*Ankistrodesmus falcatus, Chlorella vulgaris, Scenedesmus acuminatus, S. arcuatus, S. abundans, S. bijuga, S. dimorphus, S. obliquus, S. prismaticus* and *S. quadricauda*), 26 bacillariophycean (*Cyclotella meneghiniana, Cocconeis placentula, Cymbella affinis, C. cymbiformis, C. tumida, C. turgida, Epithemia gibberula* var. *producta, Gyrosigma acuminatum, G. kuetzingii, G. scalproides, Gomphonema constrictum* var. *capitatum, G. olivaceum, G. sphaerophorum, Melosira granulata, Navicula cryptocephala, N. dicephala, N. gastrum, N. lanceolata, N. salinarum, N. simplex, Nitzschia acicularis, N. palea, Pinnularia undulata, Rhopalodia gibba,*

Synedra ulna and *Surirella elegans*) and 5 cyanophycean (*Anabaena fertilissima*, *A. iyengarii* var. *tenuis*, *A. planctonica*, *Microcystis aeruginosa* and *Spirulina major*) forms were found growing in Ganga water. Such algae, especially diatoms appear to be of considerable significance as a source of fish food.

Chapter 7

Biology of Algae and Pollution

Algae are involved in water pollution in multiple significant ways. Pollution may bring about enrichment of algal nutrients in water and may selectively stimulate the growth of few species that in turn change water quality. Certain algae are able to flourish in water polluted with organic wastes and play an important role in "self-purification" of water. The selective types of algae that exist in polluted water are used as indicators of pollution. The polluted water algae may frequently include certain forms toxic to man or animals while drinking the water or getting in its contact. The Ganga water shows a very broad spectrum of algae. The topic on biology of algae and pollution has been divided for the sake of convenience into following subheads : (a) Algae as indicators of high acidity, (b) Algae causing natural softening of water, (c) Eutrophic algae, (d) Sewage algae, (e) Algae of allergenic significance, (f) Clean water algae, (g) Pollution algae, (h) Taste and odour algae, (i) Algae causing colouration of water, (j) Toxic algae. Such broad spectrum of chapter is aimed to pin-point importance of algae in biology and water pollution.

ALGAE AS INDICATORS OF HIGH ACIDITY

Changes in pH of water due to natural causes greatly modifies the algal population. Following acid tolerant algae have been reported. Such algal indicators of high acidity include 4 chlorophycean (*Cladophora, Closteriopsis, Mougeotia* and *Ulothrix zonata*) and 1 bacillariophycean *Stauroneis anceps.*

ALGAE CAUSING NATURAL SOFTENING OF WATER

Following algae have been reported to cause natural softening of Ganga water. Two chlorophycean (*Cosmarium* and *Scenedesmus*), 1 bacillariophycean (*Synedra*) and 1 *cyanophycean* genera (*Anabaena*) are recorded involved in natural softening.

EUTROPHIC ALGAE

Algae invariably cause part or most of the eutrophication. The quantitative and qualitative assessment of algae in Ganga river shows growth of following algae indicating eutrophic state of water. Twelve chlorophycean (*Actinastrum, Ankistrodesmus, Closterium, Cosmarium, Mougeotia, Oocystis, Pediastrum, Staurastrum, Scenedesmus, Spirogyra, Sphaerocystis, Ulothrix*), 7 bacillariophycean (*Asterionella, Diatoma, Fragilaria, Gyrosigma, Nitzschia, Synedra* and *Surirella*), 1 euglenophycean *Euglena* and 6 cyanophycean forms (*Anabaena, Microcystis, Calothrix, Lyngbya, Oscillatoria* and *Spirulina*) are recorded as eutrophic algae from Ganga water.

SEWAGE ALGAE

Algae of river Ganga indicate the presence of sewage discharge in the river. Out of 37 sewage algae 17 chlorophycean (*Ankistrodesmus falcatus, Actinastrum, Chlorococcum humicola, Characium, Chlorella, Cladophora, Cosmarium, Botrytis, Glenkinia radiata, Oocystis, Pediastrum, Scenedesmus dimorphus, Selenastrum, Sphaerocystis schroeteri, Stigeoclonium, Spirogyra, Tetraedron* and *Zygnema*), 11 bacillariophycean (*Cymbella, Cyclotella, Epithemia, Fragilaria, Gomphonema, Hantzschia amphioxys, Melosira, Navicula, Nitzschia, Surirella* and *Synedra*), 2 euglenophycean (*Euglena* and *Phacus*), and 7 cyanophycean forms (*Anabaena, Calothrix, Lyngbya,*

Microcystis, Oscillatoria, Phormidium and *Spirulina subtilissima*) are indicators of sewage.

ALGAE OF ALLERGENIC SIGNIFICANCE

Algae are reported to cause allergenic diseases of skin, bronchial tract and eye inflammation. Observations reveal that out of 25 species of allergenic algae 11 chlorophycean (*Chlorococcum humicola, C. infusionum, Chlorella vulgaris, Scenedesmus acuminatus, S. arcuatus, S. abundans, S. bijuga, S. dimorphus, S. obliquus, S. prismaticus* and *S. quadricauda*) and 14 cyanophycean forms (*Anabaena fertilissima, A. iyengarii* var. *tenuis, A. oscillarioides, A. planctonica, Microcystis aeruginosa, Nostoc linckia* var. *arvense, Nodularia spumigena, Oscillatoria anguina O. formosa, O. limnetica, O. princeps, O. okeni, O. pseudogeminata* var. *unigranulata* and *O. subbrevis*) are recorded.

CLEAN WATER ALGAE

Water free from sewage water or other organic enrichment due to waste discharge usually shows the presence of clean water organisms. These organisms particularly algae are typical of the oligotrophic zone where self-purification process of water or the mineralization of the waste material is completed. Following algae have been recorded from Ganga water which indicate the cleaner quality of water at places. Two chlorophycean (*Cladophora glomerata* and *Rhizoclonium hieroglyphicum*), 4 bacillariophycean (*Amphora ovalis, Cocconeis placentula, Navicula gracilis* and *Nitzschia linearis*) and 1 euglenophycean (*Euglena spirogyra*) are recorded as clean water algae.

POLLUTION ALGAE

The variation in algal population at different points or under different conditions of organic pollution constitutes one of the indices applied to any desired location in the stream to determine the presence or absence of domestic sewage or other putrescible wastes or to measure the degree of recovery from pollution with these wastes. The algae represent a conspicuous and significant group in this continuously changing population in a stream. Algae of Ganga river indicating the pollutional load in river and nature of pollution include 20 chlorophycean (*Actinastrum hantzschii, Ankistrodesmus falcatus, Chlorella vulgaris, Closterium acerosum, Coelastrum microporum, Pediastrum boryanum, P. duplex, P. simplex, P. tetras, Scenedesmus acuminatus, S. arcuatus, S. abundans, S. bijuga, S. dimorphus, S. obliquus, S. prismaticus, S. quadricauda, Stigeoclonium glomeratum* and *S. tenue*), 17 bacillariophycean (*Achnanthes minutissima, Cymbella affinis, C. cymbiformis, Cymatopleura solea, Cocconeis placentula, Cyclotella meneghiniana, Fragilaria crotonensis, Gomphonema parvulum, Hantzschia amphioxys, Melosira granulata, M. varians, Navicula cryptocephala, Nitzschia acicularis, N. palea, N. obtusa, Synedra acus* and *S. ulna*), 1 euglenophycean *Euglena gracilis* and 18 cyanophycean forms (*Anabaena fertilissima, A. iyengarii* var. *tenuis, A. planctonica, Chroococcus, turgidus, Lyngbya martensiana, L. putealis, Merismopedia punctata, M. tenuissima, Microcystis aeruginosa, M. flos-aquae, Oscillatoria chalybea, O. chlorina, O. formosa, O. limosa, O. princeps, O. tenuis, Spirulina laxissima* and *S. major*) species.

TASTE AND ODOUR ALGAE

Taste and odour occurrence is linked with the presence of algal blooms with few exceptions noted during declines of dense algal growth. A considerable proportion of all decaying vegetation is often composed of dead algal cells. The odour produced as a result of the activities of fungi and bacteria may be either from the intermediate products formed during the decomposition or special substances synthesized within the cells of the micro-organisms. Following algae are the indicators of specific taste and odour and have been observed growing in Ganga river. Such taste and odour algae include 12 chlorophycean (*Cladophora glomerata, Chlorella vulgaris, Cosmarium reniforme, Hydrodictyon reticulatum, Pediastrum boryanum, P. duplex, P. simplex, P. tetras, Staurastrum gracile, Scenedesmus abundans, S. bijuga* and *S. quadricauda*), 6 bacillariophycean (*Asterionella gracillima, Cyclotella comta, C. meneghiniana, Fragilaria construens, Melosira granulata* and *Synedra ulna*), 1 euglenophycean (*Euglena spirogyra*) and 11 cyanophycean (*Anabaena planctonica, A. spiroides, Cylindrospermum muscicola, Gloeotrichia natans, Microcystis aeruginosa, M. protocystis, Oscillatoria curviceps, O. okeni, O. princeps, O. tenuis* and *Symploca muscorum*) forms.

ALGAE OF ALLERGENIC SIGNIFICANCE

S.No.	Name of Species	Allergenic Disease Caused	Periodicity	Quantitative Abundance
1.	*Chlorococcum humicola*	Rhinitis, bronchial asthma, hypersensitivity and pneumonitis	All three seasons at Nanamau	Common
2.	*Chlorococcum infusionum*	"	Winter-summer at Nanamau and Bithoor	Rare
3.	*Chlorella vulgaris*	"	All three seasons at Nanamau and Bithoor	Common
4.	*Scenedesmus acuminatus*	Bronchial provocations	Winter-summer at Nanamau and Bithoor	Rare
5.	*Scenedesmus abundans*	"	Winter-summer at Nanamau and in winter at Bithoor	Rare
6.	*Scenedesmus bijuga*	"	Throughout the year at Nanamau and Bithoor	Major
7.	*Scenedesmus dimorphus*	"	All three seasons at Nanamau and Bithoor	Common
8.	*Scenedesmus obliquus*	"	All three seasons at Nanamau and in winter-summer at Bithoor	Common
9.	*Scenedesmus prismaticus*	"	Winter-summer at Nanamau and in summer-rain at Bithoor	Rare
10.	*Scenedesmus quadricauda*	"	All three seasons at Nanamau, winter-summer at Bithoor	Major
11.	*Scenedesmus arcuatus*	Mortality in cattle and fish. Affected animals show symptoms like partial paralysis, loss of balance, hard stool covered with blood, reduced milk yield, general weakness and photosensitization of skin.	Winter-summer at Nanamau and in summer at Bithoor	Rare
12.	*Microcystis aeruginosa*	"	All three seasons at Nanamau and Bithoor	Common
13.	*Oscillatoria subbrevis*	"	All three seasons at Nanamau	Major
14.	*Oscillatoria princeps*	"	All three seasons at Nanamau and in winter - summer at Bithoor	Major Common
15.	*Oscillatoria anguina*	"	Winter - summer at Nanamau and in summer at Bithoor	Rare

S.No.	Name of Species	Allergenic Disease Caused	Periodicity	Quantitative Abundance
16.	*Oscillatoria limnetica*	Mortality in cattle and fish. Affected animals show symptoms like partial paralysis, loss of balance, hard stool covered with blood, reduced milk yield, general weakness and photo-sensitizations of skin.	Summer - rain at Nanamau and in all three seasons at Bithoor	Rare
17.	*Oscillatoria pseudogeminata* var. *Unigranulata*	"	Summer at Nanamau and Bithoor	Rare
18.	*Oscillatoria formosa*	"	Winter - summer at Nanamau and Bithoor	Common, Major
19.	*Oscillatoria calcuttensis*	"	Summer at Bithoor	Rare
20.	*Nostoc linckia* var. *arvense*	"	Winter - summer at Nanamau, summer at Bithoor	Common
21.	*Anabaena planctonica*	Mortality in cattle and fish. Affected animals show symptoms like partial paralysis, loss of balance, hard stool covered with blood, reduced milk yield, general weakness and photosensitization of skin and contact causes dermatitis.	Winter-summer at Nanamau, and Bithoor	Common
22.	*Anabaena fertilissima*	"	All three seasons at Nanamau and Bithoor	Common, Major
23.	*Anabaena iyengarii* var. *tenuis*	"	Winter at Nanamau	Rare
24.	*Anabaena oscillarioides*	"	Winter-summer at Nanamau and in winter at Bithoor	Common
25.	*Nodularia spumigena*	Unconciousness and rigid spasm in sheep, horses, dogs and pigs.	Winter-summer at Nanamau and Bithoor	Common

ALGAE CAUSING COLOURATION OF WATER

Algae reported to impart colouration of water in Ganga water include two chlorophycean (*Chlorella* and *Cosmarium*) and only 1 cyanophycean (*Microcystis*) genera.

TOXIC ALGAE

Following algae have been recorded from Ganga river as indicators of toxicity in water. Such toxic algae include 4 cyanophycean forms (*Lyngbya contorta, Microcystis aeruginosa, Microcystis flos-aquae* and *Nodularia spumigena*).

ALGAE IMPARTING TASTE AND ODOUR

Genus	Odour		Taste
	Moderate	Abundant	
Actinastrum	-	Grassy, Musty	-
Chlorella	-	Musty	-
Chara	Skunk, garlic	Spoiled, garlic	-
Cladophora	-	Septic	-
Cosmarium	-	Grassy	-
Closterium	-	Grassy	-
Hydrodictyon	-	Septic	-
Pediastrum	-	Grassy	-
Scenedesmus	-	Grassy	-
Spirogyra	-	Grassy	-
Staurastrum	-	Grassy	-
Ulothrix	-	Grassy	-
Asterionella	Geranium, Spicy	Fishy	-
Cyclotella	Geranium	Fishy	-
Fragilaria	Geranium	Musty	-
Melosira	Geranium	Musty	-
Synedra	Grassy	Musty	-
Anabaena	Grassy, Nasturtium, Musty	Septic	-
Cylindrospermum	Grassy	Septic	-
Gloeotrichia	-	Grassy	-
Gloeocapsa	Grassy	Septic	Sweet
Microcystis	Grassy	Septic	Sweet
Nostoc	Musty	Septic	-
Oscillatoria	Grassy	Musty, Spicy	-
Rivularia	Grassy	Musty	-
Symploca	-	Musty	-
Euglena	-	Fishy	Sweet

Chapter 8

Microbes and Self Purification

The process of self-purification of lotic waters of Ganga exhibits a high degree of retrieving capacity from pollution and in the process assimilating and digesting the pollutants spilled into the river. In this context algae play a very vital role in self-purification process.

The pollutants in flowing waters are in the form of particulate matter causing turbidity. Chemical substances of dissolved or undissolved nature show toxic effects and biological organisms cause problem of deoxygenation of water, making it unhealthy for use. Certain disease-spreading organisms are interlinked with oxygen level of the water. It may be pointed out that suspended undissolved pollutants can be removed by sieving and filteration of water but dissolved substances and micro-organisms pose severe threats and problems of pollution. The group of coliform bacteria are known to cause enteric diseases in both cattle and human beings. The growth and infestation of coliforms appears linked with oxygen level of water. Therefore, healthy waters with high oxygen contents are prone to lesser coliform menace. The group of fungi is responsible for degradation of organic matter, converting it into simpler substances. Some of the fungi cause severe diseases of allergenic nature in both human beings or cattle. Some of the keratinophilic fungi cause severe diseases of hair, horns, hoofs or nales because of their keratin degradation ability. Unfortunately, as fungi are non-green, non-photosynthetic organisms, they only consume oxygen present in water during respiration. As such they are responsible for depletion of oxygen and make water unhealthy.

However, a very large number of algal species amounting 577 are recorded from Ganga. Algae show problems of pollution as well as purification. Among problems which contribute to degradation of water quality are toxic effect of algae (*Lyngbya contorta, Microcystis aeruginosa, M. flos-aquae* and *Nodularia spumigena*), allergenic disease spreading algae (*Chlorococcum humicola, C. infusionum, Chlorella vulgaris, Scenedesmus acuminatus, S. arcuatus, S. abundans, S. bijuga, S. dimorphus, S. obliquus, S. prismaticus, S. quadricauda, Anabaena fertilissima, A. iyengarii* var. *tenuis, A. planctonica, A. oscillarioides, Microcystis aeruginosa, Nostoc linckia* var. *arvense Nodularia spumigena, Oscillatoria anguina, O. formosa, O. limnetica, O. princeps, O. okeni, O. pseudogeminata* var. *unigranulata* and *O. subbrevis*), colour producing algae (*Chlorella, Cosmarium* and *Microcystis*), taste and odour producing algae (*Cladophora glomerata, Chlorella vulgaris, Cosmarium reniforme, Hydrodictyon reticulatum, Pediastrum boryanum, P. duplex, P. simplex, P. tetras, Staurastrum gracile, Scenedesmus abundans, S. bijuga, S. quadricauda, Asterionella gracillima, Cyclotella comta, C. meneghiniana, Fragilaria construens, Melosira granulata, Synedra ulna, Euglena spirogyra, Anabaena planctonica, A. spiroides, Microcystis aeruginosa, M. protocystis, Cloeotrichia natans, Cylindrospermum muscicola, Oscillatoria curviceps, O. princeps, O. okeni, O. tenuis* and *Symploca muscorum*). On the other hand algae invariably contribute to oxygenation of water. During photosynthetic activity they consume carbondioxide producing oxygen. The released oxygen dissolves in water and increases oxygen level. Algae play a vital role in oxygenation of water. Some of the dissolved chemical substances otherwise toxic and polluting water are absorbed by algae as nutrient for their growth. Substances like calcium, magnesium, nitrate, nitrite potassium and iron are consumed as nutrients by algae during their growth. Thus, these dissolved substances get eliminated from water and brought into the body of the algae and can be removed mechanically by sieving or filteration of water.

It can be said that algae perform a good deal of self-purification of water, help in removal of dissolved polluting substances and oxygenation of water. However, some of the algae do cause toxicity, degradation in water quality and impart colouration and odour. They cause disease spread yet their capacity to oxygenate river water out weighs harmful effects. Fungi cause certain human and cattle diseases but are also excellent degradors of organic matter. Ganga eco-system presents such an excellent biological fabric that its self-purification capacity exceeds that of any other river.

Chapter 9

Organisational Structure and Objectives of Ganga Action Plan

Nature-man close interweb has been part of our ancient Indian heritage. But a strong voice to express deep concern over environmental protection was of our late Prime Minister Mrs Indira Gandhi. She mobilized public concern and made people alive to the fact that survival of mankind depended on survival and continuance of plant and animal species performing yeo-man's service in ecological balance of earth. Her attachment to the river Ganga made her enquire from Dr. Nilay Chaudhury a scientist "Can we not clean the Ganga." Dr. Chaudhury's affirmative reply prompted her to ask him to prepare an Action Plan, and if need be propose an Authority to implement this plan (Yojana June 1-5, 1985 Dr. Nilay Chaudhury). In 1980 she had asked the Central Pollution Control Board to study the sources and extent of pollution in the Ganga. The Board in 1984 prepared its first comprehensive assessment of the problem under following sub-heads:

(a) Three fourth's of the pollution of the Ganga comprised of untreated municipal sewage.
(b) Out of this 88 per cent comes from 27 Class-I cities.

(c) Only a few towns have sewage treatment facilities but even they are not functioning.

(d) Industries accounted 25 per cent of the pollution but this was far more acute in areas such as Calcutta and Kanpur (Ganga Action Plan, Ministry of Environment and Forest, Govt. of India, New Delhi).

The inventory of pollution prepared by the Board like fertilizer and pesticide use, industrial pollution input load, land use pattern, hydrological aspects, formed the basis for Ganga Action Plan document.

The first five years of the attainments of Ganga Action Plan was published in October, 1984 but was not presented to the nation formally due to assassination of Smt. Indira Gandhi. The Plan was part of the legacy of Indira Gandhi to her son, Sri Rajiv Gandhi to take up the cleaning of the entire river as a singl high priority project. In his very first address to the nation he announced the intention of cleaning the Ganga. In February, 1985, a ten member Central Ganga Authority was set up with Prime Minister as its Chairman. Chief Ministers of U.P., Bihar and West Bengal were also on the Board to ensure coordination and quick action. The CGA was formed to improve the water quality of the Ganga and its tributaries to acceptable standards through joint and mobilised efforts of various agencies like State Governments, local civil bodies, voluntary agencies and other organisations. The main functions of CGA were framed to include : (a) Formulate, promote and approve appropriate policies and programmes to achieve the objective of healthier and cleaner Ganga, (b) Examine and approve Action Plan, (c) Mobilisation of financial resources, (d) Monitor and review progress of implementation giving necessary directions to the Steering Committee, (e) Achieve objectives, meet annually and take necessary measures for its success.

The CGA under the Chairmanship of the Prime Minister comprises the following members:

(i) Union Finance Minister
(ii) Chief Minister, U.P.
(iii) Chief Minister, Bihar
(iv) Chief Minister, West Bengal
(v) Deputy Chairman, Planning Commission

(vi) Minister of State, Planning, Government of India
(vii) Minister of State, Environment and Forests, Government of India.
(viii) Minister of State, Science and Technology, Government of India
(ix) Member (Science), Planning Commission Secretary, Department of Non-Conventional Energy Sources and Secretary, Department of Environment, Government of India would be permanent invitees to this Authority.

The Steering Committee was made responsible for approval of schemes, annual allocations and review of progress. Another committee under the chairmanship of the Member (Science), Planning Commission monitors progress of schemes and reports it to the CGA.

The Ganga Project Directorate is part of the Ministry of Environment and Forests to appraise and sanction schemes and to coordinate the implementation of the Action Plan in the three States of U.P., Bihar and West Bengal.

The main agency responsible for coordinating the preparation and execution of schemes in U.P. and Bihar is the Department of Urban Development and in West Bengal it is the Department of Environment. Other agencies assisting in execution of this work are Calcutta Metropolitan Development Authority, U.P. Jal Nigam, Bihar State Water Supply and Sewage Board etc. The GPD has set up Regional Offices at Allahabad and Calcutta to coordinate activities of three State Governments and to monitor progress of work.

A State level Task Force under the Chairmanship of Chief Secretary alongwith City Task Forces and local Project Managers to periodically review progress, has been set up by the State Governments to coordinate work at the State and local levels.

MONITORING THE PROGRESS

Detailed formats, progress charts and schedules have been prepared for assessment of progress and schemes are monitored in terms of physical and financial progress so that obstacles and loopholes may be identified for remedial measures. Since the river water quality in India is assessed on the basis of standards

ORGANIZATIONAL STRUCTURE FOR THE GANGA ACTION PLAN

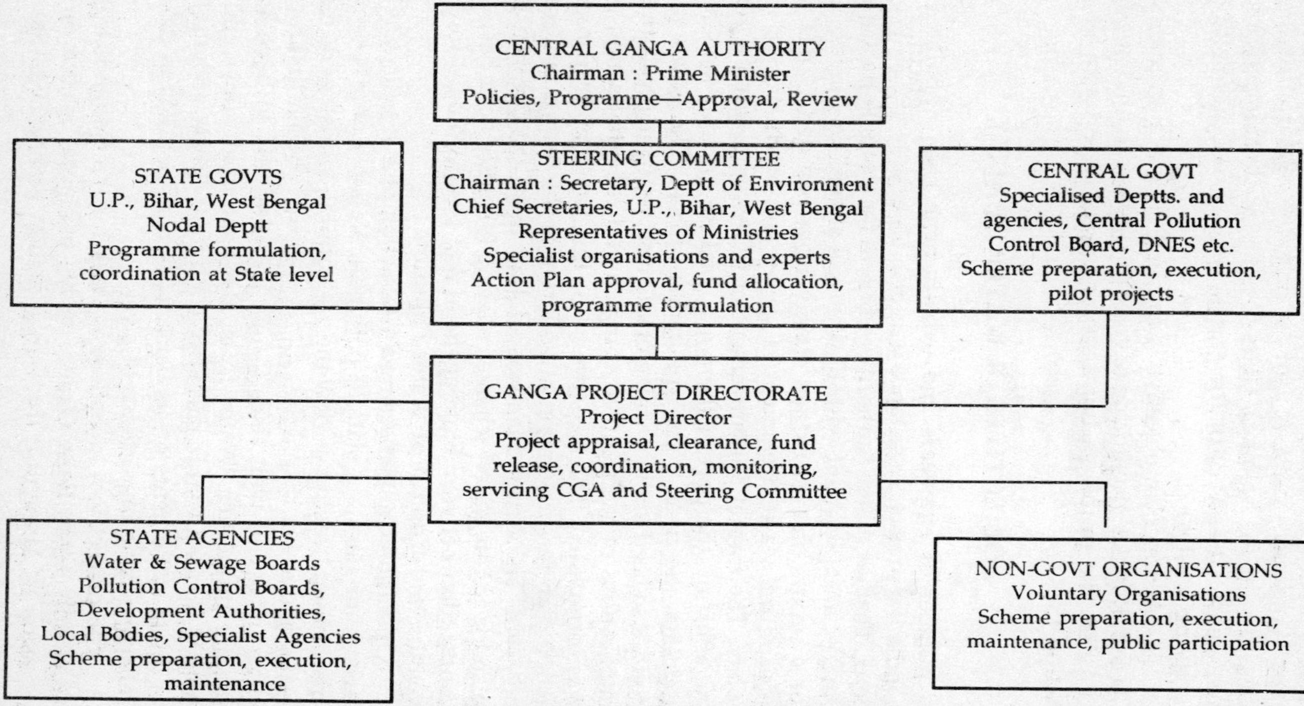

Source : Ministry of Environment Govt. of India, New Delhi (1984).

fixed for different uses the Central Pollution Control Board has fixed standards - Class-A : (fit for drinking after chlorination), Class-B : (fit for bathing), Class-C : (fit for drinking after full treatment), Class-D : (fit for aquatic life), and Class-E (fit for receiving wastes).

The Action Plan objective is to bring river water quality to Class-B. Regular monitoring of river quality is essential not only to assess the problems but also for impact assessment of the Action Plan.

OBJECTIVES OF WATER QUALITY

A number of engineers and experts have been appointed to handle the project. Assistance was also offered by U.S.A. World Bank, Thames Water Authority of Britain for cleaning the Ganga. Finally, an agreement with France was signed on June 9, 1985 for a programme to extract energy from polluted Ganga and agricultural waste, treatment of wastes and production of electricity. The formal announcement of CGA represented a genuine effort on the part of the policy makers to clean the Ganga and save the river from exploitation of man. The government decided to heed to the warning of scientists regarding hazarduous implications of the polluted Ganga and a remedial action plan was formulated. The Department of Environment prepared an action plan for prevention of pollution of the Ganga in December, 1984. The Action Plan was a follow up of the October, 1984 plan for an integrated scheme covering (a) entire river basin, (b) cooperation between State agencies and Central Board for Prevention and Control of Water Pollution.

Operation Ganga became a vital action plan for cleaning the river from alarming state of pollution. National efforts became channelised keeping in view the agricultural, industrial drinking water and sewer disposal needs. Ganga was chosen first for the clean up action plan since Ganga is considered to have some unique properties. Water from Hardwar upstream could be bottled up for a year in germ free conditions speaks about purity of urban settlements on its banks with high population density reflecting civic importance of river.

The project was a challenging task for the Ganga is India's most holy river. Rajiv Gandhi in his inaugural speech on the occasion of launching the Ganga Action Plan at Varanasi on

June 14, 1986 said:

> "The purity of the Ganga has never been in doubt. Yet we have allowed the pollution of this river which is the symbol of our spirituality. The felling of trees has caused severe floods, and silt and mud now flow into the Ganga making the river shallow so that boats cannot ply in it as they did before. Sewage and pollution from cities, industries and factories and dead animals are also being thrown into the Ganga. From now on, we shall put a stop to all this. We shall see that the waters of the Ganga become clean once again.
>
> The Ganga Action Plan is not just a Government plan. It has not been prepared for the PWD or government officials alone. It is a plan for all the people of India; one in which they can come forward and participate. It is up to us to clean the whole of the Ganga and refrain from polluting it.
>
> This programme, starting at Varanasi here today will reach out to every corner of our land and to all our rivers. In the years to come, not only the Ganga, but all our rivers will be clean and pure as they were thousands of years ago."

OBJECTIVES OF THE GAP

Since domestic wastes from 27 Class-I cities along the river were identified as major cause of pollution, the Ganga Action Plan sought to divert such wastes away from the river. The objective of this plan was to divert this sewage to other locations for treatment, and convert it into valuable energy source. The purpose was to reduce pollution by at least 75 per cent under various schemes proposed by GAP. The 2621 schemes under the proposed plan treated sewage for use in irrigation, bio-electricity, fish ponds and aquaculture. Rs. 240 crores were earmarked under the Seventh Plan for the purpose. The towns covered under the GAP are:

U.P.	*Bihar*	*West Bengal*
1. Hardwar-Rishikesh	1. Patna	1. Berhampore
2. Farukhabad-Fatehgarh	2. Chapra	2. Nabadwip
3. Kanpur	3. Monghyr	3. Calcutta Howrah and otherClass-I towns of the Calcutta Metropolitan district
4. Allahabad	4. Bhagalpur	
5. Mirzapur		
6. Varanasi		

SALIENT FEATURES OF GANGA ACTION PLAN

While the Ganga of the past was a purifier of sins, cured various diseases and main water source forming life-line of the people, the Ganga of today is a symbol of gross pollution, unhealthy water, destroyed plant and aquatic life, carrier of various germs and diseases, cause of human and animal health hazards using its water.

The main sources of pollution identified have been mainly the free-flowing and non-biodegradable effluents discharged in the form of industrial wastes, urban wastes, chemical and fertilizer surface run offs, dumped animal carcasses and of dead bodies, large scale wallowing of cattles and construction of dams, on the river and its tributaries.

Cleaning the river was essential. The Ganga Action Plan began with this stupendous task of cleaning the main river course. The Ganga in its 2,525 km length starting from Gangotri in the Himalayas to its meeting point with the sea at the Ganga Sagar in the Bay of Bengal, passes along 29 Class-I cities (population over a lakh), 23 Class-II cities (population between 50,000 and 1,00,000) and about 48 towns having less than 50,000 population.

The level of pollution also depends on the degree of dilution and velocity of the flow of water so it is necessary to maintain a minimum discharge in the river, specially at critical points where waste discharge is heavy. Due to diversion of water for irrigation purposes this objective could not be fulfilled and the entire dry weather flow is diverted to the Upper Ganga Canal at Hardwar and the rest at Lower Ganga Canal near Aligarh making the pollutant inload heavier in the region from Kannauj to Kanpur. Sixty per cent of its water comes from Yamuna, Ghaghra, Kosi and Gandak, all joining the main river at or points below Allahabad. The Hardwar-Allahabad stretch is, therefore, particularly vulnerable. The main features of the Ganga Action Plan are as under :

SEWAGE TREATMENT PLANTS

The bulk of the CGA allocation goes to the Sewage Treatment Plants which eventually not only covers its own cost but also be a profitable venture. Sewage is a very rich source of energy and manurial matter. The production of energy can be used to ensure a regular supply of power to run the pumping and treatment plants while the manurial matter including Nitrogen, Phosphorus and Potassium are marketable products. They are a good substitute for fertilizers. Thus, human sewage can be transformed into five by-products:

(a) Bio-gas
(b) Manure
(c) Poultry feed in form of Algae
(d) Aquaculture
(e) Irrigation water

The first step involves separating solids from liquids by conventional settling process. The solids then go to digester, and produce methane gas and digested solid. The gas is fed into the gas turbine to generate power. The digested solid is sold as manure. The liquid is biologically treated and is clean enough to be used for aquaculture where algae and fishes can be cultivated. Water from aquaculture flows out as irrigation water. Thus each sewage treatment plant serves as a resource recycling unit where various products are produced ready to be marketed.

Other projects include the renovation, cleaning and repairing of existing sewer systems to prevent the overflow of sewage into the Ganga. Renovation of all existing sewage treatment plants in towns along the river and installation of new plants to recover maximum possible resources to churn out maximum revenue and recover maintenance and operational costs of the plants is a must. A total of 10 sewage treatment plants are to be renovated and another 21 new plants to be constructed in different locations. Interceptors to divert the flow of sewage and other liquid wastes into Ganga are also to be constructed. Arrangement for bringing human and animal wastes from locations close to the sewage/sullage digester for treatment are to be built. Provision of sewage pumping stations at the outfall points of open drains to divert discharge from the river into the nearest sewers and treatment plants is needed. In 8 locations pisciculture facilities are to be provided. Bio-energy component is located in 5 locations and provision for bio-energy has been kept alive in 6 more locations. Treated effluents as irrigants to be improved and expanded in 24 locations are earmarked.

LOW COST SANITATION

Low cost sanitation schemes in areas adjoining the river to reduce or prevent the flow of human wastes into the river are necessary. In many towns sewerage network is not extensive. The public toilet complexes, construction of new pour flush latrines and conversion of dry latrines has been taken up under the Ganga Action Plan. Construction of 3,026 public toilet complexes and conversion and construction of more than one lakh new pour flush latrines is also to be taken up. In cities with unsewered areas or areas where extension of sewage would take a lot of time and cost, low cost sanitation schemes particularly in specific parts of the city are to be constructed to help reduce BOD in the river.

RIVER BANK FACILITIES

Development of the river front areas for prevention of pollution in the river also forms part of the Ganga Action Plan. Arrangements to prevent discharge of animal and human wastes directly from cattlesheds located on the river banks have been made. Schemes have been taken up in various towns. In Varanasi schemes for river front development, protection of slopes of ghats, construction of community toilets, improvement of lanes and by-lanes, street lighting, etc., have been taken up.

BUILDING ELECTRIC CREMATORIA

The tradition of throwing unburnt, partly burnt, ashes and animal carcasses into the river is considered to be a grave cause of pollution in specific locations. Under the Ganga Action Plan a total 29 crematoria have been sanctioned for different locations.

The sanitary conditions near the river at these locations have to be improved and the use of electric crematoria will help in saving fuelwood and will also be cost effective. The local agencies will be responsible for the operation and maintenance of these facilities. In Varanasi a scheme has also been identified as part of the city development package under the Ganga Action Plan. Arrangements for improving river traffic police to supervise and patrol area will also help in preventing disposal of animal carcasses, and other pollutant input load.

COST ESTIMATES AND PHASING, CO-FINANCING AND POOLING OF RESOURCES

The amount for execution in the Seventh Five-Year Plan period amounts Rs. 250 crores. These estimates are for Class-I cities situated on the bank of the river. The cost of certain schemes connected with the prevention of the pollution of the river is shared with other concerned agencies. Resources have been pooled from different sources like Railway Ministry, Ministry of Surface Transport, Ministry of Welfare, U.P. Government and Ganga Action Plan. Depending on the site, place and scheme various departments and ministries help in co-financing and pooling resources.

EFFLUENT AND POLLUTION FROM INDUSTRIES

Apart from civic and municipal waste, industries have also contributed to pollution of the river and its gross abuse. Industrial effluents contaminate the water and the Action Plan was directed towards industrial pollution in specific locations. An expert Group reviewed the position of 264 industrial units located along the river bank identified 68 units as gross polluters. Out of them 34 are in U.P., 5 in Bihar and 29 in West Bengal. Under laws of the country the responsibility for treating industrial effluents rests upon the industry. While some industries have been asked to establish treatment plants such as tanneries in Jajmau, Kanpur, small industries located near the river causing direct pollution in the river require plans for sewerage and treatment plants to tackle industrial effluents directly to city sewer for treatment.

TECHNOLOGY DEVELOPMENT

The Action Plan provides for establishment of pilot projects to assess feasibility of technology applications and development for the treatment of wastes and energy recovery. National Laboratories and Scientific Organisations like NERI, Nagpur; IARI,

Delhi; Central Leather Research Institute, Madras; Central Soil Conservation Institute, Dehradun and Industrial Toxicology Research Centre, Lucknow help in devising low cost technology in treatment of waste water, resource recovery through treated effluents and for training personnel for monitoring heavy metals and pesticides etc.

CONTINUOUS RESEARCH IN RELEVANT AREAS

A research committee has been constituted to aid and adivise the Ganga Project Directorate towards promotion of research to formulate holistic policies for conservation of the river resources. State level Committees, Steering Committee, etc., have all a role to play in implementation of Ganga Action Plan through Scientific Research. Several workshops relating to areas relevant in the Ganga Action Plan have been held from time to time where scientists and technologists interacted for assisting the plan. Despite all this the Ganga Authority Project could never evolve detailed objective research areas and the physico-chemical and biological profile investigated could not be used for applied exploitation of river. Hetrogenity of research projects, non-implementation of grass-root recommendations of First Phase coupled with allocations to certain institutions and politics in Ganga Action Plan eludes applied exploitation and restoration of water purity.

SCIENTIFIC CONTENT OF THE PLAN

The plan aims to secure a better knowledge of the river, its physico-chemical and biological profile, technology development and application in waste water treatment, and research in resource recovery mechanisms. Interaction was made possible among scientists and technologists from 14 Universities under University Research Plan of GPD in area of hydro-biological profile of the river, yet its applied exploitation is yet to be accomplished.

PROGRESS OF WORK

Under the first phase of the GAP as many as 261 schemes were sanctioned at a cost of Rs. 304.85 crores. This cost was later revised keeping in view the State-wise break up, the GPD expenses and expenditure on research amounted approximately Rs. 358 crores. The categorisation of schemes mentioned earlier led the plan into action. By the end of the first phase 159 schemes were completed—69 in Uttar Pradesh, 31 in Bihar and 59 in West Bengal. The remaining 102 schemes in various stages of completion are likely to be completed by 1993.

IMPACT OF WATER QUALITY

The main objective of GAP has been improvement of water quality of Ganga by interception, diversion and treatment of about 873 mld of sewage of 25 Class-I towns on its banks. The purpose was to reduce industrial pollution along the Ganga by enforcing pollution control laws and implementing low-cost sanitation measures and establishing electric crematoria.

LEVEL OF SEWAGE

Sewage flow into Ganga is about 1340 mld, and the flow of industrial effluents is estimated at 260 mld. Other polluting acts like mass bathing, washing clothes, pesticide and agricultural run off and carcass dumping add to both organic and bacteriological pollution. The first phase of the GAP does not take into consideration the other aspects of pollution abatement, along with pollution inload from tributaries as well as Class-II towns along the river Ganga.

RESTORATION OF THE BIO-DIVERSITY AND PRODUCTIVITY

The acts of riverine population has been the main reason for reduction in river flow, water pollution, sedimentation of the river bed and habitat destruction due to gravel and sand abstraction. The GPD has constituted a Fisheries Expert Committee comprising of scientists and officials to assess and identify the status and causes of decline of aquatic life and also revive its productivity. The report of Committee has led to the establishment of two expert consultants to prepare a Model Draft of

Fisheries Bill and also to suggest measures of technical and legislative nature to conserve the genetic resources and habitats and improve productivity of fish in the river.

IMPROVED WOOD CREMATORIA

The basic difference between the ordinary wooden crematorium platform and the new design is that instead of keeping the body on the ground and stacking the entire wood on the concrete platform, the body in new design is kept in a restricted space raised above the ground on a fire/brick or cast-iron cradle in a way that there is a free air circulation ensuring better combustion, efficiency and saving of upto fifty per cent of the quantity of wood normally used with reduction in the burning time by half. The GPD sanctioned electric crematoria at 34 locations. But due to their high cost, against the proposed design cost of only Rs. 8000 per unit as well as the problem of departing from traditional method, electric crematoria could not be installed. One hundred units of the kind were to be constructed in the towns along the Ganga (35 each in U.P. and West Bengal and 30 in Bihar) amounting to expenditure of Rs. 8 lakhs to be borne by the GPD.

TRAINING PROGRAMME

Since a large number of treatment plants, and pumping stations are to be operated and maintained skilfully for proper implementation of the Ganga Action Plan it is imperative that technical training be imparted for the operation of these plants. Institutes have been selected and the training programme already initiated to the following categories of personnel :

(i) Decision makers/senior management personnel
(ii) Middle level engineers
(iii) Plant operators.

The NEERI, Nagpur has also been involved with this training. A Waste Water Training Institute is to be set up at Kanpur with the help of NEERI. Till this institute comes up, training courses have been held from time to time at Nagpur, Varanasi and Calcutta for category I, II and III personnel employed on operation and maintenance works of the implementing agencies. For the training institute funds will be provided under the Indo-Dutch Bilateral Programmes.

UASB (UPLOW ANAEROBIC SLUDGE BLANKET) TECHNOLOGY

This technology recently introduced brings about treatment of waste water through a layer of micro-organism in an air tight reactor. In this reactor bio-gas generation is ensured, apart from the sludge which can be used as manure. The other advantages are no bad odour, flies and mosquitoes, less land area required, and reduced sludge pumping and drying costs. The process is cost effective and operational with low maintenance cost. The experimental Jajmau Plant of 5 mld has been found successful while 3 plants of 14 mld capacity at Mirzapur, 36 mld capacity at Kanpur and 8 mld capacity at Chapra in Bihar are under installation.

INTERNATIONAL COOPERATION

International cooperation has been provided through various agencies like the Overseas Development Administration Programme, World Bank and assistance from Netherlands. Technical assistance has been provided by ODA Thames Water International of U.K. in the areas of water quality monitoring and modelling, sewerage and sewage treatment systems and training. The contract commenced in February, 1990 with £11,42,000 sanctioned by ODA for this purpose. The Indo-Dutch collaboration has been taken up in cities of Kanpur and Mirzapur for integral sanitation programme. Interception diversion and treatment of sewage and tannery waste water, water supply, solid waste management, community participation and health promotion have been taken up at an exchequer of 50 million Dutch Quilders (Rs. 50 crores) to be completed by 1993. The scheme of common conveyance and treatment in Kanpur is part of this sanitation programme. The assistance from World Bank constitutes Rs. 63 crores as part of the U.P. Urban Development Project and the money is utilised for pollution control works, purchase of materials and equipment and training. The Project began in

December, 1987, and is expected to be completed by March, 1996. The progress has been poor and unsatisfactory due to extended litigation regarding sewage treatment plant contracts in Allahabad and Kanpur. However, the contracts have now been accorded.

PUBLIC PARTICIPATION

Another objective of the GAP is to achieve awareness of the people and their participation in implementation of the Action Plan. Its purpose is to make the programme a truely, people's affair. Involvement of public was sought and awareness created through media, exhibitions, posters, seminars, *shramdans*, *padyatras* and hoardings, etc. This kind of programme is facilitated through voluntary organisations. NGO's, school and college students and pilgrims; youth organisations like NSS, NCC, Scouts and Guides; Nehru Yuvak Kendra, Yuva Mandalies and other organisations. Certain educational kits are designed by Centre for Environmental Education, Ahmedabad for the high school children in the form of Quiz competitions, essay contests and debates on simple water quality monitoring. Involvement of pilgrims awareness about plantation of saplings during Kumbh Melas and other religious festivals have been undertaken. Screening of films in the three States, *Seva Shivers* and *Shramdan* to keep the river banks clean and afforested are also encouraged.

The CGA has made arrangements for the operation and maintenance of the schemes of the Ganga Action Plan. The State Governments will make necessary provisions in State Budgets for the treatment plants, pumping stations and intercepting sewers. Regarding other schemes like River Front Facilities, Electric Crematorium and Community Toilets, the State Governments had agreed that they will be self-sustained or will be subsidised. However, it has been found that in spite of these decisions of defining the responsibility with regard to bearing the 0.8 M cost the State Governments have not been fulfilling their obligations. Gross irregularities of diversion of Government of India's funds meant for execution, operation and maintenance has been committed. (GAP 20th meeting of Steering Committee February, 1991, Government of India, Ministry of Environment and Forests.) As regards money utilized on completed schemes a mechanism to monitor proper functioning of the schemes has to be devised.

POLLUTION LOAD SURVEYS

As per the decision of the 18th Steering Committee Meeting, it was decided that the State Governments would carry out short term, one time pollution load field surveys for total outfall into the river Ganga. The quantity of waste-water flows and sewage characteristics are to be measured in these surveys. CMDA in West Bengal, Bihar and Uttar Pradesh entrusted this work, to the NEERI and on the basis of these reports the Eighth Five-Year Plan priorities have been fixed, finalised and reviewed.

RESOURCE RECOVERY

Resource recovery constitutes a major component of the GAP as it is expected to meet a substantial part of the 0.8 M expenditure. It has been found that presently biogas is being generated at four STPs at DLW, BHW (Varanasi), Jajmau (Kanpur) and Saipur (Patna). While bio-gas can be usd as an alternative to save natural fuel, the Chairman of CPWB stated that other means of resources recovery like pisciculture and algal culture in the low cost treatment plants is not receiving the desired attention.

INDUSTRIAL POLLUTION

With a view to take measures for pollution abatement along the river from industrial effluents many polluting industries have been brought under the purview of Environment (Protection) Act, which takes serious action against industries not setting up Effluent Treatment Plants (ETPs). But on the other hand many industries are still not under its purview due to corporate lobbying, political and vested interests. Such industries are continue to still pollute as they have obtained stay order from High Courts. The cases of may tanners in Jajmau (Kanpur) may be cited in this regard. Many tanneries have also been closed down by the UPPCB for not abiding by pollution control acts. A proposal for clusters of small scale polluting units for the consideration of joint funding based on the principle of "polluter pays" is also being expedited by State Governments. There were 68 gross polluting industries as on December, 1990 and 41 has set up effluent treatment plants. Legal action was initiated against defaulting units by CP Division of Ministry of Environment, the Regional Directors of GPD and the respective State Pollution Control Boards. The SPCB has also been directed to submit data on industries polluting in the Ganga Basin including the tributaries. In June, 1990 the Ministry sanctioned a scheme for promoting common treatment plants for clusters of small scale polluting industrial units where individual ETPs are not techno-economically feasible. Central assistance upto 25 per cent of the capital cost or upto Rs. 25 lakhs and in special cases Rs. 50 lakhs will be given for each treatment plant provided the concerned State Government gives its matching share (GAP 20th meeting of Steering Committee, Government of India, Ministry of Environment and Forests).

The organisational structure and objectives of Ganga Action Plan seem well conceived to restore sanctity and purity of the holy rivers. However, in a massive plan of this kind administrators, bureaucrats, scientists, technologists, social workers and common man are involved weaknesses and caveats of the plan have many pathways and entries. The success and failures of organisation is conditioned by the fact how sincerely and honestly various groups and agencies involved work in unison.

Chapter 10

The Ganga Action Plan: An Overview

ACTION PLAN

Involvement of State Governments divided the interest and responsibility and was a set back to project. The plan suffers from a number of flaws largely due to the haste in drafting and conceiving the plan. It appears not the CGA, but State Governments have been entrusted with the responsibility of treating effluents accounting for the major share of pollution in Ganga which, therefore, logically figures as major component in project to clean the river.

Apart from this other foundational factors not envisaged and considered in plan were other reasons of CGA failure like deforestation in the Himalayas from where the river originates and resultant consequent siltation. Due to continuous deforestation billion tonnes of top soil is washed away into the sea by our river systems leading to rise of the river bed, thus, enhancing flood areas. In areas of heavy rainfall in the North-East like Cherrapunji a place of heaviest rainfall in the world there is acute scarcity of water for months. The devastation of natural resources is most in Himalayan region. The main contribution of the Himalayan eco-system to the Ganga Basin has been its perennial water resources. The mountains through orographic influence induce precipitation of water from the atmosphere and in forest cover converts seasonal rainfall into perennial water resources. However, the hydrological role of mountains has been ignored as they are considered sources of unexploited raw materials. The most widely known people's movement against this kind of ecological destruction has been in the Doon Valley and Mussoorie hills against limestone quarrying and commercial exploitation of timber leading to total denudation of hills.

Water is the essence of earth and plants are the essence of water said the Chandogya Upanishad. To save this ecological havoc of forest plunder GAP should have also considered of replenishing forest belts through afforestation to save the whole Ganga basin turn into a polluted drain leading to desertification. The priorities of afforestation have not yet been allocated by States so far. It has been recommended that the Agricultural Finance Consultants Ltd. should draw an Action Plan for all the three States for afforestation along the river Ganga to reduce soil erosion and siltation. In view to sustain the flow of Ganga the Himalayan eco-system has to be preserved. A concern for Ganga also demands rethinking on development projects such as the Tehri Dam, proposed Thermal Plants on the river banks at Uchhalar and Kahalgaon, Atomic Power Plant at Narora each having potential to pour loads of poison in the river.

There appears to be a credibility gap between performance and the action plan punctuated with instances of confused thinking. During the Kumbh Mela in Hardwar, diversion of sewage to a suitable location without simultaneous installation of treatment plant for resource recovery, utilisation of effluents for pisciculture, algae culture, bio-gas was a crushing disappointment in spite of governments plan for ecological safeguards. Such diversion of toxic effluents without adequate treatment can cause seepage into ground water and has disastrous after effects in the region.

The fundamental question today is that the Action Plan should pay due attention to various issues like augmentation of flow of water during lean season, elimination of pollution at source, coordination in totality so that sub-plans are feasible and merged into an integrated programme. The exclusive emphasis presently is on domestic and industrial sewage and sullage and not phasing out or shifting polluting industries or reducing it at source.

The plan needs rationalisation, and selling water to industries at remunerative rates to recover costs. Green belts on both banks in urbanised areas to prevent pollution needs to be incorporated in the plan. Execution of the plan depends upon five major development projects to be constructed on the banks of the river.

OVERVIEW

The efforts to clean up the Ganga is an acid test for the water pollution control strategy in India. The work is of a gigantic nature because of constraints on technological capability and financial resources. Problems regarding maintenance of a minimum flow, especially, at critical points along the river with large industrial and urban settlements will determine the degree and velocity of flow of water.

Resource recycling and the clean up operation involves coordinated efforts of various agencies. The CGA with its skeleton technical staff can ensure quick decisions but it's ultimate implementation as well as monitoring depends on concerned State Water Pollution Control Boards responsible to supervise and watch with vigilance the biggest clean up operation undertaken in India.

Another problem that the CGA needs to consider is the working of the pumping plants to stop discharge of sewage and sullage into the river. It is believed that this stoppage can reduce pollution of the river to a tune of more than 80 per cent. But due to improper maintenance and operation the sewage overflows into the river.

The Ganga Action Plan envisages improvement in sanitation and lay out of domestic urban drainage system in a way to reduce pollution load into the river using alternative technology and sanitary methods for disposal of human excreta. The treatment methods have also been suggested to harness recovery of utility products from wastes and generate energy to recover the cost of treatment plant. As an alternative, bio-gas production for use as domestic fuel and energy for lights, fertilizers and wastewater supplement enriched with nutrients for irrigation of croplands and use of recycled waste water for pisciculture supplicate for better utilization of river water.

Purity of water is interlinked with nature of rocks, forests, and rainfall and nature of eco-system acting jointly to provide a complexion to the river water. But this is a complex issue and diverse parameters form the canvas of river water. Despite scientific literature in area of river water quality, there has been a lack of information, data and effort on part of CGA to fill this vacuum of utmost importance which shapes the complexion of river as it comes into being in the hilly terrains of Himalayas. Significantly, it may be added that the nature of water as the river originates perpetuates itself in the subsequent flow unless otherwise changes are made due to negative stresses.

While there are several kinds of pollutants drained into the river in the form of liquid wastes, industrial wastes, solid wastes, agricultural washouts and mixture of dissolved gases, the GAP focuses only to the major parameters of pollution, visualised as sewage and sludge. However, it needs to be remarked that other kinds of pollutants should not be ignored because they together form a complex syndrome likely to recoil with greater vengeance than single sewage factor. Admittedly, the old belief that Ganga waters are immune to extraneous pollution and maintain purity has been partly belied because the external pollutant input encompasses the assimilative limit of river.

Corporate interests invariably deepen our doubts about the implementation of the GAP. The meeting organised by the Central Pollution Board to discuss industrial pollution have always been marked by absentees and poor representation both from private and public sector undertakings. The apathy of the industry is evident and its only interest lies in the incentives of government for anti-pollution measures. It should be alternatively suggested that supply of water to industries should be purely on payment basis and the amount of money generated could be used for supporting ways and means to undertake maintenance of standard of water quality. Non-payment should be made punishable. Incidentally, it may be mentioned that industry receives water on nominal payment of taxes but illegal supply of water and utilization of other sources of ground water remain still untaxed as water is a cumulative commodity with a system of existence in the form of ground water, river water and rain water. It is indicated that State Governments have been advised to review their water tariffs to help augment the local resources, promote recycling, minimise wastes and pollution and a methodology for pricing of water for industry be evolved as fixed by the Bureau of Industrial Costs and Prices Study. Over utilization of any of these kinds of water effects the river waters. As such there is a need to apply controls on use and abuse of water by industries.

The CGA Plan of Ganga cleaning is based on pattern of Thames river but problems of Thames and Ganga are entirely different. Thames suffers from pollution spillage mainly from steam ships and sewage whereas problems of Ganga are more of domestic and industrial spillage. The eco-systems operative in the two rivers are also different. One flows through temperate, while the other tropical land. The socio-economic backdrop and cultural practices are also variant. Industrial belts are mostly situated away from the Thames river. The river has an evenly distributed rainfall throughout the year without any leap and lean season. The tropical system is highly fragile due to human interference. The Thames is very near the sea and open to tidal waves and enables cleaning easier. It is difficult to discern how problems of such diverse nature in the two rivers could be tackled at par.

States face the problems of acquisition of land for the project as land owners have moved to the court and land acquisition is pending on account of delay in verdicts of courts. But in public interest it becomes imperative to obtain such land and acceleration in disposal of court cases to acquire the land needs attention.

Apart from this the importance of sewerage and drainage in large towns has to be taken care of because schemes for expansion of water supply finds 85 per cent of the water back in the form of sewage. Mostly existing drainage facilities are choked in almost all the towns and need immediate attention. The State Governments are entrusted with this task but improvement in drainage appears to be moving at a snail's pace.

The maintenance of minimum flow of Ganga also needs to be revamping. A river management committee for both the Ganga and the Yamuna has been set up by the Ministry of Water Resources. The CGA agreed in principle to the construction of a gated barrage at Kanpur to maintain minimum flow in the river and also to divert requisite quantum of water towards the bank on city side. This plan has not been able to make much headway as Ministries of Environment and Forests, Resources and Urban Development have expressed inabilities to provide Central assistance of Rs. 50 crores sought by U.P. Government for this project despite the U.P. Government's offer to spend Rs. 100 crores out of its own funds. The cost escalation was not committed upon. The project has now escalated over Rs. 200 crores. This project also has other benefits like a road bridge, reclamation of land, water supply and control of floods damages. Yet project cost has not been geared up and this important plan is still pending implementation and execution.

The greenbelts on both the banks from Josimath to the Gangasagar island is treated as a preserve of the forest department. This entails only piecemeal jobs or patch work and causes ecological damage. This damage should not be inflicted on nature as it will result in ecological death. Pollution Control Boards in India have not been able to enforce the existing penal provisions.

Growing urbanisation needs to be contained. Degraded land resources have also been ejecting a large number of people out of the rural areas adding to the urban population pressures. These factors have a major contribution in keeping the Ganga clean. The Himalayan forests, need restoration. The muddier water without phyto planktons, suffers from oxygen deficiency and renders the river biologically dead. Siltation caused continuous floods, filth and public health hazards. The diversion of Ganga waters at Hardwar and near Aligarh have reduced the flow of the river to the extent that during the lean season it can be crossed walking at various places. Such processes need to reversed to keep the river free from pollution. Incentives like award for meritorious work under GAP for officers needs encouragement.

Chapter 11

Fiscal Analysis of Ganga

The literature documents the significance of Ganga and its linkage with the economy of Indo Gangetic Plain. Admittedly, the fact points out the economic value of Ganga as a river. Due to the exploitation the river Ganga was reduced to a completely deteriorated flowing water and attempts to restore the water quality have tantamounted to an expenditure of a colossal amount of investment in the Government sponsored GAP. Like all business trade and industrial endeavours national projects of GAP dimension should be evaluated for the economic give and take. The nation must know what is the fiscal accounting involved in the Ganga Action Plan. It may be remarked that the Government must function as an efficient accountant to keep a balance of expenditure and income of any national project.

Swami Chidananda of Divine Life Society rightly pointed out "We should not think of Holy and Divine Mother Ganga in terms of kilowatts, in terms of hectares of irrigated land, in terms of drinking water facility for any city, whatsoever it be. It is a very narrow way of understanding. Ganga is not only a holy river but is most important for national integrity and cultural unity and the oneness of our nation." In addition to this, should the Ganga be allowed to flow polluted? The public health hazards it causes would also mean enormous amount of money being spent on control and remedial measures of human and cattle diseases. It becomes, therefore, apparent that the expenditure on Ganga cleaning operations churns out dividends in the form of saving on public health expenditure as well as feeds water requirements and a host of biological products of the river. Above all the pains, sufferings, displeasures which the river would cause because of its low level water quality cannot be calculated in terms of money. Human sufferings are difficult to be quantified fiscally. Yet they have paramount significance in social upgradation. This multifaceted canvas of Ganga is beaded with financial units on both expenditure and income sides, while calculating economic give and take of the river. However, it may be remarked that despite fiscal analysis of environmental protection being in a nascent phase of development, attempts to work out fiscal analysis of Ganga seems of utmost priority (Asthana and Shukla, 1992).

DEPLETION AND DEGRADATION OF WATER RESOURCES

The economic growth and population multiplication constantly strains our water resource base. The growth and developmental activity is admittedly, doing enormous loss to the river waters and the external cost of the water quality degradation will have to be borne by humanity now and in future. Historically natural water resource is assumed as a gift of nature. This erroneous thinking pre-empts economic assessment of natural resource depletion. Instead of considering depletion and degradation of natural resources equivalent to man-made assets that can be regenerated and made again, natural resources include certain assets which if exploited cannot be regenerated back.

A shift in the Ganga river from Kanpur towards Unnao due to excessive utilisation of water, or change of river bed area or its drying up and depletion in water quality may tantamount to extermination of this water resource in the region. Accounting of costs and benefits in exactness should include income and drawing down capital assets for resource degradation.

While the Government has made efforts in protecting our environment and natural resources through policies and conservation projects the acceptable rates of return in the context of stock of renewable natural resources is rarely considered in a systematic and comprehensive way at the planning level.

QUANTUM OF INCOME AND EXPENDITURE IN GAP

The GAP was projected and planned for its grand period of growth and activity during Seventh Plan of Government of India. This period stands as the baseline for futuristic harvesting of dividends from Ganga Action Plan. Seventh Plan period (1985-1990) showed an estimated total cost of works amounting to about Rs. 250 crores spread over as detailed below:

Year	*Amount (Rs.)*
1985-86	30 Crores
1986-87	50 Crores
1987-88	60 Crores
1988-89	60 Crores
1989-90	50 Crores

Incidentally this time frame coincided with the second half of United Nations sponsored International Water Supply decade and Sanitation decade. It got support through the UNDP

	Annual Per Capita Expenditure on Pollution Control	Rs.		Annual Per Capita Revenue	Rs.
1.	Power for Sewage Pumping	0.10	1.	Sale of electricity generated	12.00
2.	Supervising, operating staff and establishment charge	1.00	2.	Sale of manures	3.00
3.	Cost of diesel for operating generators	16.80	3.	Algae and additional fish produced	12.00
4.	Annual repairs and replacement of mechanical components	3.30	4.	Water hyacinth revenue	5.00
5.	Annual repairs and replacement of Civil work	0.70	5.	Final effluents sold	5.00
6.	Depreciation on mechanical components	1.10			
	Total:	Rs. 23.00		Total :	Rs. 37.00

Source : Data Based on Executive Summary of Action Plan for Prevention of Pollution of Ganga 1984.

Global resource recovery project of the World Bank. This data shows exponential increase in expenditure with passage of time during Seventh Plan. It has to be borne in mind that with this increase in expenditure it was expected that some income will start generating due to development of Ganga as a water resource for generation of output in multiple ways. The amount of Rs. 250 crores got inflated while creation to Rs. 262 crores upto 1990 in the Seventh Plan. Another 77 crores is anticipated till 1993 end (GAP, 1990). The per capita annual revenue and expenditure lay out is given on previous page.

Above figures exclude amounts spent on public health with special reference to medication, control of diseases or remedial thereaupetic treatment of patients. It also does not reflect the quantity of increased pleasure, happiness and longevity of life. Nevertheless these facts are ultimate while analysing the fiscal give and take in the maintenance and exploitation of Ganga waters. Increased pollution raises the cost of disposal of waste. It harms public health and depletes enjoyment of nature's endowments. Blades (1989) differentiates pollution into output of pollutants, the damage of pollution, the costs of abatement and benefits from adoption of abatement. In view to measure market value of benefits of pollution abatement there are practical difficulties and their quantification in national accounts difficult. Hueting (1989) discusses pros and cons of various options and alternatives to standardise quantifications of these values in SNA framework but the correct depletion, degradation or natural input to sustain industrialisation, growth and economic activities are still beyond callibration. While the UN-SNA framework was basically desired to calculate National Accounts on Input-Output structure as undermining feedback for National Economic Accounting, Peskin (1989) advocates four sectors with introduction of "Nature Account" in addition to standard accounts for household, industry and government. There does arise a difficulty on how natural capital is to be treated and attaching value to environmental services and damages.

In the backdrop of riverine waters and a fiscal accounting of our natural capital water resource—the Ganga it may be pointed out that pains, pleasures, sufferings, disease spread, longevity of life cannot be quantified in absolute terms. The socio economic and public health problems cannot be quantified in terms of money but expenditure before and after upgradation of Ganga waters could be worked out in terms of amount spent on public health. This would depict in close approximation the fiscal accounting of Ganga water quality maintenance. Since water is elixir of life even if higher amounts are involved in cleaning than net dividends, public health improvement should be taken as enough dividend to undertake proper cleaning of the river.

It may be pointed out that cost analysis procedure of environment mentioned above with special reference to the UN-SNA system and Peskin model are only pointers for framing a system for fiscal analysis of river waters.

PROCEDURE FOR FISCAL ANALYSIS OF RIVER WATERS (Shukla and Vandana model)

The natural account of Peskin's model has been modified in the backdrop of Ganga waters for analysis of fiscal accounting as under:

GANGA NATURAL ACCOUNT

Input	*Output*
1. Power for sewage pumping	1. Sale of Electrcity
2. Supervising, operating services and establishment charge	2. Sale of manure
3. Expenditure on pollution control measures	3. Algal production
4. Expenditure on disease spread	4. Bio mass production (Hyacinth, fish etc.)
5. Expenditure on medication of mass disease	5. Effluents sale
6. Expenses on drinking and industrial water supply.	6. Less expenditure on disease control and medication
	7. Taxes received from domestic and industrial water supply.

Net Ganga Water Benefit
Sum (1 to 7) – (1 tp 6)

Above model need be applied after collection of above feedback of data collected by GAP for arriving at fiscal analysis of Ganga waters, but entire data is not available in the reports of GAP.

A developing country like India cannot ignore environmental fiscal accounting in their policy making yet does not have the means to work out such calculations. The accounting of such assets leading to computation of GNP and GDP through integration of environmental depletion and degradation has to be formalised. Unless the accounting of pains and pleasures due to disease spread, public health hazards, depletion in scenic beauty are established (whether through indirect methods), unless regeneration of water cost on its depreciation value is calliberated and accounted in the depreciation of man-made assets at the expense of water resources fiscal analysis of Ganga water will be a farce dream.

Aforesaid clearly pinpoints to the fact that fiscal accounting apart, cost and effectiveness of control measures, risk-cost comparison, industrial impact analysis, economic implication analysis, cost benefit ratio and Ganga Water impact analysis are the focal areas which need special emphasis *vis-a-vis* fiscal accounting of this riverine ecosystem. The analysis would explicit the areas of money input and output for unquantified parameters. This would bring fiscal analysis translating apparent fiscal account in near absolute terms. The analysis of riverine eco-systems may be discussed briefly as given below -

RISKS AND BENEFITS

Protection, maintenance and upgradation of riverine waters accrues technical analysis as to how far benefits could be derived while exploiting the water without involvement of any risk. It may be mentioned that dealing with one kind of risk leads to manifestation of other risks in the society. For *e.g.*, it could be said that decision to prevent growth of certain fungi or algae to check biological pollution may lead to an adverse effect and risk to fish productivity. Likewise the decision to restrict impregnation of certain chemicals like calcium or nitrogen compounds may effect adversely the micro-organism growth and thus impair the self-purification of water. The experience of developed countries in the regulatory operations of riverine pollution may provide answers to several such questions yet there are specific problems of Ganga Eco-system needing study to devise measures for deriving benefits with minimal risks.

COST AND EFFECTIVENESS

There is seldom use of regulatory measures, based on sound cost and effectiveness. Usually available control techniques are used without calliberation of risks but with a special emphasis to the cheap control device. However, technical effectiveness and cost viability should be the guideline of approach while dealing with abatement of pollution.

RISK COST COMPARISON

While cost effectiveness is ascertained risk cost ratio becomes vital in improvement of water quality. The cost usually tends to be universely proportional to risk because new advanced technologies are costlier but lesser risky. While following methodology for reducing risk it should be the endeavour to adopt the cheapest methodology available. Alternative technology without hazards implications or technology friendly to water quality need be adopted. There are numerous variations that can reduce economic impact and can have lesser deleterious risk involvement.

ECONOMIC IMPACT ANALYSIS

Quality control of river waters in many countries is more of a financial burden on its economy, but sophisticated macromodels of economy are used for calliberation of such programmes. Countries like France, Japan, Netherlands and the USA have shown little impact because they can afford to invest for cleaner supply of water for domestic and industrial use. However, developing countries like India can ill-afford only an expenditure greater than 2 per cent of GDP. It becomes, therefore, important that while dealing with project like GAP cooperation of Planning and Finance Ministries should be sought to keep alive technical feasibility and economic viability of the project.

BENEFIT - COST ANALYSIS

Gains and expenditure form usual units for deriving cost benefit ratios in economic analysis and this is by far adopted in most of the methods for fiscal accounting. But still unquantified parameters of social costs and public health paradigms reflect the drawback in the system in fiscal accounting of Ganga waters.

DISTRIBUTIONAL ANALYSIS

This analysis dwells in the area of quantum of investment which could be shared or allocated to different sections, trade units, industries, individuals or population of individuals. The distribution of benefits of health and water protection programmes should be related to risk assessment. Estimation of distribution of costs and monetary benefits or dividends of biological, physical or chemical productivity is difficult to predict and needs a trained economist who is capable of calliberating distributional patterns of cost benefit parameters.

ENVIRONMENTAL IMPACT ANALYSIS

The EIA is an assessment of the short term and long term effects of any proposed action or inaction covering policy planning, legislative proposals, projects like dam contruction, waste disposal, sewer pipe construction, and other operational practices. Environmental impact assessment tries to identify the probable undesirable events that can occur and ways of avoiding it. It has to be based on considerations of environmental, economic, social and policy planning and then evaluated.

Thus problem of fiscal analysis of river waters in developing countries needs cumulative consideration of the above techniques before evolving a final model for analysis.

THE POLITICS OF ECONOMICS

Economists could do precisely little towards incorporation of wants and needs while maintaining and harnessing quality of Ganga waters. Ganga Water Accounting could not attain perfection in absence of quantification of public health hazards, pains, pleasures, happiness and joys on one hand and expenditure on providing such public amenities like personal expenditure of population on remedial measures or medication of common man on the other hand. Ganga fiscal accounting as such has been reduced merely to an academic exercise as firm conclusions on health effects were obscured. Economists and environmentalists could never accept objectivity, subjectivity and rationality of economic analysis. While economists laid stress and emphasis on objective numbers, common man wanted to provide credence to the personal experience, choice and control. It is in this diverse thinking that political manoeuvring of Ganga eco-system has cropped up, thrived and prospers to meet the interests of politicians for political attainments.

There is some amount of internalising of polluting cost by some industries along Ganga such as those of Tanneries, Jute and Cotton industry and Chemical plants. Instead of passing it on to others this kind of recovery of pollution damage rests on purchase of products of polluting industry. Taxing pollutor has also been an alternative to Ganga Pollution Abatement.

Ganga Public Costs of eco-development coupled with economic development were frequently regarded unjustified. It is difficult to calculate at first and project distant, indirect or long term effects. Economists must understand these processes of interactions in human and physical world. Quantification in terms of as many variables as possible will help in assessments and projections. Given the complexity of the various physical and behavioural linkages in water resource management there is massive uncertainty and economists must recognise their technical limits of profession and adopt a multi-disciplinary approach.

The politics of economics has been playing a major role in the development of Ganga eco-system. There are two vital factors along involvements of politics in the Indo-Gangetic Plain. Directly the development in the region is looked upon to influence the vote banks of political parties and individual leaders. This partially is responsible for formulation of developmental plans. But development and industrialisation is directly proportional to environmental degradation in general and deteriorated water quality in particular. Local environmentalists, NGO's and people nursing a 'close to nature' philosophy of life and propounders of cleaner environment with healthier waters constantly oppose such developmental plans that are instrumental in damaging the riverine system. The two blocks constantly pressurise each other and give vent to politics in the region. There is another set of public which is for construction of Ganga Barrage while all those who have grabbed the lands along the river bank and derive colossal amount of benefits are constantly opposed to construction of barrage. The barrage is likely to solve the problems of energy, floods and change in the course of river. There is good deal of politics among two diverse groups involved with construction of the barrage.

Heterogeneous policy planning through various ministries concerned stands as a barrier in the pollution abatement. The bureaucratic administration and leadership is unfortunately prone to be influenced by vested interests and politics involved, negating integrated development of Ganga Basin at large, and pollution abatement of the river in particular with proper utilisation of government funds.

Chapter 12

Role of NGO's in Ganga Water Purity

The challenges of Ganga water purity maintenance are broadly two folds. On one hand there are governmental agencies and on the other the citizen's. Admittedly, the governmental agencies are products of citizens, yet after their creation they adorn a different character—a trait to sustain governmental seat and in this course they have no alternative but to seek mass popular vote banks. It is well known that this populist approach blends governmental actions working for development *vis-a-vis* economic and industrial growth. It is an irony of fate that despite populist moves there is greater urgency for man—nature political and religious firmament form the basic tenets of Ganga Water purity. The role of NGO organisation is more than the importance ascribed to them. No amount of Governmental decisions or policy planning could turn out a tide in upgradation of Ganga but deeply entrenched people's movement through the aegis of NGO's could dramatically change people's perception leading ultimately to emancipation among Ganga dwellers.

The Ganga water was regarded to be endowed with magical curative powers but of late cities like Kanpur, Barauni, Hooghly, Allahabad, Varanasi have grown immensely polluted and millions of Ganga bathers contact deadly diseases like cholera, hepatitis, dysentry, polio, typhoid and paratyphoid. There are fungal diseases causing severe damage to nails, hair, horns and hooves among human beings and animals.

This situation is a product of industrialisation and excessive population. At Ghazipur effluents from the opium industry have turned Ganga Waters to 'Opium Waters'. In Patna about 40 million litres of sewage water is flown everyday into the river. The effluents of shoe factory at Barauni, distilleries and oil refinery coupled with thermal power station influx, fertilizer factories and overflow of molasis of sugar factories have attracted the attention of non-governmental organisations. The establishment of a nuclear power station at Narora has led to nuclearisation of Ganga Basin. But all this has happened at the helm of governmental planning. Who could stall this Ganga Water degradation? The answer is in the development of a people's movement of a larger dimension and people's involvement than ever before.

The history of People's movement dates back to 1914, launched to oppose construction of a dam at Hardwar upstream to Har Ki Pauri. The aim of this dam was to divert the water to the upper Ganga Canal. The construction of dam was opposed by Purohits on the grounds that religious rituals like '*Shraddha*' cannot be performed in captive water. When their protests could not be heard they approached Pandit Madan Mohan Malviya whose call attracted citizens and Dharam Gurus like Shankaracharyas of Puri and Sringeri Maths, the Maharajas of Gwalior, Jaipur, Bikaner, Patiala, Banaras, Alwar, Darbhanga, Kasimbazar and others who contributed to the first ever people's movement to 'Save Ganga'. Herein it would be pertinent to remark that though the movement was for religious performance of rituals it amounted to maintaining the sanctity of Ganga and its saviour from impeaching artificially. The movement emphasised that natural current of the river should be maintained as it descends from the Himalayas. The movement culminated into withdrawal of British Government's decision and instead of a dam, 'Kutcha Bund' was made.

The Bhoodan footmarch of Acharya Vinobha Bhave in 1952 was to appeal for voluntary action for cleaning of Ghats. The Chipko movement of Sunderlal Bahugana, sparked in hills of Uttar Pradesh in the early seventies was against deforestation, and protection of local hill resources. Deforestation in the region leads to rise in river beds due to silting and soil erosion thereby increasing the possibility of floods and land slides, disturbing the water balance and drawing the Ganga basin to the fury of the river. The movement is now involved in 'Save Ganga Campaign' with active participation in the Anti-Tehri dam movement, the dam if constructed, will disturb the ecological balance of the Himalayan region. The 'Clean Ganga' movement launched in the early eighties at Varanasi also focused attention on the pollution of Ganga. Local NGO Units in cities along the banks of Ganga take up localised problems and contribute to Save Ganga in their limited spheres.

Apart from taking up localised issues or cleaning Ganga as a whole, NGO's have other important roles where they contribute directly and indirectly to Ganga water purity. Acting as a catalytic agent between government departments and the people for promoting the cause of Ganga water quality they can also champion the cause of Clean Ganga Operation through education and awakening.

The importance of Ganga as a natural resource which can solve our problems of poverty with its vast fertile basin is not known to many of us. Drinking water scarcity level can be minimised when pollution level in the river is checked. Each one of us needs to contribute our own share in maintaining the purity. Marches, *Padyatras*, Posters, Paintings, audio visuals should be used by NGO's to educate and awaken these masses regarding the river philosophy, restoration of ecology and maintenance of riverine water quality. There is a constant duo between exploiters and restorers of Ganga water purity. The role of Non-governmental Organisations (NGOs) is delicately poised to maintain a semblance between the government and citizens, between ecology and development and above all cleaner water and waste water. The last of these, cleaner water and waste water needs spatiation. Cleaner water stands for natural healthier water, ecological purity and waste water symbolises products of industrial waste and effluents. Herein, a dimension which could not find mention is that of agriculture. Agriculture is now rated as an industry and the amount of pollution impregnated in the rivers as a result of washouts of agricultural pesticides, fertilizers and insecticides cause another kind of pollution. The role of non governmental organisations is not anti-development or anti-industrialisation. But they strive for better environment development and industrialisation friendly to environment.

It is a matter of satisfaction that citizen's environment and development movement has flowed in great diversity and now reaches all corners of Indo-Gangetic Plain. It has brought national concern to the fore and has aroused Ganga Consciousness. Historically, Ganga Seva has been enshrined in Indian religion and mythology. The ecological integrity and interdependence of Ganga Basin weighs heavily on quality, biological diversity and chemical complexity of the river Ganga.

It may be anticipated that the NGO movement in the Ganga basin would assume greater dimension and on one hand work for social and political change of thinking to stall river pollution and on the other hand fan such problems for cleaner Ganga before the government. The role of NGO's working as a link between government and citizens is likely to herald an era of Ganga water purity, sanctity and ecological prosperity.

Chapter 13

Management and Development of Ganga Eco-system

Anthropogenic source spilled into natural riverine system causes a change. As is explicitly implied, such a change from natural condition is deleterious and therefore, termed pollutional (Campbell, 1981). It is widely accepted that natural waters possess digestive, assimilative or regenerative capacity. Within this scheme of assimilative capacity of water, operate changes that are not deleterious, with some notable exceptions like radioactive materials. The regenerative capacity of Ganga waters is much more pronounced than other rivers. This character of Ganga tantamounts to its holy nature. However, it seems that assimilative capacity of Ganga waters is based on two factors—time and space. Unfortunately, ruthless exploitation of river water has stressed Ganga eco-system to a breaking point. It is a common assumption that introduction of anthropogenic material beyond a certain limit imparts deleterious effects. The threshold beyond deleterious level ascribed to Ganga eco-system is no better than thresholds of artefacts of the inability to measure certain responses. It may be also ascribed to the ignorance of the functions of the system. However, there is another concept persuasive of the presence of multiple thresholds in the Ganga Eco-system. Possibly, both these concepts may be correct in their own way. Odum, Finn and Frang opined a new aspect to the foregoing and proposed that anthropogenic materials may be viewed under certain conditions as a subsidy to the eco-system causing diversion from normative state and viewed by most ecologists as beneficial. The catch in this thinking is that what may be beneficial to one group of organisms may be harmful to another. The beneficial or harmful nature depends upon a human value judgement. How best it could be ascertained what anthropogenic material is beneficial or harmful for a riverine eco-system is intriguing the ecologists world over. The biological complexities of Ganga waters perforce negate any clear cut dictate on beneficial or harmful roles ascribed to a given pollutant, a group of pollutants and their extent. Perhaps despite national efforts on this front under Ganga Action Plan and beyond, there seems a need for studying in-depth this aspect for sustainable development and exploitation of the river water.

What should form the strategy for development of Ganga eco-system conducive to protection and enhancement of water quality? The starting point seems from the worst possible case assumption—no response thresholds exist at the eco-system level and subtle effects unmeasurable. The perceptible changes could only be determined. The most common assumption seems a drastic curtailment in population, in face of industrial or agricultural impacts but this is a myth and an impossibility to conjure. As a policy in the Ganga Basin of drastic population curtailment is likely to espouse reduction in amenities of civilization. Even protoganists of society and industry restrained in their views of environment would feel anguished if such drastic measures are planned. Population curtailment is not possible as it is not enshrined in Ganga civilization. Therefore, to conjure a society without harm or risk to natural Ganga eco-system is just an impossibility. However, a redeeming feature for Ganga eco-system development seems plausible from benefits derived out of technological advance in the area of enjoying quality eco-systems. It is easier to imagine this ideal situation in the wake of Rene Dubos's optimistic view, but how this desirable objective could be achieved has to pass the acid test of social, economic, political, scientific and technological paradigms expedient in a developing country like India. Herein development of sustainable Ganga eco-system and maintenance of water quality of the river is discussed for futuristic exploitation of the river water.

REGULATING HUMAN IMPACTS

Pollutant stress could be regulated through following basic options:

(a) *Pipe Standards:* The regulation of concentration and/or quality of potentially hazardous material that leave a pipe by chemical measurement at the end of pipe or somewhere in the system discharging into the environment.

(b) *Technologically based Standards:* This assumes installation of best technology for waste treatment to protect Ganga water.

(c) *Receiving System Standards:* The use of the responses of natural systems or surrogates of natural systems to determine loading capacity of natural waters.

Although the first of the three options, former two only provide a verbal option but no direct measurements of response receiving system standards have been made successfully so far. There are pitfalls and caveats in the three options described *op. cit.* It seems that chemical and physico-mechanical methods are not sufficient to administer maintenance of Ganga water quality. The biological control is more than necessary. However, biological control technology is still in nascent stage of development and whatever data is available conforms for application in specific riverine systems. The diversity of Ganga eco-system is so different and complexity of its water changes in time and space that it requires in-depth study before a sustained effort could be launched.

Such a prospect needs a constant spirit of enquiry and investigation alive in frontiers of knowledge, planning futuristic strategy for development of Ganga eco-system. A new method and new theory has to be developed on the edifice of knowledge of other riverine systems. The faults need be identified and caveats removed. The responsibility of biologists, therefore, becomes wider than just monitoring biological pollution. But unfortunately, ideal river waters are so scarce that negate working prospects of biologists to study pollutional change and device even the crudest biological methods for sustainable water quality improvement.

It is the irony of fate that Ganga Action Plan has laid greater emphasis on engineering skills to cleanse Ganga water of its pollution but the research and approach for biological control of pollution could never assume the dimension of importance it deserves. The first phase of GAP has been devoted for research on monitoring physico-chemical pollution in Ganga and making a biological profile. This constitutes the basic data but it seems Pollution Control Boards are just confined to a routine monitoring work. The applied study for Ganga water resource development which should have formed the second phase seems to have lost priority in the corridors of bias of Ganga Authority and its leaning towards routine agencies and institutions. It seems apathetic that institutions of higher learning associated with first phase could not be utilised for second phase for which they preferred themselves. The policy of restricted reliance on academic community and seeking alternative in employment of regulatory agencies, industry and environmental firms and the like is reflected in the success or failure of Ganga Action Plan.

A cohesive, integrated approach has to be resorted, physico-chemical methods need be clubbed with biological devices. Theoretical ecology and biology must be used in the development of new technology for regular adoption and management.

WATER QUALITY STANDARDS

The water quality standards set by WHO for river waters are as under:

However, such rigid limitations and regulations on Ganga water quality are impossible to achieve in India. Range finding tests and establishing Ganga water standards is required and technical feasibility and economic viability of such standards tested from time to time.

ASCERTAINING UTILITY OF TECHNOLOGY

The predictive utility of technology should be ascertained taking into consideration interpretation and closeness of laboratory results and actual water hazard; sensitivity of the method for false negatives; variability to detect changes in impacts; and finally replicability of the method for adoption as routine time and again.

DECISION CRITERIA ON WATER QUALITY

The decision of criteria on water quality standards should be based on social, technical, legal and economic relevance.

REGENERATION FROM ECO-ACCIDENTS

While developing a sustained plan of action proper safeguards against risk hazards should be ensured. Management options include (a) Doing nothing (b) Restoring original condition (c) Rehabilitating to ecologically improved and socially acceptable condition. Converting the damaged eco-system to an alternative eco-system bearing little or no resemblance to the original. The first option is only for a minor hazard which could be automatically corrected in time by self-regenerative capacity of Ganga but other options are meeting emergencies needing corrective measures. Such options for protection against accidents are detailed below :

Restoration of original condition

But for practical difficulties restoration of original condition is the best option but there are scientific, social or economic impediments in the path of the attainment of this option. The option selects out itself when any ecologically unique eco-system has been damaged or damaged area is part of larger system with aesthetic, recreational or ecological value. The option could be tested for its feasibility taking into consideration (a) Feasibility of restoration of original condition (b) Availability of ecological data base about original condition (c) Reference accessibility to a side where restoration has to be achieved (d) Availability of original species or physiological races to help restoration (e) Responsibility taken by the managements for restoration (f) Importance of the damaged eco-system in local and national perspective (g) Social utility of restoration. If the answers are in affirmative this option is valid for adoption.

SUMMARY OF WHO WATER QUALITY GUIDELINE VALUES

Values recommended in these guidelines are for total concentrations (*i.e.*, all forms of substances present).

1. **Bacteriological quality**

Piped supplies	Number of organisms per 100 ml
(*a*) treated water entering the distribution system	faecal coliforms : 0 coliform organisms : 0
(*b*) untreated water entering the distribution system	faecal coliforms : 0 coliform organisms: 0 in 98% of samples over the year; 3 in an occasional sample, but not in consecutive samples.
(*c*) water in the distribution system	faecal coliforms: 0 coliform organisms : 0 in 95% of samples over the year; 3 in an occasional sample, but not in consecutive samples.
unpiped supplies	faecal coliforms : 0 coliform organisms: 0
bottled drinking-water	faecal coliforms : 0 coliform organisms: 0
emergency supplies of drinking water	faecal coliforms : 0 coliform organisms : 0

2. **Inorganic constituents of health significance**

	Guideline value (mg/litre)
arsenic	0.05
cadmium	0.005
chromium	0.05
cyanide	0.1
fluoride	1.5
lead	0.05
mercury	0.001
nitrate (as N)	10.00
selenium	0.01

3. **Organic constituents of health significance**

	Guideline value (μg/litre)
benzene	10
chlorinated alkanes and alkenes	
carbon tetrachloride	3[a]
1,2-dichloroethane	10

[a]Further information can be obtained from ; WHO. *Guidelines for drinking-water quality*. Geneva, World Health Organization Vol. 1-3, 1984/85.

1,1-dichloroethene	0.3
tetrachloroethene	10[a]
trichloroethene	30[a]
chlorophenols	
pentachlorophenol	10
2,4,6-trichlorophenol	10 (odour threshold concentration, 0,1 µg/litre)
polynuclear aromatic hydrocarbons	
benzo [a] pyrene	0.01
trihalomethanes	
chloroform	30
pesticides	
aldrin/dieldrin	0.03
chlordane	0.3
2,4-D	100
DDT	1
heptachlor and heptachlor epoxide	0.1
hexachlorobenzene	0.01
lindane	3
methoxychlor	30

4. **Radioactive materials**

gross alpha activity	0.1 Bq/litre
gross beta activity	1 Bq/litre

5. **Aesthetic quality**

	Guideline value (mg/litre)
aluminium	0.2
chloride	250
copper	1.0
hardness (as $CaCO_3$)	500
iron	0.3
manganese	0.1
sodium	200
sulfate	400
total dissolved solids	1000
zinc	5.0
colour	15 true colour units (TCU)
taste and odour	not offensive to most of the consumers
turbidity	5 nephelometric turbidity units. Preferably < 1 for effective disinfection.
pH	6.5-8.5

[a]Tentative guideline values. When available carcinogenicity data could not support a guideline value, but the compounds were judged to be of importance in drinking-water and guidance was considered essential, a tentative guideline value was set on the basis of available health-related data.

Rehabilitation of Certain Characteristics

There is circumstancial evidence that desirable characteristics could be restored even though original condition cannot be achieved. This rehabilitation serves as an alternative to the perfect restoration of the damaged water. The following test should be applied to strike adoption of this option:

(a) Desirable ecological parameters be known.
(b) Feasibility of restoration of these characteristics without restoration of entire eco-system.
(c) Sustainability of these characteristics in face of otherwise changed water complexion.
(d) Prospect of creeping any undesirable side effects.
(e) Significance of choosing this option against restoration option.

Establishing Alternative Eco-system

The option is of little relevance as another Ganga Eco-system cannot be developed. But part of the damaged riverine system can be refurnished and transplanted elsewhere in the Ganga for its continuance.

Doing Nothing

The regenerative capacity of Ganga waters is tremendous. In the circumstances of a normal environmental hazard or an accidental spill the Ganga water restores itself in time and space. A flow of approximately 35 to 40 kms restores the water quality but any such hazard like spill of radioactive material or a major spill negates any activation in the part of the restoration process and rehabilitation options are required to be adopted.

Recommendations for Biological Upgradation

In view of vast variance of the river water eco-system and its implications in biology and pollution there is a strong need to monitor and record biological profile of Ganga in time and space. Implications of coliforms in spread of enteric diseases and their response to pollutants needs be studied for effective control of diseases and improvement in water quality. Algal resources of Ganga could be used as a genetic stock for use in basic and applied research and could be harnessed for extraction of antibiotic substances, and extracts used to boost productivity of crops and as a source of protein. Role of algae in fish productivity and as fish food needs attention. Algae could be used as a fair indicator of water quality and indicator organisms need greater attention for monitoring of water quality. Algae and fungi concerned with spread of allergenic and Keratinophilic diseases need be studied to combat problems of public health. Cellulosic and Keratin degradators need be studied for its implications in maintenance and improvement of water quality. Utilization of tannery effluents and sludge in agriculture needs urgent attention in areas like Kanpur, supporting a well-established tannery industry. Use of macrophytes in treatment of tannery and other industrial wastes needs attention. Response of microbes to major pollutants in isolation and mixture simulating conditions prevailing in river or as predicted for future on the basis of past experience need be studied.

HUMAN AND SCIENCE FACTORS

The problem of maintenance of clarity and sanctity of Ganga waters is shrouded with a mysterious admixture of social, religious, political, scientific and technological factors. Keeping Ganga waters clean is by no means a technological aspect but is mixed with social and given problems. The religious factor also negates adoption of many options. There are references to the success stories of improvement in water quality in rivers like Thames and Clyde in Britain and Connecticut river in America. The successes in these rivers were instrumented through technology and methods without a pinch of financial sacrifices of the industry or Government but in the two countries there is existence of a harmonious link between industry, regulatory agencies, academic institutions and society. The technological and economic affluence added to the success story but such prospects are bleak in Ganga eco-system because of poverty ridden country like India with a mosaic of varied religions and heterogeneous community blended in Indian polity. Perhaps there is the need for greater thrust among people in the domains of industry, trade, leadership and religion in the regulatory approach. Financial incentives for upgradation of water quality stressing the individual, groups or industrial units, economically be provided for them to survive and make up for the losses in course of regulatory mechanisms.

A magnanimous approach of various religions in performance of rituals unhealthy for maintenance of water quality be resorted to. Such rituals which perpetually damage water should be abandoned.

PLANNING GANGA ECO-SYSTEM DEVELOPMENT

The objective of pollution control in the river is upgradation and better resource exploitation. Naturally, Ganga water resources include products of flora and fauna, which serve as a source of feed, food, medicine, energy and industrial substances. Above all, Ganga serves as a source of domestic water for drinking and bathing purposes as well as water required for irrigation and industrial use. The water policy should strive to weigh the normative implications of objectives and economics. Besides economic lineage, the human aspects of Ganga waters should not be lost sight of. The social and religious importance ascribed to Ganga cannot be justified unless the sanctity of the water becomes the character of the river. It is this religious fervour and fulfilment of thousands and millions of people's faith in the river that signifies importance and specialised place in Indian polity while planning Ganga cleaning operation. Admittedly, pollution control planning of any kind is for the good of life and pollution is regarded as theft from nature's treasure. Pollution is regarded as producers profit and consumers low price while non-polluters bear the cost of pollution without any kind of dividend to them. Planning policy should ensure elimination of such injustices. Concept of policy and planning should not only confine to the imposition of penalties to polluters but also compensate victims from pollutional damage. However, unfortunately in India citizens victim to Ganga water pollution *vis-a-vis* spread of diseases, public health problems, pains and displeasures of polluted waters are in no way. Problems of pollution control policy and planning raise such basic questions as fairness of increased equality of real incomes forcing pollution control at its lowest efficiency. The cost and benefit distribution is vital while considering Ganga eco-system development. Pollution control policy of the river and planning should be such that provides for only proportionate shares of costs commensurate to the benefit, thus, taxing rich and poor according to their contribution to the Ganga water damage. An agenda for planning Ganga environment and its resource should include:

(a) Assessment of the existing natural resource, patterns of resource utilization and prospects for future
(b) Impact Analysis of resource depletion on Gross National Product
(c) Government policies to keep in mind long term benefits instead of political and electoral considerations
(d) Policy interventions in matters of subsidies, taxes, proper water pricing, all of which have an indirect impact on resource use.
(e) Such concerns like sanctity, holiness, pleasures and religious devotions should find a place while determining policy and planning.

Policy making is by and large a domain of political set up and the government. As such self-interest of policy makers comes above basic environmental issues, especially, in the matters of financial dividends, taking care that their own position in the decision making is not eroded. Since politics is a game of numbers the riverine and water management policy is framed to capture the majority interest and votes. However, in implementation of such policy, vested interests are realised. Deferring the policy implementation for shifting the responsibility to another place, the political system tends to pass decisions from Central to State and State legislatures to bureaucratic agencies. Thus the resolution of political conflicts results in piecemeal fragmented decisions. Consequently, planning suffers due to involvement of fiscal politics. Ganga water protection is not a technological problem but also politics of power and diplomacy for financial gains.

DECISION MAKING

It is difficult to describe a uniform decision making process to deal with riverine pollution abatement, utilisation and exploitation of riverine water attributes. There are multiple factors involved in the decision making that a judicious divergence from the routine process catering to the specific needs is required. The physico-chemical and biological diversity of Ganga eco-system is such, that little permits a homogenous approach. However, the main objectives of Ganga water management in general include:

(a) To control a particular pollutant or set of pollutants causing public health problems and deterioration of water quality.

(b) To establish suitable regulatory plan, instrument legal laws and device financial planning to provide development of healthier riverine system.

(c) To exploit the water resource, use and reuse it for the benefit of the common man.

(d) A constant survey and impact analysis should be made and decisions taken to modulate the pitfalls in the process.

(e) The bureaucratic administrative or political interest should not be allowed to compromise the upgradation programme and the decisions in the interest of Ganga eco-system development should be taken in the interest of upgradation.

(f) The decision should not be taken in isolation but a holistic approach should be kept in view. The groups involved (Government departments, politicians, social workers, religious leaders, industrialists, journalists and consumers) should be consulted and their interests incorporated for the success of the plan.

(g) Financial subsidies and incentives to the aggrieved parties as well as industries requiring funds for treatment of their wastes be ensured.

The complexity of decision making is subject to organisational and institutional expanse. The Ganga Action Plan is not only executed through water resource departments, pollution control boards but there are other governmental agencies like Ministry of Finance, Ministry of Environment and Wild Life, Ministry of Industry and Commerce who play their part in execution. The legal expanse is also part of the abatement process wherein role of judiciary and special courts becomes specific. The social implication of the issue makes the decisions more difficult at times. Such a broad canvas of development programme has to cater national will and intent for a cohesive plan of Action.

There is another option to this holistic approach. The approach of this sectoral procedure wherein a draft proposal may be prepared by GAP in consultation with departments of health, water resources, public works, labour, industry and social welfare and necessary environmental legislations or regulations adopted after a debate on the issue, but this kind of procedure is more cumbersome and prolonged.

SCIENCE IN PUBLIC POLICY

The importance of Science and its involvement in decision making has increased in consonance with scientific and technological change. There is a changed process of policy decision making which is now based on input of information and recommendations from scientific experts. The technicalities of abatement programme emphasise involvement of scientists and technologists in the decision making process.

The social scientist working the areas of economics, sociology, political science and other domains also have a role in the decision making process for programming social aspects of abatement. Admittedly, involvement of Science and Social Science should be harmoniously wedded in the decision making process.

THE RESPONSE MECHANISM

In a developing country like India there exists a stiff competition for resources between economic and health interests. The Ganga eco-system should be developed in such a manner that an optimum of the two interests should reign supreme. Competing and conflicting health and sustainable resource mobilisation should be taken as criterion to gauge response of the abatement programme. The characteristics of standard Ganga eco-system development programme should involve various units of community. A debate on various aspects and values should be organised to churn out a plan of action. The mechanism should permit technical and policy analysis and modulation from time to time in accordance. The mechanism should provide room for hearing and discussion with conflicting interests in a bid to accommodate them and work for the success of the plan. The policy decision should be open to objectivity of the goals and subjectivity of the responses.

A STRATEGY FOR THE FUTURE

Noted environmentalist Sundarlal Bahugana opined that isolated efforts in a society, where economics has become the religion, cannot succeed till the whole concept of development is changed. We have copied a development system which is based on two lies, *i.e.*, nature is a commodity and the society is only of human beings.

This system was born and developed in the west, where the State had vast colonies to exploit and less population with a big area. Its application in densely populated countries has created a host of problems and the crisis of the Ganga is one of them.

In June 1982, the United Nations Environment Programme (UNEP) organised a public hearing on environment in London to celebrate the tenth anniversary of the Stockholm conference. A bottle of the holy Ganga water was presented to the chairman of the conference, saying that "Ganga is the representative of all the rivers and these rivers are in danger. This is the gift of the materialistic civilisation. If you want to protect the environment, protect the rivers."

Technological solutions to environmental problems have not succeeded and some people now realise that we need a technology to solve the problems created by technology. That technology was developed in countries like India, where, besides searching for the truths of the physical world, the truths of the inner were also unfolded. That is spirituality.

Its message is very simple and is the basis of our culture. Our culture sees life in all creation-human being, birds and beasts, trees and plants, rivers and mountains. Our behaviour with these should be worshipful. This can be practised only in a system where austerity is cherished.

Thus, the basis of development should be cultural values and its objective should be achievement of peace, happiness and fulfilment instead of affluence. The production system should be decentralised. Energy priorities should be human, animal, bio, solar, wind, tidal and hydel (from the run of the river). This will help in saving the Ganga from pollution created by big industries, big dams and thermal and nuclear power stations.

Big irrigation systems have created problems of salinity and water-logging, besides flooding the best lands. We should learn lessons from the USSR, where big dams have created so many problems including that of food scarcity.

Ganga, as she has been since time immemorial, can solve our problems of pollution and poverty. Dense tree cover in its hill catchments should be restored by reviving the natural vegetation to conserve the water. Trees giving food, fodder and fibre, besides medicinal herbs should be cultivated. This will check the occurrence of landslides, soil erosion and drying up of springs, which feed the Ganga.

The difference between lean and peak season river flow, which at present is 1:70 at Hardwar will be minimised. There will be enough water in the river throughout the year, which will also keep the river clean.

In no case should big dams be built, because these when filled with silt become useless. Besides getting polluted, water is lost in evaporation when impounded. A dam is the short-term solution of a permanent problem and as such dams will stand as monuments of stupidity.

The ultimate and permanent solution lies with the masses. We have a tradition of worshipping the Ganga. This should be strengthened with scientific support and the message should be taken to the masses.

Chapter 14

Still an endangered River

Rivers are the most extraordinary part of the land, and explicitly so, Ganga is of pivotal significance materially, sentimentally and environmentally. Let alone think of the probability that only fragments of its inheritance remains unexploited, but it is more valuable than ever before. People have worked to protect the river—to leave it as it was pure before—and yet the onslaught of millions perforce has endangered the holy river. Despite governmental action, love of common man and status of 'Mother Ganga' what pre-empts saving the river catelogued as the still endangered river of India, remains a million dollar question.

The manipulation of flowing water since many decades has been considered useful and good universally with some notable exceptions. While George Washington believed Western Settlement making of 350 miles of a canal little short of madness (Palmer, 1986), President John Tyler put the corps back to work on River Ohio for river development and assigned its importance as subject to posterity. In India, all importance was showered to development of Ganga with utmost fanfare. Ganga Action Plan first of its kind was launched and devoted to river development in India. The governmental will and the Plan was churned out of conservationists support for maintenance of Ganga water quality. The nation was faced with powerful arguments based on science, economics, social and political ethos. This Indian river, sustaining more than 300 million people has held an endless fascination for its people. Its importance cannot be quantified but it is a celestial river personifying the spirit of India and despite all pollution being witnessed has lost none of its glory and remains a living presence for the people of India. Ganga protection groups thus proposed a strong case for exploiting water resources and meeting needs without destruction and waste prevalent in the past.

Let us consider the water requirement statistics. A person needs only 5 to 6 pints of water a day but amounts used in agriculture and manufacture of goods are enormous. Production of one pound of cotton requires 3,100 gallons, one pound of rice 500 gallons, a gallon of milk 932 gallons and an egg 120 gallons of water. Figures project how much water is required to meet daily human need of multiple nature. In this scheme of things strategies of Ganga Water Management seem very complicated. To look at construction of barrages and dams or hydroelectric power generation instead of other alternatives needs crucial evaluation.

Protection and conservation of water and stretching what we have should guide riverine development. In India more than 70 per cent of water is consumed in agriculture and the next largest consumption of water is in manufacturing and energy. Coupled to this is the problem of growing urban needs, an ever growing population and waste from public water systems. Efficient irrigation management, recycled industrial supplies, reduced consumption by individuals and awareness to water needs may help conservation but the specter of water shortage will remain.

Even with the movement to save Ganga, the national rivers system, the beginnings of political reform and the knowledge of alternatives many more rivers can be still categorised as endangered. Despite the National SAVE GANGA MOVEMENT there has been little headway in Ganga Water maintenance. Actually, the movement should be closer to simple goal of saving the river Ganga for river's sake on lines of New Englanders working for St. John River in Maine. There is a magical moment on every St. John trip. When you push off from shore for the first time you know that nothing lies ahead but paddling, floating, watching the sun and clouds, hearing the wind and river and that each day your senses will open like flowers to the sun and absorb more of what is around you. The St. John is the only free flowing river in New England that is long enough for this to happen.

Perhaps it was astute planning and intend to curb the intention of construction of dams. Alternatively conservationists were called upon to organise Friends of St. John's in 1974. It was Aronld who single handedly lobbied and proved that even a single person can make a difference in saving the rivers. The argument that river was worth saving, that the dams were not worth the cost, and that energy conservation cheaper is impregnated with relevance. Flood control benefits were reduced when new corps levee protected the town of Fort Kent. This augments the fact that structural constructions or treatment plants on the rivers can do precious little as compared to the protection of Ganga in its natural—wild state. Beyond recreation, ecology, economics and science, river Ganga needs to be recognised as valuable for its own sake. Moreover, a spiritual side to the river surfaces in the protection movement a dimension nowhere seen except on the Stanislaus River in California U.S.A where people fought to save the river which was a sacred and holy place to them. Ganga also brings out feelings of reverence to the natural world where rivers are a source of life and inspiration and therefore, need to be saved. Let us remember Anthony Netboy's writing, "If society in its wisdom some day decides that food and game fish are more valuable to the nation than electricity . . . . . ." precisely explicits what is the importance of natural river as compared to impounded water, bears lot of substance. Flowing waters are symbolic of renewable water resources which should not be impounded.

There have been controversies over the construction of dams and barrage in the development world. Tehri Dam and Ganga Barrage construction has met with constant opposition. Recognition of river values is crucial while taking anti-Dam or barrage decision, otherwise the builders will return when politics change.

Yet there is example of Tocks Island Dam in between New York city and Philadelphia proposed as a model of contemporary planning. The Dam is known as "Grand Coulee Dam" of Delaware. The Delaware Dam was mainly for flood control but later on recreational benefits were gradually added up for nearly half of its justification. But conservationists under the banner of

Delaware Valley Conservation Association filed a suit to stop construction of dam but the case was dismissed. The story of dam opposition did not stop here. The demolition of two hundred year old stone farm houses to rubble enraged people who cared little for ecology. They demanded protection under Historic Preservation Act of 1974. The futility of dam because of farm and poultry run off, pollution and entrophication due to over fertilization, climaxed into deauthorization of dam in 1975. Gradually Delaware turned out to be a gentle flowing river between woods. The dam was beaten through technical studies of water quality and water supply alternatives and protection gained through the national and international wild scenic river system. There has been a belief that we can bring all water we want by invoking technology and money. This was a kind of religious and patriotic approach of people in the past but that era is over due to recession of money and undercurrent of religion. The importance of the word "natural" used as a prefix to a river worth keeping was recognised.

The Delaware experiment serves as an example to serve people without a dam or barrage. Does this not provide a fillip to work on alternatives to save Ganga with thinking of constructing Tehri Dam. The Indo-Dutch collaboration, commissioning treatment plants and construction of Ganga barrage would not help unless waste on water and natural Ganga water eco-system is maintained. It is not possible to recreate in nature that man once destroys. Let alone destruction be checked so that restoration is not required. This invites the movement to save Ganga. The power of NGO's and environment should prosper and work in close affiliation with the masses. This can only lead to fulfilment of the cherished dream of Ganga purity restoration.

Despite river protection, changing politics, involvement of hundreds of Non-Governmental organisations and religious institutions the problem of Ganga water has assumed greater dimension. The fact should be recognised that Ganga water resource is limited and the contribution of ground water is also depleting. The water supply shortage is a major problem in the third world. Through conservation it can be improved to a great extent but the need to explore new water resources remains as ever before. This is precisely the reason for water disputes of several national rivers traversing various States. Thank God river Ganga has no such problem. The construction of Ganga Barrage besides other objectives is to bring the river closer to the Ghats of Kanpur. The frontal questions facing change of the course of river is the long term future of migration of people towards water and in that case the Indo-Gangetic Plain will hold the golden belt of that part of civilization which was turned into the thickly populated and polluted belt. But either way importance of Indo-Gangetic Plains was, is and will be in future. The problems and dimensions will change face from positive to negative prosperity—life and happiness to poverty—struggle for life and unhappiness, in the absence of timely action.

Then is it too much of an asking, too much of work, too much to expect that a river like Ganga—the best natural homeland for people in the country can be saved to save people from killing boredom, stiffling artificiality, loneliness of life without wild creatures and the dread of money making world. If Americans needed the enthusiasm of John Muir, forethought of John Craighead, prose of Wallace Stegner, dedication of Mark Dubois, calculations of Jerry Meral, drive of David Brower, public attention of I.N.O. Douglas, old political style of John Saylor and new style of Bob Edgar, then to save Ganga Indians need another Bhagiratha to make the river metaphysical and free flowing, a Rajeev to restore its sanctity, a Sunder Lal Bahugana to maintain its natural environment, and a Medha to sustain habitats in the lap of Ganga homeland not once in their life time but for all times to come. Alas! these mortals remained larger than their life in deeds but they have been captives in the life spans of their own. Perhaps multiple projects like that of Ganga Action Plans strengthened with weaves of devotion, selflessness, purity and zeal working for the river can alone save Ganga. But these qualities exist in dreams—practicality is juxtaposed, Ganga remains, still an endangered river.

Bibliography and Further Reading

Abdel Fattah, H.M., Moubasher, A.H. and S.M. Maghazy (1982). Keratinophilic fungi in Egyptial soils. Mycopathologia 79 : 48-53.

Abhijit Sen (1993). Disappointing effort on ecological economics, Down to earth (2) 5 : 53-54

Abraham, J.G. (1962). A survey of the hydrobiology and fisheries of the Cooum river. Madras J. Fish 1 : 50.

Acheson, M.A. (1983). Water pollution control in India. Water Pollution and Management, 215-227.

Agrawal, G.P. (1955). Cultural and Pathological studies of some fungi imperfecti. D. Phil., Thesis, University of Allahabad.

Ahluwalia, P.S. Maj. The Eternal Himalayas (1982) Inter Print, New Delhi.

Ahmad, M.R. and Winter, A (1968a). Studies on the harmonal relationships of algae in pure culture. The effect of Indol-3-acetic acid on the growth of blue-green and green algae. Planta (Berl.) 78 : 277.

Ahmad, M.R. and Winter, A. (197). The effect of weak auxins on the growth of blue-green algae. Hydrobiologia 36 (2) : 305-316.

Ajello, L. (1953). The dermatophyte, Microsporum gypseum, as a sparophytes and parasites. J. Invest. Dermatol. 21 : 157-171.

Ajello, L. (1960). Geographic distribution and prevalence of dermatophytes. Ann. N.V. Acad. Sci. 89 : 30-38.

Ajello, L. (1968). A taxonomic review of the dermatophytes and related species. Sabouraudia 6 : 14-159.

Ajello, L. and E.M. Alpert (1972). Survey of eastern island soils for keratinophilic fungi. Mycologia 64 : 161-166.

Ajello, L. and A.A. Padhye (1974). Keratinophilic fungi of the Galopagos Islands. Mykosen 17 : 239-243.

Akhtar, Yasmeen and A. Gaffer (1987). The occurance and distribution of fungi in a waste stabilization pond system. Pak. J. Bot. 19 (1) : 110-111.

Al-Delauny, K.S. and S.S. Ali (1970). Antibacterial action of vegitable extracts on growth of pathogenic bacteria. J. Sci. Food Agric. 21 : 110-111.

Aleksandrova, I.V. (1951). The process of humus formation in several primitive mountain soils. Pochvovedenie, 604-616.

Alikunhi, K.H. (1958). Observations of the feeding habits of young carp fry. Ind. J. Fish 5 : 95-106.

Allen, N.B. and Arnon, D.I. (1955a). Studies on nitrogen fixing blue-green algae I. Growth and nitrogen fixation by Anabaena Cylindrica. Physiol. plant. 30 : 366-372.

Allen, N.B. and Arnon, D.I. (1955b). Studies on nitrogen fixing blue-green algae II. The Sodium requirement of Anabaena Cylindrica. Physiol. Plant. 31 : 653-660.

Amhurst, R.S. (1988). A text book of plant ecology. Students Friends & Co. Varanasi, India.

Amos William H., The Infinite River 1970. New York Bathmore.

Anita Gordon and David Suzuki. It's a matter of survival. Cambridge, M.A. : Harvard University Press, (1991).

Apha-Awwa-Wpcf (1970). Standard methods for examination of waste water (15 Ed.). American Public Health Association, Washington DC.

Asthana, V. and Shukla, A.C. (1992). Politics of Environmental Protection : Fiscal Analysis, ISA Convention Atlanta, U.S.A.

Atkinson, T. Edwin. The Himalaya Gazetteer (1981). Cosmo Publications, New Delhi.

Atlas, R.M.; Parmer, D. and R. Bartha (1978). Assessment of pesticides effects on non target soil microorganisms. Soil Biol. Biochem. 10 : 231-239.

Awasthi, A.K. and A.C. Shukla (1988). Keratinophilic fungi and dermatophytes from Ganga. Current Science.

Azegami, K.; Nishiyama and H. Tabei (1988). Suppression of bacterial seedling blight of rice by iron and its mechanism. 5th International Congress of Plant Pathology, Kyoto Japan : 109.

Aziz, K.M.S. (1974). Diarrhoea toxins obtained from water-bloom producing species, Microcystis aeruginosa Kutz. Sci 20 : 461-466.

Azmal, M. and Raziuddin (1988). Studies on the pollution of Hindon river and Kali Nadi (India). Ecology and Pollution of Indian Rivers I (Ed. R.K. Trivedy), Ashish Publishing House, New Delhi.

Bahl, N. and J.S. Grewal (1975). Influence of vitamins on growth and population of Operculate padwicki. Indian Phytopathol. 28 : 429-430.

Ballal, V.N. and B.P. Wangikar (1985). Effect of fungicides on growth and sporulation of Alternaria and Pestalotia species. Proc. 72nd. Ind. Sci. Cong., Part III : 108.

Bannerjee, B.N. Can the Ganges be cleaned ? (1989). B.R. Publication Corporation, Delhi.

Bannerjee S. and Moteani M.P. (1960). Some observations on Pollution of Savoun stream by the effluent of sugar factory, Balrampur (U.P.) Ind. J. Fish 7 : 107-128.

Bannerjee, S.; Motwani, M.P. and Karamchandani, S.J. (1956). A case study of heavy fish mortality in the river sone at Dehri-on-Sone, Bihar caused by the wastes. Ind. J. Fish. 3 : 186-196.

Barry Commoner. Making Peace with the Planet, New York : Pantheon Books, (1990)

Basham, A.L. The wonder that was India (1981). Rupa and Co. New Delhi.

Bednarz, Teresa and Helena (1985). Toxicity of zinc, cadmium, lead, copper and their mixture for chlorella pyrenoidosa. Warko wska-oratin-Al. Acta. Hydrobiol. 25/26 (3/4) : 389-400.

Bertrv, Georges, Marc Le Saout and Yvon Gu Eune (1986) (recd. 1987). Toxicity of copper to Microcystis aeruginosa Kutz. and Aphanizomenon flosaque Ralfs. Tntrev Gesamten Hydrobiol. 71 (6) : 841-849

Bhandari, J.K.S. and R.S. Singh (1976). Effect of carbon and nitrogen sources on the growth of Alternaria triticina. Indian Phytapath. 29 : 88-89

Bhargava, D.S. (1981). Pollution control strategy for Ganga. Proc. Int. Sym on Water Resources, Conservation, Pollution and Abatement (Uday Chitranshi), Roorkee, 31-38.

Bhargava D.S. (1981a). Ganga the most self purifying river. Proc. Int. Sym. on Water Resources, Conservation, Pollution and Abatement (Ed. Uday Chitaranshi) Roorkee : 11-13.

Bhatt, S.D.: Bisht, Y. and Negi, U. (1984). Hydrology and Phytoplankton populations in the river Koshi of the Kumanun Himalaya. Ind. J. Ecol. 12 : 141-146.

Bilgrami, K.S. (1986). Study of river Ganga (Munger-Farakka). Report from 1st May 1985 to Sept. 1986 Bhagalpur University.

Bilgrami, K.S. and J.S. Dutta Munsi (1985). Ecology of river Ganges. Impact of human activities and conservation of aquatic biota (Patna to Farakka') allied press, Bhagalpur, India.

Bilgrami, K.S. and Dutta Munshi, J.S. (1979). Limnological survey and impact of human activities on the river Ganges (Barauni to Farakka range). Technical Report prepared under Man and Biosphere.

Bimachar, B.S. and David, A. (1946). A study on the effect of factory effluents on the Bhadra river fisheries at Bhadrawati. Proc. 33rd Ind. Sci. Cong. III : 130.

Bollen, W.B. (1961). Interactions between pesticides and soil microorganisms. Ann. Rev. Microbiol. 15 : 69-92.

Brattberg and Gunilla (1986). Decreased phosphorus loading changes phytoplankton composition and biomass in the Stockholm (Sweden) archipelago. Vatten 42 (2) : 141-153.

Bunker, Janet, C. and M. Jam Arthur (1946). Microcalorimetric studies on the effects of media and environmental conditions on the growth of bacteria. Microbio. 47 (192/193) : 177-188.

Cabrita, J. and M.M. Figueiredo (1973). Dermatophytes in Portugal, Sabouraudia 11 : 21-29.

Cairns, J.J, Dickson, K.L. and Crossman, J.S. (1973). The response of aquatic communities to spills of hazardous materials. Fresh Water Pollution, Bacteriological and Chemical Pollutants, M.S.S.

Calvo, A., Vidal, M. and J. Guarro (1984). Keratinophilic fungi from urbansoils of Barcelona, Spain. Mycopathologia 85 : 145-147.

Canabaeus, L. (1929). Uber die heterocysten and gasvakaolen der blaualegen zueinander. Pflanzenforschung, Jena.

Caretta, G. and E. Piontelli (1975). Isolation of keratinophilic fungi from soilin Pavia, Italy. Sabouraudia 13 : 33-37.

Carmichael, W.W. (1981). Fresh water blue-green algal (cyanobacteria) toxins. A review 'The Water Environment (Ed. W.W. Carmichael), Plenum Press, New York & London, 1-13.

Carpenter, K.E. (1924). A study of fauna of river polluted by lead mining in Aberyst Wyth district, Cardinganshire. Ann. Appl. Biol. II, 1-23.

Carpenter, K.E. (1925). On the biological factors involved in the destruction of river fisheries by pollution due to lead mining. Ann. Appl. Biol. 92 : 1-13.

Chandra, K.; Singh, B.; Srivastava, G.H. and Mehrotra, S.M. (1981). Pollution of wastes of industries manufacturing nitrogenous fertilizers–a case study from river Ganga near Allahabad (India).

Chandrakala, G. and Vidyavati (1987). Influence of leaf extract of Azadirachta indica (L) on Cladophora crispata (Roth) Kuetz. 2nd All India Applied Phycological Congress : 3.

Chandrakala G. and Vidyavati (1988). Effect of bark extract of Azadirachta indica on the growth of Cladophora crispata (Roth Kuetz. Res. J. Pl. Environ. 4 (2) : 27-30.

Chadrakala, G. and Vidyavati (1989). Influence of leaf extract of Azadirachta indica on Cladophora crispata (Roth) Kuetz. Advances in Applied Phycology II : 109-112.

Chaphekar, S.B. (1979). Studies of the Kalu river eco-system. A report on an Interdisciplinary Project carried out in the Institute of Science, Bombay-32, Jan.-June, 1978.

Chaphekar, S.B. and Mhatre, G.N. (1986). Human impact on Ganga Eco-system. Concept Publishing Company, New Delhi.

Chattopadhyay, S.N. and Chandra, H. (1966). Pollution studies of river Ganges near Kanpur. Ind. J. Environ. Hlth. 8 : 270-285.

Chaudhri, H. and G.S. Sachar (1934). A study of the fungus flora the Punjab soils. Ann. Mycol. 32 : 90-100.

Chaudhuri, S. (1986). Integrated ecodevelopment of the river Ganga basin projects (Consolidated progress report) Bidhanchandra Krishi Vishwavidhyalaya, Kalyan.

Chaurasia, S.C.; Vyas, K.M. and A.K. Pathak (1977). In vitro, inhibitor effect of certain fungicides on the growth of Phytopthora parasitica var. piperina causing foot rot disease of Pan (Piper Betle)

Cheryl Simon Silver with Ruth S. De-Fries, One Earth, One Future: Our Changing Global Environment Washington, DC : National Academy Press, (1990).

Chinnaswami, R. and R.J. Patel (1983). Herbicides induced growth in the blue green alga Anabena flos-aquae. All India Applied Phycological Congress, Kanpur : 270-273.

Cholnoky, B.J. (1958). Hydrobiologische Unter Suchungen in Transvaal II. Substreinung in Juskii-Crocodile Fluss system. Hydrobiologia 11 (3-4) : 205-266.

Cooke, W.B. (1954 a). Fungi in polluted water and sewage. I. Literature review. Sewage Ind. Wastes 26 : 538-549.

Cooke, W.B. (1954 b). Fungi in polluted water and sewage. II. Isolation Technique Sewage Ind. Wastes 26 661-674.

Cooke, W.B. (1954 c). Fungi in polluted water and sewage. III. Fungi in polluted water and Sewage Ind. Wastes 26 : 790-794.

Cooke, W.B. (1970). Fungi in the Liebanon Creek sewage treatment plants and in Turtle Creek, Warren Co., Ohio, Mycopathol. Mycol. Appl 51 : 199-205.

Cooke, W.B. (1977). Fungi in streams, lakes adjacent soils and sewage treatment systems in the flathead river basin, Montana Northwest Sci. 51 : 172-182.

Cooke, W.B. and G. Matsuura (1966). A study of yeast populations in a waste stabilization pond system. Protoplasma 57 : 163-187.

Cooke, W.B. and G, Matsuura (1969). The distribution of fungi in a waste stabilization pond system. Ecology 50 : 689-694.

Couey, H.M. (1965). Inhibition of germination of Alternaria spores by sulfur dioxide under various moisture conditions. Phytopathology 55 : 525-527.

Couey, H.M. and M. Uota (1961). Effect of concentration, exposure time, time and relative humidity on the toxicity of sulfer-dioxide to the spores of Botrytis cinerea. Phytopathology 51 : 739-814.

D'Mello, Anthony and William M. Yotes (1987). The action of sodium deoxychlorate on Escherichia coli. Appl. Environ. Microbiol. 53 (8) : 1944-1946.

Daniels, G. (1954). Isolation of Keratinomyces ajelloi from soils in Great Britain. Nature 174 : 224-226.

Das, O.A.K., Shukla, A.C. and Gupta, A.B. (1967). The utilization of hydrophytes as a source of subsidiary protein food. Ichthyologia 4 (1-2) : 50-52.

David, A. (1956). Studies on pollution of Bhadra river fisheries (Mysore) with industrial effluents. Proc. Nat. Inst. Sci. Ind. 22 (3) : 131-162.

David, A. (1959). Effects of Calcutta sewage upon the fisheries of Kutti estuary and the connected cultivable fisheries. J. Asiat. Soc. New. Bot. 1 (4) : 339-363.

David, A. and Ray, P. (1966). Studies on the pollution river Daha (N. Bihar) by sugar and distillary wastes. Ind. J. Environ. Hlth. 8 (1) : 6-35.

Davis, I. (1961). Microscopies studies with ozone quantitative lethality of ozone for Escherichia coli. U.S. Air Force Aerosp. Med. Cent. : 51-54.

De Vrles, G.A. (1962). Keratinophilic fungi and their action. Antonic Von Leeuwenhoek 28 : 121-133.

De, P.K. (1939). The role of blue-green algae in nitrogen fixation in rice fields. Proc. Roy. Soc. London B 127 : 121-139.

De, S.B. and Mukherjee, G.C. (1971). A case study of river pollution by rayon industry. J. Inst. Engineer 51 (10) : 1-4.

Dekker, J. 1973. Selectivity and resistance against systematic fungicides. Eppo. Bull. 10 : 47.

Demeter, O (1956). Uber modification bei Cyanophycean. Arch. Mikrobiol. 24 : 165-173.

Deshmukh, S.K. and S.C. Agarwal (1983). Prevalence of dermatophytes and other keratinophilic fungi soil of Madhya Pradesh Mykosen 26 : 574-577.

Desikachary, T.V. (1959). Cynophyta. I.C.A.R., New Delhi.

Deshmukh, S.K. and S.C. Agarwal (1982). In vitro degradation of human hair by some keratinophilic fungi. Mykosen 25 : 454-458.

Dey, N.C. and L.M. Kakoti (1953). Microsporum gypseum in India. J. Indian Med. Assoc. 25 : 160-164.

Digamber Rao, B. and Vidyavati (1983). Maleic hydrazide induced variations in Euastrum verrucosum. All India Applied Phycological Congress, Kanpur : 274-277.

Dixit, S.N. and S.B. Misra (1977). Antifungal activity of plant extracts-I against orange dust. In symposium on physiology of micro-organisms. Today and Tomorrow Printers and Publishers : 221-224.

Durie, E.B. and Frey (1955). Isolation of Microsprum gypseum and of Keratinomyces ajellio from Australian soil. Nature 176 : 936.

Elford, W.J. and J. Van den Ende (1942). An investigation of the merits of ozone as an aerial disinfectant. J. Hyg. 42 : 240-265.

Ellis, E. (1940). Fungi isolated from pinewood soil. Trans. Brit. Mycol. Soc. 24 : 87-97.

Ellish, M.M., Westfall, B.A. and Ellish, M.D. (1946). Determination of water quality. U.S. Fish Wild Ser. Res. Rpt. 9 : 122.

Ellof, J.N. and Van Der Westhuizen, A.J. (1981). Toxilogical studies on Microcystis. A review The Water Environment (W.W. Carmich-ael), Plenum Press, New York and London, 343-363.

Evans, J.H. (1960). "A biological studies" Ann. Rept. of East African Fresh Water Fish. Res. Organisation. Jinga, Uganda.

Executive summary of Ganga Action Plan for prevention of Pollution of Ganga 1984, Ministry of Environment and Forests, New Delhi.

Fish, G.R. (1951). Digestion in Tilapia asculatta. Nature, 167 : 900-901.

Flier, C.B. (1973). Inhibition of Histoplasma capsulatum by the garlic. Mycopathol. Mycol. Appl. 50 : 227-273.

Fogg, G.E. (1944). Growth and heterocyte production in Anabaena cylindrica Lemm. New Pytol. 43 : 164-175.

Fogg, G.E. (1951). Studies on nitrogen fixation by blue-green algae II. Nitrogen fixation by Mastigocladus laminosus. Coh. J. Exp. Bot. 2 : 177-180.

Fogg, G.E. (1960). Recent advances in our knowledge of nitrogen fixation by blue green algae. Proc. Sym. Algo., 115-118.

Fogg, G.E. (1961). In physiology and biochemistry of algae. (Ed. R.A. Lewin), Academic. Ed. R.A. Lewin), Academic Press, New York, 479-486.

Fogg, G.E. (1969). Proc. Roy. Soc. London, 173B : 175.

Foog, G.E. (1949). Growth and heterocyte production in Anabaena cylindrica Lemm. In relation to carbon and nitrogen metabolism. Ann. Bot. Lond. 13 : 241-259.

Frederick, R. Steiner, Soil Conservation in the United States : Policy and Planning Baltimore, MD : Johns Hopkins University Press, (1990).

Freeman, M.C. (1986). The role of nitrogen and phosphorus into development of Cladophora glomarata in the Manawatu river, New Zealand. Hydrobiologia 131 (1) : 23-30.

Fritsch, F.E. (1951). Heterocyst a botanical enigma. Proc. Linn. Soc. London 162 : 194-211.

Fritsch, F.E. (19 7). A general consideration of the sub-aerial and fresh water algal flora of Ceylon. Proc. Roy. Soc. Bll : 79-197.

Fujiki, M. (1972). The transitional condition of Minamata bay neighbouring sea polluted by factory waste water containing mercury. 6th Int. Conf. Wat. Poll. Res. paper : 42.

Garg, A.K. (1966). Isolation of dermatophytes and other keratinophilic fungi from soils in India. Sabouraudia 4 : 259-264.

Gauthier, Michel, J. Patrick, M. Monro and Soussan Mohajer (1987). Influence of salts and sodium chloride on the recovery of Escherichia coli from sea water. Curr. Microbiol. 15 (1) : 5-10.

Geitler, L. (1960) Schizophycen, Encyclopedia of Plant Anatomy. Berlin Band. 4 Teil. I.

Ghanvat, H.K. (1983). Studies on the Phycoplankton ecology of river Krishna, Maharashtra with reference to pollution. M.Sc. Thesis, Shivaji University, Kolhapur.

Gibbs, R.J. (1973). Mechanism of trace metals transport in rivers. Sci., 180 : 71-73.

Gilman, Joseph C. (1957). A manual of Soil Fungi. 1-450

Global Tomorrow Coalition, The Global Ecology Handwork : What you can do about the Environmental Crisis. Boston : Beacon Press, (1990).

Goddard, H.N. (1911). A preliminary report of fungi found in agricultural soil. 13th Rept. Mich. Acad. Sci. : 208-213.

Gonzalves, E.A. and Vaidya, B.S. (1963). On the Larvicidal properties of charophytes. Hydrobiol. 21 188-192.

Gorham, P.R. and Carmichael, W.W. (1979). Phycotoxin from blue-green algae. Pure Appl. Chem. 52 : 165-174.

Greaves, M.A., Davies, H.A., Marsh, J.A.P. and G.I. Wingfield (1976). Herbicides and soil microorganisms. Crit. Rev. Microbiol. 5 : 1-38.

Guarro, J.; Punsola, L. and M.A. Calvo (1981) Keratinophilic fungi from soil of Tarragona catalunya. Mycapathologia 76 : 69-71.

Gunale, V.R. and Balakrishnan, M.S. (1981). Biomonitoring at Panama, Mula and Mutha rivers. Ind. J. Environ. Hlth. 23 (4): 306-312.

Gupta, R.S. and Verma, Usha (1989). Growth responses of prokaryotic algae towards heavy metals. Advances in Applied Phycology-II : 69-72.

Gupta, A.B. (1957). The algal flora of some paddy fields and its importance in soil economy. J. Res. 3(1) : 1-24.

Gupta, A.B. and Ahmad, M.R. (1966). Studies on the effect of feeding some fresh water fishes with scenedesmus obliquus (Turpin) Kutz. Hydrobiol. 28 (1) : 42-48.

Gupta, A.B. and Bahel, K.K. (1981). Effect of hormidium foveolarum and productivity of Helianthus annus L. Abs. Int. Bot. Cong. Sydney : 218.

Gupta, A.B. and Shukla, A.C. (1964). Effect of algal hormones on the growth and development of rice seedlings. Labdev. J. Sci. Tech. 2 (3) : 205

Gupta, A.B. and Shukla, A.C. (1969). The effect of algal extracts of Phormidium species on growth and development of rice seedlings. Hydrobiol. 34 (1) : 77-84.

Gupta, A.B. and Srivastava, G.C. (1962). Production of antibacterial substances by Hydrodicyton reticulatum. Curr. Sci. 32 : 19--20

Gupta, A.B. and Srivastava, G.C. (1964). On antibiotic properties of some fresh water algae. Hydrobiol. 25 (1-2) : 285-288.

Gupta, D.C. and G.N. Pandey, (1980). Environmental pollution monitoring of river Pandu at Kanpur. J. Inst. Engineers (India) 60 : 25-42.

Gupta, Pratibha and Shukla, A.C. (1988). The genus Scenedesmus Meyen. in Ganga water at Nanamau and Bithoor. Res. J. Pl. Environ. 4 (2) : 74-77.

Gupta, Pratibha and Shukla, A.C. (1990). Cyanophyceae and pollution interweb in Ganga water at Bithoor. River Pollution in India (Ed. R.K. Trivedy), Ashish Publishing House, New Delhi, 177-200.

Gupta, R.S. and Bajaj, Asha (1983). Halo-tolerant blue-green algae as a source of fertilizer. Abst. All. Ind. Appl. Phy. Cong. : 6.

Haines, R.B. (1936). The effect of pure ozone on bacteria. Gt. Brit. Dep. Sci. Ind. Res., food Invest. Bd. Spec. Rep. No. 1935 : 30-31.

Harding, J.D.C. and Whitton, B.A. (1978). Zinc, Cadmium and lead in water sediment and submerged plants of the Daewent reservoir, Northern England. Wat. Res. 12 : 305-316.

Hawkins K., Government and Enforcement (1984). Clarendon Press, Oxford.

Heagle, A.S. and A. Strickland (1972). Reaction of Erisiphe graminis F. sp. hordei to lower levels of ozone. Phytopathology 62 : 1144-1148.

Heagle, A.S. and L. Key (1973). Effect of ozone on the wheat stem rust fungus. Phytopathology 62 : 397-400.

Heremith, A.S. (1984). Studies on the role of fungi in the sewage stabilization pond ecosystem. Ph.D. Thesis, Gulbarga University, Gulbarga.

Heririckon, A. and Wright, R.F. (1978). Concentration of heavy metals in small Narweighian lakes. Wat. Res. 12 : 101-112.

Heurck, H.V. (1896), A treatise on the diatomaceae. W.W. and Sons, Publications, W.C.

Hibben, C.R. (1966), Sensitivity of fungal sporesto sulfer dioxide and ozone. Phytopathology 56 : 880.

Hibben, C.R. and G. stotsky (1969). Effects of ozone on the germination of fungus spores. Can. J. Microbiol. 15 : 1187-1196.

Hicks, L.E. (1937). The lemmaceae of Indiana. Amer. Mid. Nat. 18: 777-789.

Hillman, W.S. (1961). The lemnaceae, or duckweed. A review of the descriptive and experimental literature. Bot. Rev. 27 (2): 221-287.

Hora, S.L. (1942). A short note on the pollution of stream in India and its likely to effect of fisheries. Rept. Fish. Commtt. I.C.A.R., New Delhi.

Hosmani, S.P. and Bharti S.G. (1980). Algae as indicators of organic pollution. Phykos 19 (1) : 23-26.

Hughes, E.O.; Gorham, P.R. and Jehnder (1958). Toxicity of unialgal culture of Microcystis aeruginosa. Can, J. Micro. 4 : 225-236.

Hussain, M.A. (1976). Preliminary observations on pollution of lake Hussain sagar caused by industrial effluents. Ind. J. Environ. Hlth. 18 : 227-232.

Husted, F. (1930). Bacillariphyta (Diatomaceae) In A Pascher's Die Susswasser-Flora Mitteleuropas. 10.

Hutchinson, T.C. and P. Stokes (1975). In water quality parameters: 303-343. Philadelphia : American Society for testing of materials, Special Technical Publication. 573.

Hynes, W.B.N. (1960). The biology of polluted waters. Liverpool University Press.

Hynes, W.B.N. (1970). The ecology of running waters. Liverpool University Press.

Ingram, W.N. and Palmer C.M. (1952). Simplified procedures for collecting, examining and recording plankton in water. J. Amer. W.M. Ass. 44 : 617-624.

Iyengar, M.O.P. and H.G. Venkataram (1951). The ecology and seasonal succession of the algal flora of river Cooumat, Madras with special reference to diatomaceae. J. Madras Univ. 21 (b): 634-655.

Jain, P.C. and S.C. Agarwal (1977). Keratinophilic fungi from soil of Mount Abu (India) Geobios. 4 : 136-138.

Jain, M. (1986). Investigations on Keratinophilic fungi from Lucknow soils. Ph.D. Thesis, Kanpur University.

Jain, M. Shukla, P.K. and O.P. Srivastava (1985). Keratinophilic fungi and dermatophytes in Lucknow soils with global distribution. Mykosen 28 : 148-153.

James R. McGoodwin, Crisis in the world's fisheries : People, Problems and Policies (Standford, C.A. : Standford University Press, (1990).

James Robertson, Future Wealth New York : Bootstrap Press (1990).

James, A. and Evison, L. (1979). Biological indicators of water quality. John Wiley and Sons, London.

Jensen, C.W. (1912). Fungus flora of the soil. Soil Sci. 31 : 123-158.

Jessica Tuchman Mathews, ed., Preserving the Global Environment : The challenge of shared leadership, New York : Norton, American Assembly, (1990).

Jhingran, V.G. (1977). Fisheries of the Ganga river system (MS) from Jhingran, 1977.

Jhingran, V.G. (1977). Fish and fisheries of India, Hindustan Publishing Corporation (India), New Delhi.

Jhingran, V.G. (1983). Fisheries and water pollution. Water Pollution and Management. 166-186.

Jim MacNeill; Pieter Winsemius and Tiazo Yakushiji. Beyond Interdependence : The Meshing of the Worlds Economy and the Earth's Ecology, New York : Oxford University Press, Trilateral Commission (1991).

Jindal Krishan K. (1988). Chemical control of bacterial leaf spot and fruit gummosis of almond. 5th International Congress of Plant Pathology, Kyoto Japan : 109.

Kamal Sonjoy and S.N. Srivastava (1989). Studies on the effect of vitamins on the growth and sporulation of Colletotrichum (Condra) sp. parasitizing tropical fruits. Res. J. Pl. Environ. 5 (1) : 29-32.

Kamal, S. and S.N. Srivastava (1988). Physiological studies on two species of Colletotrichum (Condra) parasitizing tropical fruits. Res. J. Pl. Environ. 4 (2) : 71-75.

Kanswal, M.M. (1986). Ecological studies on some food fishes of land river Bhilangana of Garhwal Himalaya, Ph.D. Thesis, University of Garhwal.

Kant, Shashi (1983). Algae as indicators of organic pollution. Adv. Appl. Phyco. II, 79-86.

Kaushi, B.D. (1983). Amelioration of salt effected soils with blue-green algae. Abst. All. Ind. Appl. Phyco. Cong., 13.

Kedesia, V.P. and Sharma, C.B. (1981). Detection of trace elements in the Kali Nadi water in Meerut. Tech. Ann. Ind. Assoc. Wat. Poll. Conf. 8 : 168-170.

Keeney, W.L.; Breck, W.G.; Van Loon, G.W. and Page, J.A. (1976). The determination of trace metals in Cladophora glomerata–C. glomerata as a potential biological monitor. Wat. Res. 10 : 981-984.

Kettle W. Ocen and Frank Denovelles Jr. (1986). Effects of Calcium stress on the plankton communities of experimental ponds. Fresh water Ecol. 3 (4) : 433-444.

Khan, K.R. (1983). Biological indicators and indices of water quality. Water Pollution and Management, 198-208.

Khan, M. (1983). Algae today. Siva Printers, Dehradun.

Klein, A.E. (1977). A study of heavy metals in lake Abbaya, Ethopia and incidence of nonparasitic elephantiasis. Wat. Res. 2 : 323-325.

Klein, L. (1957). Aspects of river pollution. Butterworth Scientific Pollution, London.

Klopping, H.L. (1951). Chemical constitution and antifungal action of sulfur compounds. Utrecht. Schotanous and Jens. : 142.

Krauss, R.W. (1962). Inhibitors in biology and biochemistry of algae. (Ed. R.A. Lewin) Academic Press, New York and London : 673-685.

Kulikova, G.M. (1960). The feasibility of using algae for the purification of radioactivity contaminated water. C.R. Acad. Sci. U.S.S.R. 135 : 978-980.

Kulkarni, A.H. and J.D. Nimbalkar (1983). The effect of hypo and hyper salinity on the growth and organic constituents in Gracilaria contricata J. Ag. All India Applied Phycological Congress : 237- 239.

Kulkarni, A.H., Dixit, G.B. and J.D. Nimbalkar (1989). The effect of hypo- and hyper- salinity on the inorganic constituents in Gracilaria contricata J. AG. Advances in Applied Phycology-II : 169-176.

Kumar, A. and S. Singh (1989). Inhibitory effect of trimethoprim sulphamethooxale on growth and Medel 6 (2) : 233-236.

Kumar, D., and H.D. Kumar (1989). Clonal variability and nature of chlorophenicol – resistance in the cyanobacterium Nostoc Linckia. Advances in Applied Phycology – II : 177-183.

Kumert, J. (1972). The digestion of human hair by the dermatophyte Microsporum gypseum in a submerged culture. Mykosen 15 : 59-71.

Kushwaha, R.K.S. (1976). Studies on keratinophilic fungi from soils. Ph.D. Thesis, Sagar University, Sagar India.

Kushwaha, R.K.S. and S.C. Agarwal (1976). Some keratinophilic fungi and related drmatophytes from soil. Proc. Indian Nat. Sci. Acad. 42 : 102-110.

Kushwaha, R.K.S. and S.C. Agarwal (1977). Efficacy of some antibiotics against spore germination of keratinophilic fungi. In vitro. Acta Botanica 5 : 174-177.

Kushwaha, R.K.S.; Srivastava, J.N. and Neeta Nigam (1985). Dermatophytes and keratinophilic fungi in soils of India. IX ISHAM Congress, Georgia : 3-3.

Kuss, F.R. (1950). The effects of ozone on fungus sporulation. M.Sc. Thesis, University of New Hampshire.

Kuthubutheen, A.J. (1977). The effects of fungicides on soil and leaf surface fungi. Ph.D. Thesis, University of Aston in Birmingham.

Kuthubutheen, A.J. and G.J.F. Pugh (1978). Effects of fungicides on the physiology of phyllplane fungi. Trans. Br. Mycol. Soc. II : 261-269.

Kuthubutheen, A.J. and G.J.F. Pugh (1979). The effect of fungicides on soil fungal population. Soil. Biol. Biochem. 11 : 297-303.

Lauren Kenworthy and Eric Schaeffer. A citizens guide to Toxic Waste Reduction New York : INFORM, (1990).

Lebedev, V.S. (1986). Features of the effect of heavy metals on the membrane of the Escherichia coil. 12 V Akadnauk SSR Biol. 0 (3) : 370-377.

Lester R. Brower *et. al.*, State of the World 1992 : A Worldwatch Institute Report on Progress towards a sustainable society. New York : Norton, (1992). This is the ninth in an annual series.

Lester W. Milbrath, Envisioning a Sustainable Society : Learning Our Way Out Albany : State University of New York Press, (1989).

Linn Proc. Nat. Acad. Sci., India. 47 : 141-144

Ma, R.M. (1933). A study of the soil fungi of the Peking district. Lingnan. Sci. Jour. Supp. 12 : 115-118.

Mahajan Jagmohan, Ganga Trail (1984), Clarion Books, New Delhi.

Managing Planet Earth, Special issue of Scientific American, 261 (3): 46-175 (Sept. 1989)

Mathur, R.P. (1986). Pollution Modelling of Upper Ganga Basin (A Report) Deptt. of Civil Engineering, Roorkee University, Roorkee.

McCallon, S.E.A. and F.R. Weedon (1940). Toxicity of ammonia, chlorine, hydrogen cyanide, hydrogen sulphide and sulphur dioxide gases. 11, Fungi and bacteria. Contrib. Boyce. Thomp. Inst. 11.

Mercantini, R.; Marsella, R.; Lambiase, L. and F. Fulvi (1983). Isolation of keratinophilic fungi from floors in Roman primary schools. Mycopathologia 82 : 115-120.

Meredith, C.H. (1938). Phycomycetes in Iowa soil. Phytopath. 28 : 15-16.

Michael Silverstein, The Environmental Factor : Its Impact on the future of the World Economy and your Investments Chicago: Longman Financial Services, (1990).

Michael Williams, ed., Wetlands : A Threatened Landscape Cambridge, M.A. Basil Blackwell, (1990).

Michel-Brun, J. and M.F. Monier (1981). Keratinophilic fungi from soil in Lyon's region between 1966 and 1980, their evolution. Lyon. Med. 245 : 87-91.

Micne, Curran, D.P.J.C. and L. Wilson (1986) (secel 1987). Effects of sedimentation on removal of fecal coliform bacteria from, effluents in estuarine water. Water Res. 20 (12) : 1493-1496.

Mishan E.J., Cost-Benefit Analysis New York Praeger, (1976).

Mishra, A.P. and M. Mahmood (1961). Effect of vitamins and harmones on growth and sporulation of Colletotrichum capsici (Syd.) Butleo and Bisby. Indian Phytopathol. 14 : 20-24.

Mishra, J.K. (1986). Fungi from mangrove muds of Andaman Nicobar Islands. Indian J. Mar. Sci. 15 (3) : 185-186.

Mitchell, Ralph (1978). Water Pollution Microbiology. London John Willey and Sons : 431.

Mitra, A.K. (1947). The algal flora of certain Indian soils Ph.D. Thesis, The University, London.

Mohan, N., Kumar, N. and R.P. Katiyar (1989). Influence of tannery effluent on Anabaena. Advances in Applied Phycology II : 213-217.

Moor, G.S. and R.D. Atkins (1977). The fungicidal and fungistatic effects on aqueous garlic extract on medically important yeast like fungi. Mycopathologia 69 : 341-348.

Morozzi, Goenci G.G.; Scardazza F. and M. Pitzurra (1986). Cadmium uptake by growing cells of gram negative bacteria. Microbios 48 (194) : 27-36.

Mukerji, K.G.; Agarwal, M.K. and A.S. Saxena (1969). Where from allergenic fungal spores are coming ? In "aspects of allergic and immunology." New Heights Publishers, New Delhi 2 : 181-189.

Murthy, Shiva Rama Krishna, Ganga (1976). Orient Longman, New Delhi.

National Research Council, Board on Agriculture, Alternative Agriculture, Washington DC : National Academy Press, (1989).

Nelson, K.E. (1958). Some studies of the action of sulfer dioxide in the control of Botrytis rot of Tokay grapes. Proc. Amer. Soc. Hort. Sci. 71 : 183-189.

Newell Frederick Haynes, Water Resources : Present and Future Uses, (1920), New Haven : Yale University Press.

Nigam, Neeta and R.K.S. Kushwaha (1984). Keratinophilic fungi and related dermatophytes from Kanpur. 7th Ind. Bot. Conf. Part II : 13.

Odell Rice, Environmental Awakening 1980 Cambridge Mass : Ballinger.

Dernose, L.J. and W. Koffman (1977). Tolerance of Schlerotina fruiti-cola to benzimidazol fungicides and control of the fungus. Pytopathnol. 88 : 153-164.

OHO Eckstein, Water Resource Development, The Economics of Project Evaluation, Cambridge : Harward University Press, (1965).

Padhy, A.A. (1968). Trichophyton terrestre from poultry from soil in India. Mykosen 11 : 725-726.

Padhye, A.A. and J.W. Charmichael. (1971). The genus Anthroderma Berkeley. Cand. J. Bot. 49 : 1525-1540.

Padhye, A.A. and M.J. Thirumalachar (1962). Isolation of Trichophyton ajelloi from soils in India. Curr. Sci. 31 : 100-101.

Padhye, A.A.; Mishra, S.P. and M.J. Trirumalachar (1966). Occurrence of soil inhabiting dermatophytes and keratinophilic fungi from soil in Poona. Hindustan Antibiot. Bull. 9 : 90-93.

Padmaja, C., and Vidyavati (1989). Effects of insecticides (Dimecrons and Nuvacron) the karyology of Rhioclonium hieroglyphicum. (AG) kg. Advances in Applied Phycology II : 223-235.

Pal, R. and P. Chatterjee (1989). An assessment of the pollution effect of chromium containing factory waste on the growth potential of Spirulina platensis. Advances in Applied Phycology II : 227-235.

Palmer, C.M. (1962). Algae in water supplies. U.S. Deptt of Health Education and Welfare, Public Health Services Publn. : 657.

Palmer, T. 1986. Endangered Rivers. University of California Press Ltd., London : 1-316.

Pandey, S.N. and A.K. Tripathi (1983). Studies on the effect of zinc on the growth of algae. All India Applied Phycological Congress, Kanpur : 206-210.

Pandey, S.N. and S. Agnihotri (1989). Studies on effect of some macro-elements on spore formation in Pithophora varia. Advances in Applied Phycology II : 237-238.

Pant Pushpesh, Ganga, Origin and Descent of River Eternal 1987. Frank Bros. & Co., New Delhi.

Panse, V.G. and P.V. Sukhatme (1967). Statistical methods for agricultural workers I.C.A.R., New Delhi.

Patil, H.S. (1979). Studies on the comparative ecology of various species of algae with a view to select the best for increasing the efficiency of sewage oxidation ponds. Ph.D. Thesis, Karnataka.

Pederson, Karsten, Lars Branstrom and Anna-Kerstin olsson (1986). So some tubing materials favour attachment and growth of bacteria in drinking water systems. Vatten 42 (1) : 21-24.

Programme, Project 5 UNESCO, Bhagalpur University Bhagalpur during June, 1976 to May, 1979.

Pugh, G.J.F. and J.I. Williams (1971). Effect of an organo-mercury fungicide on saprophytic fungi and on little decomposition. Trans. Br. Mycol. Soc. 57 : 164-166.

Pugh, G.J.F. and S.C. Agarwal (1983). Sensitivity of Trichophyton ajelloi to some common agrochemicals. Mykcopathologia 81 : 117-121.

Rai, R.T. (1981). Fungal flora of raw sewage. Indian J. Microbiol. 21: 84-85.

Raillo, A.I. (1928). Microflora of the soil. Jour. All. Russ. Cong. Bot. Leningrad : 181-182.

Rama Ratnam N. Valmiki Ramayan Balkand (171) 1985, Madras.

Randhawa, H.S. and R.S. Sandu (1963). On the occurrence of two new species of Trichophyton in Indian soils. Mycopatologia. (Dee Haag) 20 : 225-238.

Randhawa, H.S. and R.S. Sandhu (1965). A survey of soil inhabiting dermatophytes and related keratinophilic fungi of India. Sabouraudia 4 : 71-79.

Randhawa, H.S., Amarnath and R. Vishwanathan (1959). Isolation of M. gypseum (bodin) Guiart and Grigorakis from soil in Delhi. Sci. and Cult. 25 : 326-327.

Randhawa, H.S., Khan, J.U. and S.N. Gaur (1983). Blastomyces dermatidis in India, First report of its isolation from clinical material. Sabouraudia 21 : 215-221.

Ray, P. and A. David (1966). Effects of industrial wastes and sewage upon the chemical composition and fisheries of the river Ganga at Kanpur (UP) Environ. Hlth. 8 : 307-339.

Reddy, T.R.K. and G. Israel (1983). Effect of BHC on blue-green algae Synecococcus aeruginosus. All India Applied Phycological Congress, Kanpur : 213-220.

Rich, S. and H. Tomlinson (1968). Effect of ozone on conidiophores and conidia of Alternaria solani. Phytopathology 58 : 444-446.

Richard A. Denison and John Ruston, eds., Recycling the choices, Washington DC : Island Press, (1990).

Richard Tobin, The Expendable Future : U.S. Politic and the protection of Biological Diversity Durham, NC : Duke University Press, (1990).

Richards, M.C. (1949). Ozone as a stimulant for fungus sporulation. Phytopathology 39 : 20.

Robbins, W.J. (1938). Thiamine and growth of species of Phytophthora. Bull. Torrey Bot. Club 65 : 265-276.

Roy, K.; Gouri, R. Ghosh, and S.K. Dutt (1972). Keratinophilic fungi and the prevalence of dermatophytes in Orissa, India. Sabouraudia 10 : 218-229.

Rushworth M. Kidder, Reinventing the Future : Global Goals for the 21st Century, Cambridge : MIT Press, (1989).

Sabet, Y.S. (1935). A preliminary study of the Egyptian soil fungi. Egyptian Univ. Bul Faculty Sci. 5 : 1-29.

Santre, S. and B. Nandi (1975). Effect of different vitamins on the myclial growth of tree strains of F. durissimus. Indian Phytopath. 28 : 366-371.

Satry, Vir and J.S. Grewal (1973). Vitamins and trace elements requirements of Fusarium caerulam te causal organism in guar wilt. Indian Phytopathol. 26 : 274-278.

Satyanarayan, T. and D.P.C. Rao (1977). Activity of some medicinal plant extracts against keratinophilic fungi. Indian Drugs and Pharmaceutical Industry : 60-61.

Saxena, K.L.; Chakraborty, R.N.; Khan, A.Q.; Chattopadhya, S.N. and H.P. Chandra (1966). Pollution studies of the river Ganges near Kanpur. Environ. Hlth. 8 : 270-285.

Saxena, S.B. (1953). A new species of Paecilomyces from soil. Jour. Indian Bot. Soc. 32 : 186-189.

Sehgal, J.R. and R.N. Siddique (1969). Characterization of waste water Kanpur city. Environ. Health 11 : 95-107.

Serat, W.F.; Budinger, F.E.; and P.K. Mueller (1966). Toxicity evaluation of air pollutants by use of luminescent bacteria, Atmos. Enviro. 1 : 21-32.

Shankar, V. (1986). Integrated study of river Ganga. A brief report (1984-86). Gurukul Kangri University Haridwar.

Shanker, V. and G. Prasad (1983). Mycoflora of Ganges water in Rishikesh (India). Kanpur Univ. Res. J. Sci. 4 : 65-68.

Sharan, R.K. (1986). The integrated ecodevelopment projects on the river Ganga basin (Buxar to Basti). Progress report submitted to Dept. of Environ. Govt. of India, Patna University.

Sharma, M.M., Nandan Kanan. The valley of flowers, 1985. Vision Books, Delhi.

Sharma R., Goswami, A., Roych, O., Wdhry, A. and Jacob, N. (1993). The world in his pocket Down to Earth. (2) 5 : 5-11.

Shukla, A.C. (1989). Biological studies of Ganga eco-system between Kannauj and Shuklaganj. Technical report of Ganga Authority Project 1-229.

Shukla, A.C. and A. Khan (1988). 5th International Congress of Plant Pathology Kyoto, Japan : 437.

Sierra Club Legal Defence Fund, The Poisoned Well : New Strategies for ground water protection. Washington DC : Island Press, (1989).

Singh, A. (1973). Effect of vitamins and carbon and nitrogen sources on growth of Helminthosporium solani. Indian Phytopathol. 26 : 510-516.

Singh Gurdeep. Global Environmental Change and Environmental Law, (1991).

Singh, H.R. (1986). Pollution study of the upper Ganga and its tributaries. Progress report submitted to Dept. of Environ., Govt. of India. Garhwal University, Srinagar.

Singh, K.V. and S.C. Agarwal (1982). Nutritional requirements of keratinophilic fungi and dermatophytes for conidial germination. Mycopathologia 80 : 27-32.

Singh, K.V. and S.C. Agarwal (1983). Fungal colonization of human hair under fungicides treatment. Indian J. Plant Sci. 1 : 110-112.

Singh P.K. and A.L. Singh (1983). Comparative studies on azolla and blue-green algae biofertilization to rice crop. All India Applied–Phycological Congress Kanpur : 334-349.

Singh, R.H. (1968). Cultural and pathological studies of certain fungi. D.Phil. Thesis Allahabad University.

Singh, S.P. and R.P. Singh (1983). pH-dependent bisulphate toxicity in the blue green algae Aphanothece stagnina. All India Applied Phycological Congress Kanpur : 155-159.

Singh, U.P.; Pathak, K.K.; and R. Singh (1979). Effect of leaf extract of garlic on Fusarium oxysporum, F. sp. ciceri, Sclcrotina Sclerotiorum and gram seeds. Mycologia 71 : 556-563.

Singh, U.P.; Singh, H.B. and R.B. Singh (1980). The fungicidal effect of neem (Azadirachata indica Linn.) extracts on some soil borne pathogens of gram (Cicer orientium). Mycologia 72 : 1077-1093.

Sinha, A.K. (1988). A comprehensive study of Ganga and its dependents. Report submitted to Dept. of Environ. Govt. of India, Feroze Gandhi College, Rai-Bareli.

Sinha, K.K. and P.L. Singh (1988). Homeopathic drugs–inhibition of growth and aflatoxin production by Aspergillus paracisiticus. Indian Phytopath. 36 : 356-357.

Skowronski and Tadeuz (1986). Adsorption of cadmium on green micro-algae Stichococcus bacillaris. Chemosphere 15 (1) : 69-76.

Smack, R.M. and R.J. Howard (1976). Tolerance to benzimidazol derivative fungicides by Fusarium roseum on kentucky blue grass. Lurf. Pl. Dis. Reporter 60 : 91-94.

Smith Frank E, The Politics of Conservation 1966 New York, Pantheon.

Srivastava, J.N. (1985). Studies on soil inhabiting dermatophytes and keratinophilic fungi. Ph.D. Thesis, Kanpur University : 1-254.

Srivastava, A.K. (1986). Coordinated study of Ganga ecosystem. A report submitted to Dept. of Environment, Govt. of India, Rajendra Agricultural University, Bihar.

Srivastava, J.P. (1955). Taxonomical and physiological studies of some pathogenic fungi, D.Phil. Thesis, University of Allahabad.

Srivastava, O.P. (1959). Antifungal activity of some sulphur containing compounds against dermatophytes and yeast like fungi. J. Sci. Industr. Res. 18 : 25-28.

Srivastava, S.N.S. (1966). The systematic control of brown spot disease of rice with sulphanilamide and griseo fulvin. Indian Phytopath. 19 : 272-275.

Srivastava, V. and D.N. Tiwari (1983). Effect of 2, 4-D on nitrogen fixation and sporulation in a cyanobacterium Anabaena doliolum. All India Applied Phycological Congress, Kanpur : 192-198.

Steven Dariain, A Ganges of the mind (1988) Ratnanagar, New Delhi.

Steven Dariain, Ganges in myth and history. (1978). Hawai Press, U.S.A.

Stockdale, P.M. (1963). Microsporum gypseum complex (Nannizzia incurvata stockd., N gypsea (Nann.) comb. Nov., N. fulva sp. (nov.) Sabouraudia 3 : 114-126.

Subramanian, C.V. (1952). Fungi isolated and recorded from Indian soils. Jour. Madras Univ. 22B : 206-222.

Symposium on Physiology of micro-organisms, Today and Tomorrow Printers : 11-16.

Takahashi, R. (1919). On the fungus flora of the soil. Ann. Phytopath. Soc. Japan 1 : 17-22.

Tansey, M.R. and J.A. Appleton (1975). Inhibition of fungal growth by garlic extract. Micologia 67 : 409-413.

Tarar, J.L. and S.A. Akre (1983). Studies on the controlling effect of some selected chemicals on algae in water supplies. All India Applied Phycological Congress, Kanpur : 126-133.

Thind, K.S. and S.C. Agarwal (1973). In vitro effect of two chemotherapeutic agents on some keratinophilic fungi. Bull. Bot. Univ. Sagar 20 : 13-16.

Thirumalachar, M.J., Menon, S.K. and V.V. Bhatt (1961). Hamyan, a new antifungal antibiotics. Hindustan Antibiotics. Bull. 3 : 136-138.

Tomkins, R.G. (1932). The action of certain volatile substances and gases on the growth of mould fungi. Proc. Roy. Soc. Ser. B III: 210-226.

Trainor, F.R. (1964). The effect of composition of the medium on morphology in Scendesmus obliquus Can. J. Bot. 42 : 515518.

Treshow, M. (1965). Response of some pathogenic fungi to sodium fluoride. Mycologia 57 : 216-221.

Treshow, M., Harner, F.M. Price, H.E. and J.R. Kormerlink (1969). Effects of ozone on growth, lipid metabolism and sporulation of fungi. Phytopathology 59 : 1223-1225.

Tripathi, A.K. and S.N. Pandey (1989). Studies on algae of polluted ponds in Kanpur (India) IX. Effect of chloride on algal growth. Advances in Applied Phycology–II : 333-337.

Tripathi, S.N. (1954). Studies on the effects of chemicals on some green algae with particular references to their cytology. Ph.D. Thesis, Banaras Hindu University.

Truzuki, Y. and T. Yamada (1979). Inhibitory action of mercury compounds against glucose-6-phosphate dehydrogenase from yeast. J. Toxicol. Sci. 4 : 105-114.

Tujak, Zbigniew (1987). The effect of crude and fuel oils on the growth, chlorophyll a content and dry matter production of a green algae Scenedesmus quadricauda (Turp.) Environ. Pollut. 47 (1)

Tyneoka, A. and A. Gos (1973). The inhibitory action of garlic (Allium sativum Linn.) on growth and respiration of some micro-organisms. Acta. Microbiol. Polon. Ser. 135 : 51-62.

Ulfig and Krzysztof (1983). A preliminary study on the occurrence of dermatophytes and other keratinophilic fungi in bottom sediments of rivers and lakes. Acta Mycol. 19 (2) : 331-340.

Ulfig, K. and M. Koroz (1983). Isolation of keratinophilic fungi from sewage sludge. Sabouraudia 21 : 247-250.

United Nation, Global Outlook, 2000 : An Economic, Social and Environmental Perspective New York : United Nations Publication, (1990).

United Nations Development Programme, Human Development Report 1991, New York : Oxford University Press, (1991).

Upadhaya, P.P.; Singh, R.K. and G.P., Rao (1989). Antifungal and antiviral activity of crude leaf extract of some medicinal plants. Medle 6 (2) : 219-222.

Vandana, A. 1992. The Politics of Environment. Ashish Publishing House, New Delhi : 1-183.

Venkateswarlu, V. (1983). Taxonomy and ecology of algae in river Moosi, Hyderabad, India, II. Phycologica, 66 : 1-41.

Venkateswarlu, V. and Sampat Kumar, P.T. (1982). Chemical and biological assessment of pollution in the river Moosi, Hyderabad, (H.P.), India. Biol. Bull. Ind. 4 (7) : 23-30.

Verma, M.P. (1989) Growth responses of Chaetrophora to nitrogen and calcium concentrations in culture. Mendel 6 (3-4) : 447-452.

Verma, T.N.; Sinha, B.K. and U.L. Das (1982). Isolation of keratinophilic fungi from soil in Bihar. Mykosen 25 : 449-452.

Vidyavati (1989). Colchicine effects on algae, A resume. Advances in Applied Phycology II : 341-350.

Vittal, B.P.R. and D. Aravazhi (1981). Preliminary studies on sewage fungi. Indian J. Environ. Health 23 : 147-148.

Vyas, K.M. and S.C. Chaurasia (1977). Effects of fungicides on oxidative metabolism of Phytophthora parasitica var. piperina (Dust.) causing foot-rot and leaf-rot disease of Pan (Piper betle).

Warcup, J.H. (1950). The soil plate method for isolation of fungi from the soil. Nature 116-117.

Were, G.W.; Dee, M.K. and Cahil, W.P (1968).

Water flora as indicators of irrigation water contaminated by D.D.T. Bull. Environ. Contam. Taxicol. 3 : 333-338.

Whiton, B.A. (1975). In Biological indicators of water quality (Ed. James and L. Erison), John Wiley and Sons, New York.

Whiton, B.A. (1984). Algae as monitors of heavy metals in fresh water. Algae as Ecological Indicators (Ed. Shubert, L.F.), Academic Press, London, 257-280.

Whiton, B.A. (1975b). River ecology. (Ed.) Blackwell Scientific Publications, Oxford, 725.

Whitton, B.A. (1970). Biology of Cladophora in fresh waters. Wat. Res. 4 : 457-476.

Whitton, B.A. (1975a). Algae and higher plants as indicators of river pollution. Biological indicators of Water Quality (Ed. James and L. Erison), John Wiley and Sons. New York.

Williams, J.I. and G.J.F. Pugh (1975). Resistance of Chrysosporium pannorum to an organo-mercury fungicide. Trans. Br. Mycol. Soc. 64 : 255-256.

Wilson, M., Epton, H.A.S. and D.C. Sigee (1988). Biological versus chemical control of fire blight of hawthorn. 5th International Congress of Plant Pathology, Kyoto, Japan : 108.

Woodwell, G.M. (1967). Toxic substances and ecological cycles. Sci. Amer. 216 : 24-31.

World Commission on Environment and Development, our Common Future, New York : Oxford University Press, (1987).

World Resources Institute, World Resources 1992-93 : A Guide to the Global Environment, New York : Oxford University Press, March (1992). This is the fifth edition of a biennial series.

Wuu, K.D. and W.F. Grant (1966). Chromosomal abberations induced in somatic cells of Vicia faba by pesticides. The Nucleus 10 : 37-46.

Q

R

S

T

U

V

W

Y

Z